ROOT-KNOT NEMATODES
(*MELOIDOGYNE* SPECIES)

ROOT-KNOT NEMATODES (*MELOIDOGYNE* SPECIES)
Systematics, Biology and Control

Edited by

F. LAMBERTI

Laboratorio di Nematologia
Agraria, Bari,
Italy

C. E. TAYLOR

Scottish Horticultural Research
Institute, Dundee,
Scotland

1979

ACADEMIC PRESS

London . New York . San Francisco

A Subsidiary of Harcourt Brace Jovanovich, Publishers

ACADEMIC PRESS INC. (LONDON) LTD.
24/28 Oval Road,
London NW1

United States Edition published by
ACADEMIC PRESS INC.
111 Fifth Avenue
New York, New York 10003

Library of Congress Catalog Card Number: 78–74843
ISBN: 0–12–434850–5

Printed in Great Britain by Galliard (Printers) Ltd., Great Yarmouth

PARTICIPANTS

AMBROGIONI, LAURA *Istituto Sperimentale Zoologia agraria, Cascine del Riccio, Firenze, Italia.*
ARRIGONI, O. *Istituto di Botanica, Università di Bari, Via G. Amendola 173, Bari, Italia.*
BIRD, A.F. *Division of Horticultural Research, C.S.I.R.O., GPO Box 350, Adelaide 5001, Australia.*
BRODIE, B. *ARS, Dept. Plant Pathology, Cornell Univ., Ithaca, NY 14853, U.S.A.*
CAVENESS, F.E. *International Institute of Tropical Agriculture, PMB 5320, Ibadan, Nigeria.*
COOLEN, W.A. *Rijksstation voor Nematologie en Entomologie, Burg. van Gansberghelaan 96, B-9220 Merelbeke, Belgium.*
COOMANS, A. *Instituut voor Dierkunde, Rijksuniversiteit, Ledeganckstraat, 35, B-9000 Gent, Belgium.*
DALMASSO, A. *I.N.R.A., Station de Recherches sur les Nématodes, 123 Boulevard du Cap, Antibes, (F)06410, France.*
D'ERRICO, F.P. *Istituto di Entomologia agraria, Facoltà di Agraria, 80055 Portici, Napoli, Italia.*
EVANS, A.A.F. *Imperial College Field Station, Ashurst Lodge, Sunninghill, Ascot, Berks, England.*
FASSULIOTIS, G. *U.S. Vegetable Laboratory, P.O. Box 3348, Charleston, S.C. 29407, U.S.A.*
FERRIS, H. *Department of Nematology, University of California, Riverside, California 92521, U.S.A.*
FRANKLIN, MARY T. *Commonwealth Institute of Helminthology, The White House, 103 St. Peters Street, St Albans, England.*
GALLO, MARIA *Istituto di Zoologia agraria, Via G. Amendola 165/A, 70126 Bari, Italia.*
GOMMERS, F.J. *Department of Nematology, Agricultural University, Binnehaven 10, Wageningen, The Netherlands.*
GRIMALDI DE ZIO, SUSANNA *Istituto di Zoologia agraria, Via G. Amendola 165/A, 70126 Bari, Italia.*
KLEINEKE, ANNETE *Institut für Pflanzenkrankheiten und Pflanzenschutz der T.U. Hannover, Herrenhäuserstr. 2, 3000 Hannover, Fed.Rep.Germany.*
LAMBERTI, F. *Laboratorio di Nematologia agraria del C.N.R., Via*

PARTICIPANTS

G. *Amendola, 165/A, 70126 Bari, Italia.*

MAAS. P.W.Th. *Dutch Plant Protection Service, Wageningen, The Netherlands.*

MANKAU, R. *Dept. of Nematology, University of California, Riverside, CA 92521, U.S.A.*

MYERS, R.F. *Plant Pathology Department, Cook College, Rutgers University, New Brunswick, New Jersey, U.S.A.*

MORRONE DE LUCIA, MARIA *Istituto di Zoologia agraria, Via G. Amendola 165/A, 70126 Bari, Italia.*

NETSCHER, C. *Laboratoire de Nematologie de l'ORSTOM, B.P. 1386, Dakar, Senegal.*

ORION, D. *Division of Nematology, ARO, The Volcani Center, Bet-Dagan, Israel.*

RITTER, M. *Station de Recherches sur les Nématodes, 123 Bvd. du Cap, BP 78 06602 Antibes, France.*

ROMERO, DOLORES *Instituto Espanol de Entomologia, J. Gutierrez, Abascal 2, Madrid 6, Espana.*

SACCARDO, F. *Laboratorio Colture Industriali, C.N.E.N., Via Anguillarese km 1300, Casaccia, Roma, Italia.*

SASSER, J.N. *North Carolina State University, Raleigh, North Carolina 27607, U.S.A.*

SEINHORST, J.W. *I.P.O., Binnenhaven 12, Wageningen, The Netherlands.*

SIKORA, R.A. *Institut für Pflanzenkrankheiten, Universität Bonn, Nussallee 9, 53 Bonn, Germany (B.D.R.)*

STOYANOV, D. *Plant Protection Institute, P.O. Box 238, Sofia, Belgrade, Bulgaria*

TAYLOR, C.E. *Scottish Horticultural Research Institute, Invergowrie, Dundee, DD2 5DA, Scotland.*

TRIANTAPHYLLOU, A.C. *Dept. Genetics, N.C. State Univ., Raleigh, North Carolina 27607, U.S.A.*

VIGLIERCHIO, D.R. *Division of Nematology, University of California, Davis, CA 95616, U.S.A.*

WOUTS, W.M. *Entomology Division D.S.I.R., Mount Albert Research Centre, Private Bag, Auckland, New Zealand.*

List of staff members

Agostinelli, Augusta	Landriscina, S.
Basile, M.	Livorti, D.
Bleve, Teresa	Melillo, V.A.
Brandonisio, A.	Paglionico, G.
Carella, A.	Radicci, Antonia
Coiro, Maria I.	Ranieri, W.
Di Vito, M.	Roca, F.
Elia, F.	Vovlas, N.
Greco, N.	Zaccheo, G.
Inserra, R.	Zacheo, F.

Zacheo, G.

PREFACE

The root-knot nematodes belonging to the genus *Meloido-gyne* Goeldi constitute a major group of plant pathogens of outstanding economic importance. They are world wide in their distribution and species of *Meloidogyne* attack almost every type of crop, causing considerable losses of yield or affecting the quality of the produce.

Staple food crops in the tropics and virtually all vegetable, fruit and ornamental plants grown in the Mediterranean and subtropical countries are subject to attack by species such as *M. incognita*, *M. javanica*, *M. hapla* and *M. arenaria*. Even in temperate climates, crops grown under glass may suffer damage by several species, or outdoors by *M. naasi*, which particularly affects cereals. In addition to the direct damage caused by root-knot nematodes, many species have been shown to predispose plants to infection by fungal and bacterial pathogens.

The root-knot nematodes have now been known for more than a century and the vast amount of literature is testimony to their importance as crop pests and of their interest to nematologists. Despite the far ranging studies on their taxonomy, morphology and general biology, the root-knot nematodes remain as an outstanding problem and satisfactory and economic control measures have yet to be devised for most crop situations.

A pre-requisite for advancement towards a satisfactory solution to the problem posed by root-knot nematodes would seem to be the analysis and co-ordination of observations made on root-knot nematodes by scientists in many different countries, and the identification of the areas where further investigations could probably be pursued. This is the aim of the USAID International *Meloidogyne* project and the premise on which our conference was initiated.

The Conference was made possible by the generosity of the Cassa per il Mezzogiorno, the Assessorati all'Agricoltura delle Regioni Puglia e Basilicata and the Amministrazioni Provinciali of Bari and Lecce. We are indebted to these organisations for their support and encouragement

which made it possible to bring together over 50 research
scientists from 14 countries to discuss most aspects of
root-knot nematode biology and control. These published
Proceedings represent the considered opinions of experts
in the various specialist fields such as systematics,
taxonomy, morphology, cytology, ecology, physiology,
pathogenicity and chemical, cultural and genetic control,
but as a whole they present a general appraisal of root-
knot nematodes. Each of the invited contributions was fol-
followed by detailed discussions and exchange of ideas
and these formed a valuable part of the Conference. It is
difficult to convey the full flavour of the discussions
to the printed page but we hope that the summaries presented after each of the contributions at least indicate
the opportunities for further investigation on all aspects
of root-knot nematode biology. A general discussion at
the end of the Conference concluded that there is still
much research to be done to extend our knowledge of this
important group of plant parasitic nematodes and to provide control measures to safeguard our crops against
their depredations. These Proceedings present a comprehensive and authoritative account of current knowledge of
Meloidogyne nematodes and we hope will form a basis and
stimulus for further investigations and observations by
nematologists and plant pathologists.

The Conference was organised by the journal "Nematologia Mediterranea" and by the Laboratorio di Nematologia
Agraria, Consiglio Nazionale delle Ricerche, Bari. The
success of the Conference was due to the interest and
enthusiasm of the participants and to the many helpers
from the Laboratorio di Nematologia Agraria and the
Scottish Horticultural Research Institute who assisted
with the organisation of the meeting and with the typing,
photography and graphical illustration involved in preparing the manuscript for publication.

F. Lamberti

C.E. Taylor

CONTENTS

CONTENTS

x

CONTENTS

1 GENERAL PRINCIPLES OF SYSTEMATICS WITH PARTICULAR REFERENCE TO SPECIATION

A. Coomans

*Instituut voor Dierkunde, Rijksuniversiteit
Gent, Belgium*

Concepts of systematics

Systematics is the science of the diversity of organisms, the combination of taxonomy and classification. Taxonomy being the theory and practice of classifying organisms and classification being the system that results from this (cf. Mayr, 1969).

Systematics is synthetic since data from different biological disciplines have to be considered for establishing the relationships between the organisms. Those data are called characters.

The aims of a classification may be quite different and so will be the approach of the systematics. The purpose can be the construction of a useful system that allows more or less easy identification even for non specialists - or it can be the detection of the relationships between the different groups of organisms. It is usually difficult to reconcile both.

In searching for the relationships or affinities between organisms or groups of organisms either the criterion of "similarity" or that of "common descent" may be used. The first leads to a typological classification in which the historical background is insignificant or not considered; the second leads to a classification that tries to reflect the historical background as much as possible.

A recent and more elaborate form of typological classification has been proposed by the numerical taxonomists.
Their system is based on overall similarity and is called
the phenetic system. The pheneticists do not deny the
existence nor the importance of the historical reality,
but claim that it cannot be reconstructed without speculation. Their approach is said to be more objective and
more repeatable since it considers a large number of
characters which are treated mathematically and give an
idea of morphological relatedness expressed as "taxonomic
distances" in dendrograms. Several examples of such methods exist in nematology, e.g. Moss & Webster (1969, 1970)
for animal parasites, Bird (1967, 1971) for *Trichodorus*
(s.l.). At first all the characters used were considered
as equal, but later on different methods for "weighting"
have been developed. The basic principles however are
the same.

The merit of pheneticists is to have demonstrated that
the analysis of numerical data could be more adequate
than it usually is. This is certainly true for nematode
taxonomy where so much emphasis has been put on measurements. The sophisticated procedures of pheneticists are
however just another way to treat the same material and,
though useful at the level of methodology, are too shallow scientifically.

Most systematists are in favour of a phylogenetic
system based on evolutionary principles. Evolution leads
to change of characters and the affinities between the
different groups are reconstructed from the degree of
divergent evolution (tempo and mode of Simpson) that
occured. The basic principle is common descent. In the
phylogenetic or evolutionary school this is mainly reconstructed from morphological similarity and dissimil-
arity. The principles of this school have been amply explained in several works of Mayr (e.g. 1969) and Simpson
(e.g. 1961). A more recent approach is that of the

so-called cladistic school where "tempo and mode" of the evolution are considered less important than common descent and geologic age. The principles of this school have been proposed by Hennig (1950, 1965, 1966) and further discussed by e.g. Kiriakoff (1956, 1968), Brundin (1966), Schlee (1971) and Bonde (1975).

The pure typological or idealistic-morphologic school has few advocates among present day systematists, but the phenetic school deserves more attention. Pheneticists do not recognise types as such but their system can be considered as typological because their criteria are structural. Their system is another way to compare a (large) number of characters, but it lacks causality since the origin of the organisms is not taken into account. They consider the present result of evolution, without insight into the process itself.

The phylogenetic or evolutionary school also deals mainly with morphological characters; in principle its conclusions are based on the origin and monophyly of the groups of organisms, but the principle of evolution - as expressed by change in structure - is considered more important than that of descent. This is in contrast with the cladistic school where descent is always the most important factor. All organisms descending from each other form a phyletic series; such a series is of course monophyletic. It means that in such a group every member (individual, species, genus, etc.) is more closely related to every other member of the group than to any member classified outside that group.

Similarity is usually the result of common descent, the former is the effect, the latter the cause. Cladists claim that it is better to base the system on the cause rather than on the effect. They try to find out when a certain character has been acquired. The origin of such a character together with reproductive isolation proves the deviation from the ancestral group and allows for the

recognition of the new series as a taxonomic unit. It does
not matter whether many other characters of both groups
are similar or different. However, the lack of sufficient
information from the fossil record presents a serious
drawback and it is often impossible to ascertain the
splitting of the groups. This does not mean that phylogenetic principles should be abandoned. No system is perfect but new techniques and ideas can improve the results.

The taxonomists' raw material is formed by the characters of the organisms. They are mainly morphological ones,
but physiological, ecological, biochemical, ethological,
geographical and gentical features should also be considered whenever available. Similar characters can be subdivided in homologous and analogous ones. Homologous are
those characters which have a common origin; analogous
are those characters which are similar not because of a
common origin, but because of an adaptation to a similar
function or mode of life. Analogous characters are also
called convergent characters. A few examples will illustrate what is meant by analogous characters. The odonto-
style (or onchiostyle) of dorylaims and the stomatostyle
in tylenchs are clearly analogous characters. Overlapping
gland lobes have originated at least four times independently during the evolution of the Tylenchida:
1) in Aphelenchina;
2) in Anguinidae (cf. *Ditylenchus* → *Pseudhalenchus*);
3) in Dolichodoridae and Belonolaimidae (with the genera
Uliginotylenchus and *Trichotylenchus* forming the transition);
4) in Pratylenchidae, Hoplolaimidae, Meloidogynidae and
Heteroderidae. The saccate body shape and production of
a gelatinous matrix in sedentary females also originated
several times independently, namely in Tylenchulidae,
Pratylenchidae (Nacobbinae), Hoplolaimidae (Rotylenchu-
linae) and Meloidogynidae-Heteroderidae. Even the site
of production of the gelatinous material for the egg

sac may be analogous - the rectal glands in Meloidogynae, the excretory gland in Tylenchuldiae.

The German author Remane (1952) proposed a number of criteria to be used for detecting homologies. These criteria are often referred to in modern taxonomy and deserve some attention:

1) the criterion of the same position in an organism or a system, cf. the position of the cephalic sense organs, of the phasmids, of the oesophageal glands;

2) the criterion of the special property of the structure which allows the homologization of structures that may have quite a different general shape, e.g. oesophagus, oesophageal glands, excretory gland;

3) the criterion of intermediate forms, by comparing the ontogeny as well as the adult organisms, cf. the series of Mononchuloidea - Nygolaimoidea - Dorylamimoidea in Dorylaimida; the various juvenile and adult forms in Criconematidae; the series of Diplogasterids - *Tylopharynx*-like forms - Tylenchida; the already mentioned transition between Dolichodoridae and Belonolaimidae through the *Uliginotylenchus* - *Trichotylenchus* complex. It is better to use these criteria together instead of alone. In addition some other criteria may be used, such as

1) the more numerous are the related species that possess a common character, the more likely the homology : all Tylenchida have in principle a stomatostylet;

2) the higher the number of other common characters in related organisms, the more likely the homology on the basis that these characters are not correlated with each other;

3) the more a character occurs in non-related species, the less the probability of an homology, cf. the overlapping gland lobes in Tylenchida.

In the cladistic systematics characters are classified as plesiomorphous or apomorphous (Hennig, 1950). The first are original characters, already present in the

ancestors, that have been very little changed during evolution. The second are "new" characters, acquired during
evolution by a change of plesiomorphous characters.

The basic unit of the cladistic system is an individual at a particular moment or short period of its life,
called semaphoront by Hennig (1950). The whole set of
morphologic and other characters of an individual or
semaphoront-group is called the holomorph. In practice
the life of an individual can be subdivided in periods
wherein the changes of the holomorph are slow, e.g. egg
(embryonated or not) - juvenile (different stages) -
adult (immature, mature). Semaphoronts are linked together by ontogenetic relationships; individuals are
linked by tokogenetic relationships governed by the phenomenon of reproduction. Groups of individuals that are
interconnected by tokogenetic relationships are called
species. The genetic relationships that interconnect
species are called phylogentic relationships. The total-
ity of these relationships are called hologenetic relationships. All these relationships lie at the base of the
cladistic system.

Morphologic characters nevertheless also form the
bulk of the information for cladists, but they use them
differently. What matters is not so much the extent of
the present day differences among organisms, but the
relation of the present characters to the earlier ones.
The existing similarities have to be analysed in order
to find out whether a group is monophyletic or not. The
common presence of plesiomorphous characters (synplesio-
morphy) is no proof of monophyly and hence no proof of a
close relationship. On the other hand the common presence
of apomorphous characters (synapomorphy) points to close
relationship and proves monophyly. The position of the
dorsal gland outlet in Aphelenchina, the typical oesophagus in Criconematoidea, the sausage-shaped or saccate
3rd and 4th stage juveniles in Heteroderidae and

Meloidogynidae can be considered as such synapomorphies.
Only monophyletic groups can be used to build up the system. Apomorphous characters may be scattered throughout
the related branches, which complicates the picture. It
means that a given group may be apomorphous in a character that remained plesiomorphous in the group from which
it originated, but that in the latter group another character may become apomorphous while remaining plesiomorphous in the former group. Furthermore, an apomorphous
character may originate in different groups through parallel evolution. This is exemplified by the subterminal
position of the vulva in Meloidogynidae and Heteroderidae
or by the occurrence of saccate females in Rotylenchu-
linae and Meloidogynidae-Heteroderidae. Another problem
is that of the incongruities between different developmental stages of generations. The existence of sufficient
synapomorphies at one stage of the life cycle however
can demonstrate monophyly. This is shown by the peculiar
life cycle of some dimorphic *Deladenus*-species and by
the degenerate spearless males in some Criconematoidea.

In the cladistic system the same taxonomic level is
attributed to "sister groups". The latter are formed
when a new branch splits off from an existing one. Once
an apomorphous group has split off from a plesiomorphous
one the latter has been changed to such an extent that
it can no longer be considered the same as before the
splitting. Monophyletic groups are delimited by the
point at which they arise (splitting of the ancestral
group) and the point at which they split. A phylogenetic
tree of Meloidogynidae and their closest relatives, the
Heteroderidae, illustrates the system (Fig. 1).

Meloidogynidae (1) and Heteroderidae (2) most probably
originated from a common tylenchid ancestor. The first
family represents the apomorphous branch because of the
following characters : anterior position of the excretory
pore in females, the absence of a stylet in 3rd and 4th

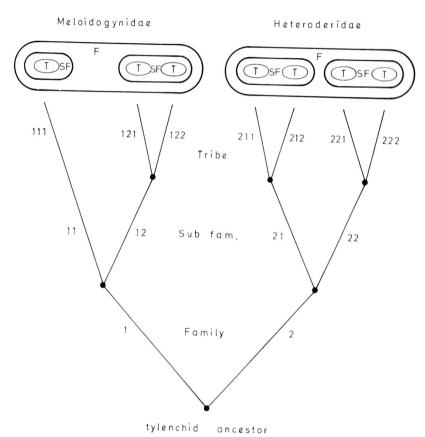

Fig. 1. Phylogenetic relationships of Meloidogynidae and Heteroderidae.

stage juveniles (as far as known) the reduced number of cephalic sensory organs, the basic chromosome number n=18, the mostly parthenogenetic reproduction and the formation of root galls.

Within Meloidogynidae two subfamilies can be recognized, Nacobboderinae (11) (with *Nacobbodera* Golden & Jensen, 1974 a junior synonym of *Meloinema* Choi & Geraert, 1974) and Meloidogyninae (12). Here the latter subfamily is the apomorphous branch because the juvenile females are saccate, the vulva is subterminal to terminal, reproduction is mostly parthenogenetic, the females are

embedded in well formed root galls. The subfamily Meloidogyninae can be subdivided into two tribes, Meloidogynini
(121) and Meloidoderellini (122), the former being the
plesiomorphous, the latter the apomorphous branch. Meloidoderellini are characterised by a cyst-like stage and
therefore are an example of parallel evolution with Heteroderinae. This trend seems also present in Meloidogynini
with the thickened cuticle in *Meloidogyne* or *Hypsoperine*
acronea.

Within Heteroderidae a first subdivison can be made
between a subfamily Meloidoderinae (21) and another subfamily Heteroderinae (22). The latter group is apomorphous in the cuticle structure of females, since annulations are lacking; and the vulva is always terminal.
Meloidoderinae comprise two different branches - a plesiomorphous one represented by Meloidoderini (211), and
an apomorphous one represented by Cryphoderini (212).
The former tribe is characterized by a subequatorial
vulva position, while the representatives of Cryphoderini
have a subterminal vulva. In the Heteroderinae the apomorphous branch is formed by the Heteroderini (222),
where the female cuticle forms a cyst after death; Atal-
oderini (221) are plesiomorphous with their non-cyst
forming cuticle.

The main difference from the classification proposed
by Wouts (1973a,b, 1974 and in these Proceedings) is the
use of tribes in the present classification and resulting
from this the different taxonomic level at which *Meloid-*
oderella on the one hand, and *Atalodera* and *Sherodera*
on the other, are linked to other genera. The use of
tribes provides an additional possibility of illustrating
relationships, namely within a subfamily e.g. Meloidoderinae. It can also prevent overevaluation of a particular character, e.g. the formation of cysts. This
character is clearly an adaptation to overcome temporary
unfavourable environmental conditions directly or

indirectly due to climatic factors that may have originated more than once (Meloidoderellini, Heteroderini).
However, for practical reasons or because of lack of information it may be decided to use a more limited number
of taxonomic categories.

In Table I it has been indicated where recent diagnoses for most of the taxa can be found. For other taxa
brief diagnoses are given below:

Meloidogyninae Skarbilovich, 1959 : Meloidogynidae.
Cuticle of female thin to thick, partly annulated, forming a cyst or not. Type genus : *Meloidogyne* Goeldi, 1887.

Meloidogynini n. tribe : Meloidogyninae. Cuticle of
female thin to moderately thick, partly annulated, not
forming a cyst. Type genus : *Meloidogyne* Goeldi, 1887.

Meloidoderellini n. tribe : Meloidogyninae. Cuticle
of female thick, partly annulated forming a brown to
black cyst containing few eggs, majority of eggs deposited in gelatinous matrix. Type and only genus : *Meloidoderella* Khan, 1972.

Meloidoderini n. tribe : Meloidoderinae. Vulva subequatorial. Type and only genus : *Meloidodera* Chitwood,
Hannon & Esser, 1956.

Cryphoderini n. tribe : Meloidoderinae. Vulva subterminal. Type genus : *Cryphodera* Colbran, 1966.

Heteroderinae Filipjev & S. Stekhoven, 1941 : Heteroderidae. Female cuticle non-annulated. Type genus :
Heterodera Schmidt, 1871.

Speciation and the species complex

Species are usually defined as actually or potentially
interbreeding communities that are reproductively isolated from other such communities. This leaves us with
the problem of the parthenogenetic organisms which then
cannot be considered as true species, but rather as
clones or groups of clones. However, Simpson (1961) has
given a definition including also the parthenogenetic

TABLE I

Outline classification of Meloidogynidae and Heteroderidae (cf. Fig. 1)

Categories		Diagnosis
1. Family	MELOIDOGYNIDAE (Skarbilovich, 1959) Wouts, 1973	Wouts (1973a), Wouts, this vol. p.30
1.1. Subfamily	Nacobboderinae Golden & Jensen, 1974 (syn. Meloineminae Husain, 1976)	Wouts, this vol. p.32
1.1.1 Tribe	Nacobboderini n. tribe	same as for subfamily
Genus	Meloinema Choi & Geraert, 1974 (syn. Nacobbodera Golden & Jensen, 1974)	
1.2. Subfamily	Meloidogyninae Skarbilovich, 1959	see text
1.2.1. Tribe	Meloidogynini n. tribe	see text
Genus	Meloidogyne Goeldi, 1887	
Genus	Hypsoperine Sledge & Golden, 1964	
1.2.2. Tribe	Meloidoderellini n. tribe	see text
Genus	Meloidoderella Khan, 1972	
2. Family	HETERODERIDAE (Filipjev & S.Stekhoven, 1941) Skarbilovich, 1947	Wouts (1973a)
2.1. Subfamily	Meloidoderinae Golden	Wouts (1973b)
2.1.1. Tribe	Meloidoderini n. tribe	see text
Genus	Meloidodera Chitwood, Hannon & Esser, 1956	
2.1.2 Tribe	Cryphoderini n. tribe	see text
Genus	Cryphodera Colbran, 1966	
Genus	Zelandodera Wouts, 1973	
2.2. Subfamily	Heteroderinae Filipjev & S.Stekhoven, 1941	see text
2.2.1. Tribe	Ataloderini n. tribe	same as for subfamily Ataloderinae proposed by Wouts (1973b)
Genus	Atalodera Wouts & Sher, 1971	
Genus	Sherodera Wouts, 1974	
2.2.2. Tribe	Heteroderini n. tribe	same as for subfamily Heteroderinae as given by Mulvey & Stone (1976)
Genus	Heterodera Schmidt, 1871	
Genus	Globodera (Skarbilovich, 1959) Mulvey & Stone, 1976	
Genus	Sarisodera Wouts & Sher, 1971	
Genus	Punctodera Mulvey & Stone, 1976	

and syngonic forms: "An evolutionary species is a lineage
evolving separately from others and with its own unitary
evolutionary role and tendencies". For example, in practice
populations of parthenogenetic *Meloidogyne* that are dis-
tinctive enough morphologically as well as biologically are
recognised as species. In amphimictic species the sexual
isolation between related forms is not absolute, while fer-
tility between widely separated populations of the same
species may be reduced.

A species is in fact composed of a number of spatially
distributed smaller, interbreeding communities, called
demes. In any such group the individuals are in intimate
spatial and temporal relation to one another. They constitute groups of genetically similar individuals which
are however usually only partially isolated from other,
adjacent populations (demes are open genetic systems and
gene flow through migration is possible). In the soil we
may expect that nematodes are not randomly distributed and
that aggregations exist. This is perhaps even more so for
slowly moving plant parasitic species. In such cases gene
flow between different demes is small and considerable
variation exists between them, e.g. in Criconematoidea
(cf. Loof, 1970).

The largest populational unit is the species itself.
Its genetic constitution changes continuously from generation to generation by the evolutionary forces acting upon
it. These evolutionary forces are mutation, natural selection and genetic drift. Mutations produce new alleles and
new gene combinations by which the genetic variability increases. Natural selections by the environment moulds the
genetic variation by favouring some gene combinations and/
or eliminating others; it brings about improvement in
adaptive relations between organisms and their environment. Genetic drift is the phenomenon of random fluctuations in the frequencies of certain alleles or gene combinations; it may be important in small populations where

it can establish non-adaptive or neutral characters. Which one of these forces is the most important depends upon the circumstances: when for some reason (e.g. environmental changes), the population increases, selection and drift will have a lesser impact and variation increases. During the next generation this variation can be considerably reduced by more pronounced selection. All these changes in genotypic variation of a given species are called sequential evolution.

The formation of new species (= speciation) requires more than that. It requires divergent evolution by which new adaptive types arise. During this process populations can be split up and genetic diversity increased. Speciation depends on reproductive isolation. Three basic modes can be distinguished: Allopatric, parapatric and sympatric speciation (cf. Bush, 1975; Table II).

Populations separated by spatial barriers preventing gene glow are called allopatric. The generation of new species can occur in two different ways:
1) by subdivision and
2) by the founder effect.

Subdivision of a widely distributed species into two or more relatively large populations by some extrinsic barrier prevents gene flow. Each subdivision evolves in its own way and if separated long enough can develop some reproductive isolating mechanism. This way of speciation is important for K-strategies, i.e. organisms with low reproductive rate, long life span and high competitive ability. In the second case a new colony is established by a small number of founders. Such populations usually exist in peripheral habitats where isolation of the parent population and invasion of an unexploited area is easy. This phenomenon is important in species with a high reproductive rate, short life span and low competitive ability (r-strategists).

Parapatric speciation occurs when species evolve as

TABLE II

The basic modes of speciation (modified after Bush, 1975)

	Extrinsic origin of barriers to gene flow		Intrinsic origin of barriers to gene flow	
	ALLOPATRIC		PARAPATRIC	SYMPATRIC
Freely interbreeding population				
Establishment of barriers to gene flow; development of reproductive isolating mechanisms (RIM)				
Expansion of range; perfection of RIM				
Species distribution after stabilization of ranges	parapatric or allopatric	broadly sympatric to allopatric	narrowly parapatric or allopatric	broadly sympatric
Reproductive strategy	K-strategists	r-strategists	r-strategists	r-strategists
Vagility	high	high	low	high to low
Initial population size	large	small	small	variable
Change in niche	no radical shift	shift or no shift	shift to new niche	shift
Evolution of RIM	result of long term adaptive genetic changes during isolation	genetic changes occur rapidly during isolation	occur during or soon after shift	occur before shift
Gene flow	none	none	some initially	some initially
Speciation rate	slow	fast	fast	fast

TIME

contiguous populations in a contiguous cline. Here speciation is accompanied by a shift to a new niche and is usually associated with organisms of low vagility. Diverging
populations are in constant contact and gene flow may occur
initially but is restricted to the contact zone. Reproductive isolating mechanisms are selected for during or soon
after the shift to the new niche.

Sympatric speciation is similar to the previous one but
premating reproductive isolation arises before a population
shifts to a new niche.

Allopatric speciation has been generally considered as
the main mode of speciation and this can be true for many
animals, yet parapatric and sympatric speciation may be
equally important and probably are even more important
than allopatric speciation in the case of parasites.

Newly acquired adaptations can be general at first and
involve a shift in ecological niche. Interspecies competition then may lead to more pronounced special adaptation
and to adaptive radiation.

Before becoming species the isolated populations may
pass through a stage of recognisable differences and are
then called races or subspecies (geographic, ecologic,
physiologic). Such races have been detected, e.g. for
Ditylenchus, Globodera, Heterodera and *Meloidogyne*. The
species are then called polytypic (as opposed to mono-
typic). Closely related species that separated only recently from each other are called sibling species; *Globodera rostochiensis* and *G. pallida* are probable examples.

The sex attraction and mating between species of *Globodera* also proves that speciation is recent (Green & Miller,
1969). The species of the genus *Heterodera* in general are
characterized by rather constant morphological characters
as well as great karyoptic uniformity, suggesting stability of the genus and close phylogenetic relationships
(Triantaphyllou, 1976). *Meloidogyne* is far more variable
than *Heterodera*, with some species already differentiated

and others segregating (Franklin, 1971). The fact that
most species are parthenogenetic complicates the story
since new mutations can give rise to new clonal populations and the impact of a changing environment can be more
drastic leading to a rather large variation between populations of the same species. The occurrence of polyploidy in
such species also leads to increased variation and evolutionary possibilities.

Some species may be polytypic, e.g. *M. incognita*, *M.
hapla* and *M. arenaria*, and some may constitute a super-
species such as the *M. incognita-M. javanica* complex.

In general speciation is faster when the founder population is small compared to the parent population. Small
populations can diverge rapidly just by chance and because
they can respond to looal selection pressures that may
differ from those encountered by the parent population.
This is even more so when distribution is limited. Active
distribution in nematodes is limited but passive distribution may be considerable. It is obvious that those *Meloidogyne* species that may be considered as primitive do not,
or not to a high degree, occur on economically important
crops, hence have not been spread so much by man. These
species either reproduce exclusively by amphimixis (*M.
carolinensis*) or by a combination of amphimixis and meiotic parthenogenesis (cf. Triantaphyllou, 1970). We ought
to know more about the environmental conditions in the
species' habitat in order to compare them with the way of
reproduction. In a given climate such conditions may still
vary a lot. A successful parasitic form must be adapted to
its life on the plant, but also to the time the plant is
in the soil and must be able to overcome adverse conditions such as temporary absence of the host. Therefore, we
need a better knowledge of the biology of the species. For
example, are the predominant species so successful because
they have shorter life cycles that cope with the relatively
short time their economically important host plants are in

the soil? Or are they, because of their higher variability, more able than other species to survive periods of absence of the typical host by substituting it with another host plant? If we can answer all such questions, we will be better able to appreciate the differences among the different species and from this gain more information about speciation in the genus. It is undoubtedly a challenging problem!

References

Bird, G.W. (1967). Numerical analysis of the genus *Trichodorus*. *Phytopathology* 57, 804.

Bird, G.W. (1971). *In* "Plant Parasitic Nematodes." (Eds. B.M. Zuckerman, W.F. Mai & R.A. Rohde). 1, 117-138. Academic Press, New York and London.

Bonde, N. (1975). *In* "Problèmes actuels de Paléontologie - Evolution des Vertébrés." 218, 293-324. Colloque International.

Brundin, L. (1966). Transantarctic relationships and their significance, as evidenced by chironomid midges. *K.Svenska.Vetensk.Akad. Handl.*, 11, 1-472.

Bush, G.L. (1975). Modes of animal speciation. *Ann.Rev.Ecol.Syst.*, 6, 339-364.

Franklin, M.T. (1971). *In* "Plant Parasitic Nematodes." (Eds. B.M. Zuckerman, W.F. Mai & R.A. Rodhe) 1, 139-162.

Green, G.D. & Miller, L.I. (1969). *Rep.Rothamsted.exp.Stn. for 1968:* 154-155.

Hennig, W. (1950). *In* "Grundzüge einer Theorie der phylogenetischen Systematik." 370. Deutscher Zentralverlag, Berlin.

Hennig, W. (1965). Phylogenetic systematics. *Ann.Rev.Entomol.*, 10, 97-116.

Hennig, W. (1966). *In* "Phylogenetic Systematics." 263. Univ. of Illinois Press, Urbana.

Kiriakoff, S.G. (1956). *In* "Beginselen der kierkundige systematiek." 167. De Sikkel, Antwerpen.

Kiriakoff, S.G. (1956). Demoderne dierkundige systematiek. *Natuurwet. Tijdschr.*, 50, 3-43.

Loof, P.A.A. (1970). Die Taxonomie der Nematoden als Hilfswissenschaft des Pflanzenschutzes. *Nachr Bl. dt. Pfl Schutzdienst, Berl.*, 24, 177-181.

Mayr, E. (1969). *In* "Principles of Animal Taxonomy." 428. McGraw-Hill, New York.

Moss, W.W. & Webster, W.A. (1969). A numerical taxonomic study of a group of selected strongylates (Nematoda). *Syst.Zool.*, 18, 423-443.

Moss, W.W. & Webster, W.A. (1970). Phenetics and numerical taxonomy applied to systematic nematology. *J.Nematol.*, 2, 16-25.

Mulvey, R.H. & Stone, A.R. (1976). Description of *Punctodera matadorensis* n.gen., n.sp. (Nematoda : Heteroderidae) from

Saskatchewan with lists of species and generic diagnosis of *Globodera* (n. rank), *Heterodera* and *Sarisodera*. *Can.J.Zool.*, 54, 772-785.

Remane, A. (1952). Die Grundlagen des natürlichen Systems der vergleichenden Anatomie an der Phylogenetik. *Akad.Verlagsgesellsch. Leipzig*, 400.

Schlee, D. (1971). *In* "Die Rekonstruktion der Phylogenese mit Hennig's Prinzip." 62. W. Kramer, Frankfurt am Main.

Simpson, G.G. (1961). *In* "Principles of Animal Taxonomy." 247. Columbia University Press, New York.

Triantaphyllou, A.C. (1970). Cytogenetic aspects of evolution of the family Heteroderidae. *J.Nematol.*, 2, 26-32.

Wouts, W.M. (1973a). A revision of the family Heteroderidae (Nematoda : Tylenchoidea). *Nematologica* 18, 439-446.

Wouts, W.M. (1973b). A revision of the family Heteroderidae (Nematoda : Tylenchoidea). *Nematologica* 19, 218-235.

Wouts, W.M. (1974). A revision of the family Heteroderidae (Nematoda : Tylenchoidea). *Nematologica* 19, 279-284.

Discussion

The discussion opened with a question whether the nematodes that have been described are to be regarded as species or demes. Coomans replied that demes are part of a species and went on to suggest that there has been far too much descriptive work but no in depth studies of characters other than morphological ones. He suggested that several populations should be described and compared, and that in general many characters of nematodes should be studied and taken into account for taxonomic work, at least until one knows what are the important (= apomorphic) ones. It should be borne in mind that morphology concerns only the phenotype and gives no direct information on the genetic constitution; more detailed study of presently described species may lead to the conclusion that some of them are demes of one species. In replying to a question about the dimensions of differences that should be established in order to separate species, Coomans replied that it is difficult to give precise rules about the importance to be attached to differences in characters. Lamberti suggested that a larger genus with many species could be split up into tribes and thus indicate relationships between genera. Coomans supported the use of tribes but said that it was important to recognise the rules of nomenclature and also to bear in mind that keys had to be established for the identification of taxa, but that the ultimate goal of systematics is to reflect as closely as possible the phylogeny.

Netscher questioned whether there should be more involvement with experimental systematics. He referred to the work of Fortuner who had studied *Meloidogyne* populations which according to Sher's criteria formed two or three species but in a study of the offspring of one individual the same so-called species differences were apparent. Coomans said that this pointed to the need for all characters of nematodes to be described and that all stages of development should be taken into account when establishing descriptions for a particular species. Discussing preferences for particular systems of classification Coomans said that Moss and Webster (see References) had described good techniques for the phenetic system and went on to suggest that anyone with access to a computer and with adequate statistical knowledge should use it. Nevertheless the phenetic system is shallow because it does not reflect evolution. The phylogenetic system, on the other hand, uses similar methods to the phenetic system and much is already known about it. The cladistic system is the most logical and deserves to be better known and used in nematode systematics. Seinhorst concurred that numerical taxonomy is indeed just a technique and that the results obtained with this method could be used with each of the systems.

2 CHARACTERIZATION OF THE FAMILY MELOIDOGYNIDAE WITH A DISCUSSION ON ITS RELATIONSHIP TO OTHER FAMILIES OF THE SUBORDER TYLENCHINA BASED ON GONAD MORPHOLOGY

W.M. Wouts

Mount Albert Research Centre, Department of Scientific and Industrial Research, Auckland, New Zealand

Introduction

"It is well recognised that a classification system developed today [including characterization of taxa] based on the relatively small, known percentage of the total nematodes, must at best be a rather transient one" (Golden, 1971). It merely represents our latest ideas on the relative significance of certain characters and serves only as a guide to an intelligent 'ordering' system. The characterization of the root-knot nematodes is no exception. Here too much of the necessary information is still lacking and there is even difference of opinion as to which super-family root-knot nematodes belong.

The present characterization of Meloidogynidae (Skarbilovich, 1959) Wouts, 1972 was achieved by considering its morphology with reference to the morphology of related families. To help understand this system the history and development of root-knot nematode taxonomy is reviewed and the systemic position of the species in the suborder Tylenchina Chitwood, 1933 as a whole is discussed. Many of the diagnostic characters used to separate families are difficult to determine, particularly in small forms. Review of existing groups suggests that the readily observable mono- and didelphic gonad morphology is important in establishing relationships between families and is used here for the classification of the suborder Tylenchina.

History

The first root-knot nematode was described by Cornu in 1879; a swollen form, inducing nodules in sainfoin (*Dodortia orientalis*) roots. From published abstracts of the literature Cornu knew that four plant parasitic nematodes had been described: *Anguillula tritici* Steinbuch, 1799; *A. dipsaci* Kuhn, 1857; *Heterodera schachtii* Schmidt, 1871; and *A. radicicola* Greeff, 1872, and that the last of these induced swellings in the host roots. He had no access to the original descriptions of the four species and could not be sure that his species was identical to *A. radicicola*. He therefore introduced it as new, provisionally naming it *Anguillula marioni*.

Müllet (1884) studied the host range of root-knot nematodes and concluded that there was only one root-knot nematode species, *A. radicicola*. Because he observed eggs in some of the swollen females of this plant parasitic species he transferred it to the genus *Heterodera* Schmidt, 1871. This move was supported by Marcinowski (1909), and for the next 50 years the accepted name for the root-knot nematode was *Heterodera radicicola* (Cornu, 1849) Müller, 1884.

In 1932 Goodey, studying Greeff's original paper, noted that the description of *Anguillula radicicola* Greeff, 1872 was in fact of a slender species belonging to a genus presently known as *Ditylenchus* Filipjev, 1936. Consequently Goodey recognised *A. marioni* Cornu, 1879 as the first and only described root-knot nematode species, and transferred it to the genus *Heterodera*. *H. marioni* (Cornu, 1849) Marcinowski, 1909 was the accepted name for the root-knot nematodes until 1949.

Goeldi (1892) was the first worker to recognize root-knot nematodes as a genus. He introduced it as *Meloidogyne* for the type of species *M. exigua*. Unaware of Goeldi's

work, Cobb (1924) introduced the name *Caconema* for the
same taxon with *C. radicicola* (Greeff, 1892) Cobb 1924 as
the type species. Neither proposal was well founded and
it was only in 1949 that Chitwood, on the basis of cuticular line patterns of the perineum, definitely identified
five different root-knot nematode species and convinc-
ingly established the root-knot nematodes as a group of
species - a genus, for which he reintroduced Goeldi's
name *Meloidogyne*.

In 1959 Skarbilovich raised the genus *Meloidogyne* to
subfamily Meloidogyninae and placed it with the subfamily
Heteroderinae Filipjev amd Schuurmans Stekhoven, 1941 in
the family Heteroderidae (Filipjev and Schuurmans Stekhoven, 1941) Scarbilovich, 1947. Golden (1971) added
Meloidoderinae as a third subfamily and Wouts in 1972
added one further subfamily,the Ataloderinae.

In redefining each of the subfamilies Wouts (1972)
observed that they are all closely related except Meloidogyninae Skarbilovich, 1958, the subfamily containing
the root-knot nematodes. These are different in several
characters; the excretory pore of the female is located
anteriorly to the level of the valve of the median bulb,
the basal plate subdivides the head into six sectors with
the two lateral sections wider than the other four;
usually the eggs are laid, occasionally some eggs may
be retained by the female; galls are produced on the roots
of the host. The only similarity between the Meloidogyninae and the other 3 subfamilies appears to be the
stylet length and the male and female body shape. These
similarities were not considered sufficiently significant
to justify retention of the Meloidogyninae in the Heteroderidae and as the subfamily could not be satisfactorily
accommodated in any of the other Tylenchoidea families
it was proposed as an independent family Meloidogynidae
(Wouts, 1972).

Classification of Tylenchina based on gonad morphology

There is a considerable difference in opinion as to
which superfamily of the Tylenchina Meloidogynidae belongs: Andrassy (1976) placed root-knot nematodes in the
superfamily Hoplolaimodea (Filipjev, 1934) Paramonov,
1967; Siddiqi (1971) in Tylenchoidea (Orley, 1880) Chitwood and Chitwood, 1937; and Golden (1971) in Heteroderoidea (Filipjev, 1934) Golden 1971. To determine the
most adequate superfamily it is necessary to consider the
next highest category, the suborder Tylenchina (Orley,
1880) Geraert, 1966.

The suborder Tylenchina is well defined and it is
generally agreed that it contains the stylet-bearing members of the Secernentea in which the dorsal oesophageal
gland opens a short distance posterior to the stylet
knobs. It is also generally accepted that within the
Tylenchina there are three basic types of oesophagus; one
which lacks a valved median bulb, as is typical for the
superfamily Neotylenchoidea (Thorne, 1941) Jairajpuri and
Siddiqi, 1969; another in which the procorpus and the median bulb are amalgamated, as is characteristic for the
members of the superfamily Criconematoidea (Taylor, 1936)
Geraert, 1966; and a third in which the median bulb is
valved, and a more or less distinct demarcation is present between procorpus and median bulb, as is expressed
in Tylenchoidea *sensu lato*. However, several different
classifications have been presented for subdivision of
the superfamily Tylenchoidea *sensu lato*. Siddiqi (1971)
accepted no further subdivision and recognised the three
basic superfamilies: Tylenchoidea, Criconematoidea and
Neotylenchoidea. Golden (1971) accepted 5 superfamilies
and proposed that species of the superfamily Tylenchoidea
with swollen females, and those with cephalic setae, be
placed in separate superfamilies: Heteroderoidea and
Atylenchoidea (Scarbilovich, 1959) Golden, 1971. Andrássy
(1976) combined all Tylenchoidea with a pronounced

stylet into a separate superfamily and therefore recognised four superfamilies. Raski and Siddiqi (1975), besides accepting Golden's proposal on Tylenchoidea *sensu
lato*, also subdivided the superfamily Criconematoidea,
accepting a total of seven superfamilies.

It is recognised that the superfamily Tylenchoidea
sensu lato is a conglomerate of Tylenchids that do not
fit in the other two superfamilies and it is to be expected that further subdivision of this taxon will be made
at some future time. However, until such subdivision can
be based on convincing characters I propose that the
taxon should be recognised as one unit, therefore accepting Siddiqi's (1971, 1976) system.

Of the three superfamilies in Siddiqi's system, two,
the Neotylenchoidea and the Criconematoidea, encompass
only monodelphic forms. Only the amorphous superfamily
Tylenchoidea contains both didelphic and monodelphic
forms. The absence of didelphic forms in two of the
three higher categories of the suborder Tylenchina could
suggest that within this taxon mono- and didelphic forms
do not evolve readily from each other. Although this is
not immediately apparent in Siddiqi's classification of
the superfamily Tylenchoidea it can be observed in the
classifications proposed by Golden (1971) and Andrássy
(1976), where within individual families almost always
either one or the other of the sexual forms exist. In
Siddiqi's (1971, 1976) classification it will appear
only if some slight changes are made within the family
Tylenchidae Orley, 1880 similar to the recent transfer
of the didelphic genus *Tetylenchus* Filipjev 1936 (Tety-
lenchinae Siddiqi, 1971) to the didelphic families Tylen-
chorhynchinae Eliava 1964 (Golden, 1971) or Leptotylen-
chinae Sher, 1973 (Sher 1973). Likewise, *Psilenchus* de
Man 1821, the only didelphic genus in the Tylenchidae,
could be transferred to the didelphic family Tylodoridae
(Paramonov, 1967) Siddiqi, 1976 and monodelphic forms

in Tylodoridae could be transferred to Tylenchidae. Such changes seem justified as the only differences between the families are the body length and the stylet length. These changes would make the separation of Tylenchidae (monodelphic) and Psilenchidae (Paramonov, 1967) Andrássy 1976 (didelphic) more apparent (Table I).

In Belonolaimidae Whitehead, 1959 and Dolichodoridae (Chitwood and Chitwood, 1950) Skarbilovich, 1959 the females are didelphic. *Trophurus* Loof, 1958 although monodelphic, has the vulva placed equatorially as if it were didelphic and inclusion of this genus in the family Dolichodoridae is therefore felt to be justified.

The Hoplolaimidae (Filipjev, 1934) Wieser, 1953 in Siddiqi's system are didelphic except for *Hoplotylus* s'Jacob, 1959 and *Acontylus* Meagher, 1968 in the Hoplotylinae Khan, 1969 and *Rotylenchoides* Whitehead, 1958 in the Rotylenchoidinae Whitehead, 1958. *Hoplotylus* and *Acontylus* closely resemble the species of the family Pratylenchidae (Thorne, 1949) Siddiqi, 1963 and were placed in the Pratylenchinae Thorne, 1949 by the original authors (s'Jacob, 1959; Meagher, 1968). *Hoplotylus* resembles forms with strong sexual dimorphism; *Acontylus* resembles *Nacobbus* Thorne and Allen, 1944 with females swelling only slightly. In erecting the genus *Rotylenchoides* Whitehead (1958) considered it to belong between Hoplolaiminae Filipjev, 1934 and Pratylenchinae. Andrássy (1976) placed the genus in Radopholinae Allen and Sher 1967, and Golden (1971) placed it close to *Acontylus*. It is thus apparent that *Rotylenchoides* may be more closely related to genera in the Pratylenchidae than to genera in the Hoplolaimidae. *Hoplotylus, Acontylus* and *Rotylenchoides* are therefore placed here in the family Hoplolaimidae. Andrássy (1976) went one step further. By including all didelphic Pratylenchidae (Radopholinae) in Hoplolaimidae he proposed a fully didelphic family Hoplolaimidae and a fully monodelphic family Pratylenchidae.

TABLE I

Classification of the suborder Tylenchina as proposed by Siddiqi
(1971, 1976) modified to a classification system based on the number
of ovaries

Neotylenchoidea	– – – – – – – – – –	Neotylenchoidea (m)
Criconematoidea	– – – – – – – – – – –	Criconematoidea (m)
Tylenchoidea	– – – – – – – – – – –	Tylenchoidea (m+d)
Anguinidae	– – – – – – – – – –	Anguinidae (m)
Tylenchidae	– – – – – – – – – –	Tylenchidae (m)
Tylenchinae	– – – – – – – – – –	Tylenchinae
Psilenchinae		Tylodorinae
Basiria etc		
Psilenchus	– – – – – – – –	Psilenchidae (d)
Leipotylenchinae		Psilenchinae
Tylodoridae		
Tylodorinae		Leipotylenchinae
Antarctenchinae	– – – – – – – – – –	Antarctenchinae
Dolichodoridae	– – – – – – – – – –	Dolichodoridae (d)
Belonolaimidae	– – – – – – – – – –	Belonolaimidae (d)
Pratylenchidae	– – – – – – – – – –	Pratylenchidae (m+d)
Nacobbinae	– – – – – – – – – –	Nacobbinae (m)
Pratylenchinae	– – – – – – – – – –	Pratylenchinae (m)
Pratylenchus	– – – – – – – – – –	*Pratylenchus*
Zygotylenchus		*Radopholoides*
Hirschmanniella		
Radopholinae	– – – – – –	Radopholinae (d)
Radopholus	– – – – – –	*Radopholus*
Pratylenchoides	– – – – – – – –	*Pratylenchoides*
Radopholoides		*Zygotylenchus*
Hoplolaimidae		*Hirschmanniella*
Hoplotylinae	– – – – – – – – –	Rotylenchoidinae (m)
Rotylenchoidinae		Hopoloaimidae (d)
Rotylenchoides		Hoplolaiminae
Helicotylenchus	– – – – – – – – – –	incl. *Helicotylenchus*
Hoplolaiminae		
Rotylenchulinae	– – – – – – – – – –	Rotylenchulinae
Heteroderidae	– – – – – – – – – –	Heteroderidae (d)
	– – – – – – – – – –	Meloidogynidae (d)

m = monodelphic   d = didelphic

Unfortunately such a classification fails to indicate the
close relationship between Pratylenchinae and Radopholinae and cannot be accepted. Pratylenchidae is therefore
recognised here as the only family of the Tylenchina containing both monodelphic and didelphic forms, with the
didelphic forms combined in the subfamily Radopholinae
while the monodelphic forms comprise the subfamilies
Pratylenchinae, Nacobbinae Chitwood and Chitwood, 1950
and Rotylenchoidinae.

Except for members of the Pratylenchidae, all other
families of the Tylenchoidea suborder Tylenchina are
either mono- or didelphic. Didelphic forms in the otherwise monodelphic family Pratylenchidae are the species
of the subfamily Radopholinae.

Classification of Meloidogynidae

Accepting the stability in the type of genital tract
within the taxonomic units of the suborder Tylenchina,
only one type of ovary can be expected in the family
Meloidogynidae; the didelphic type. Since the family's
two closest relatives the Heteroderidae and Hoplolaimidae
are didelphic, it seems that the Meloidogynidae developed
from a stable line of didelphic forms. Golden's (1971)
proposal of combining this didelphic taxon and the Heteroderidae with the monodelphic genus *Nacobbus* in a superfamily Heteroderoidea must therefore be rejected. *Nacobbus*
resembles Meloidogynidae only in female shape; in all
other characters it resembles Pratylenchidae, the family
to which it clearly belongs (Siddiqi, 1971; Andrássy,
1976). *Meloidoderita* Poghossian, 1966 another monodelphic
genus invariably associated with root-knot nematodes
(Siddiqi, 1971; Golden, 1971; Wouts, 1972; Andrássy, 1976)
was correctly transferred to the superfamily Criconematoidea by Kirjanova and Poghossian (1975). Characters of
the recently described male of this taxon (Poghossian,
1975) support this transfer.

Recognised in the family Meloidogynidae are: the type genus *Meloidogyne*; and the genera *Hypsoperine* Sledge and Golden, 1964, *Meloidoderella* Khan, 1972 and *Meloinema* Choi and Geraert, 1973.

The genus *Hypsoperine* was established for root-knot nematodes with a protruding vulval area (Sledge & Golden, 1964). Subsequently it has been accepted and rejected by several workers and its taxonomic status is not clear. A detailed comparative morphological study is required to establish the validity of this genus. Here it is considered a synonym of *Meloidogyne*.

In members of the genus *Meloidoderella* the cuticle of the female is not annulated and after death transforms into a hardened protective layer around a few unlaid eggs (Khan, 1972). The larvae possess longitudinal lines. These are unusual characters for root-knot nematodes, and therefore Husain (1976) proposed that forms with such characters be included in the subfamily Meloidoderellinae.

The genus *Meloinema* is identical to the genus *Nacobbodera* of the subfamily Nacobboderinae Golden and Jensen, 1974. The genus name *Meloinema* has priority and must be accepted as valid, with *Nacobbodera* as junior synonym. Members of this genus possess two ovaries and an anteriorly located excretory pore and are therefore not related to species of the genus *Nacobbus* but belong to the family Meloidodgynidae. Because of its different life-cycle in which the female is initially free-living and slender and subsequently swells up to the typical root-knot nematode shape, and because of the absence of a typical root-knot nematode perineal pattern and the presence of a large larval stylet, *Nacobbus* was placed in a new subfamily Nacobboderinae.

The classification of the Meloidogynidae is as follows:

Family Meloidogynidae (Skarbilovich, 1959) Wouts, 1972
    Subfamily Meloidogyninae Skarbilovich, 1959
        Genus *Meloidogyne* Goeldi, 1892
            syn. *Caconema* Cobb, 1924
            syn. *Hypsoperine* Sledge and Golden, 1964
    Subfamily Meloidoderellinae Husain, 1976
        Genus *Meloidoderella* Khan, 1972
    Subfamily Nacobboderinae Golden and Jensen, 1974
        Genus *Meloinema* Choi and Geraert, 1973
            syn. *Nacobbodera* Golden and Jensen, 1974

Characterization of Meloidogynidae (Skarbilovich, 1959)
Wouts, 1972

Skarbilovich, Acta Parasitol. Pol. 1959, 7: 122-1?7
(Meloidogyninae). Wouts, Nematologica 1972, 18: 441.

Diagnosis emended: Superfamily Tylenchoidea (Oerley,
1880) Chitwood & Chitwood, 1937. *Fully developed female* -
swollen to subspherical, sedentary. Tail absent or reduced
to short stump. Anus located terminally. Excretory pore
opposite or anterior to median bulb. Basal plate subdivides head into six sectors with the two lateral sectors
of same size or wider than the other four. Stylet well
developed. Median bulb ovate to spherical with well developed musculature and crescentic valve plates. Oesophageal
glands overlap intestine ventrally. Ovaries two. Most eggs
laid. *Second-stage juvenile* - vermiform, migratory stage.
Male - vermiform, migratory. Body twisted as much as half
a turn (180°) from head to tail. Tail short. Caudal alae
absent. Stylet and cephalic framework well developed.

    Type genus: *Meloidogyne* Goeldi, 1892
Wouts, Nematologica, 1972, 18: 441
    Families included:
            Melodogyninae Skarbilovich, 1959
            Meloidoderellinae Husain, 1976
            Nacobboderinae Golden, 1974
    Differential diagnosis: The family Meloidogynidae can

be distinguished from the other families of the superfamily Tylenchoidea by the anterior position of the excretory pore of the swollen female.

Subfamily Meloidogyninae Skarbilovich, 1959

Diagnosis emended: Family Meloidogynidae (Skarbilovich, 1959) Wouts, 1972. *Young female* - swollen to subspherical, sedentary. Cuticle annulated. Tail absent. Anus located terminally. Stylet well developed, usually less than 20 μm long. *Fully developed female* - never transforming into brown to black hard-walled cyst, cuticle remains annulated. *Second-stage juvenile* - infective stage. Stylet moderately developed, usually less than 15 μm long. Cuticle without longitudinal line except in lateral field. *Third-stage juvenile* - swollen, sedentary, with short blunt tail. Stylet absent. *Male* - testis one (or two where sex reversal occurs).

Type genus: *Meloidogyne* Goeldi, 1892.

Differential diagnosis: The subfamily Meloidogyninae can be recognized by the swollen young females and swollen third and fourth stage juveniles of which the latter do not possess a stylet; and by the annulated non cyst-forming cuticle of the female; and the second-stage juvenile without longitudinal lines.

Type genus Meloidogyne Goeldi, 1892 Diagnosis: As this is the only genus in the subfamily the definition of the genus is the same as for the subfamily.

Type species: *Meloidogyne exigua* Goeldi, 1892

Subfamily Meloidoderellinae Husain, 1976

Diagnosis: Family Meloidogynidae (Skarbilovich, 1959) Wouts, 1972. *Young female* - swollen to subspherical, sedentary. Cuticle at midbody not annulated but with zig-zag pattern. Tail absent. Anus located terminally. Stylet well developed, less than 20 μm long. *Fully developed female* - developing into black, hard-walled cyst, cuticle with

zig-zag pattern. *Second-stage juvenile* - infective stage.
Stylet moderately developed, usually less than 15 μm long.
Cuticle with longitudinal lines. *Third-and fourth-stage
juvenile* - development unknown. *Male* - testis one.

Type genus: *Meloidoderella* Khan, 1972.
Khan, Final Technical Report. Aligarh Muslim University,
Aligarh, India, 1972: 21.

Differential diagnosis: The subfamily Meloidoderellinae can be recognized by the swollen females that trans-
form into dark brown cysts with zig-zag pattern on the
cuticle; and the second-stage juveniles having longitudinal lines.

Type genus Meloidoderella Khan, 1972. Diagnosis: As this
is the only genus in the subfamily the definition of the
genus is the same as for the subfamily.

Type species: *Meloidoderella indica* Khan, 1972.

Subfamily Nacobboderinae Golden & Jensen, 1974

Golden & Jensen, J. Nematology, 1974, 6: 31.

Diagnosis: Family Meloidogynidae (Skarbilovich, 1959)
Wouts, 1972. *Young female* - slender, mobile. Cuticle at
midbody annulated. Tail more than one body width long.
Stylet well developed, usually more than 30 μm long.
Fully developed female - developing into swollen annulated
form. *Second-stage juvenile* - stylet strongly developed,
more than 30 μm long. Cuticle without longitudinal lines.
Third-and fourth-stage juvenile - development unknown.
Probably slender with stylet. *Male* - testis one.

Type genus: *Meloinema* Choi & Geraert, 1973
Choi & Geraert, Nematologica, 1973, 19: 334.

Differential diagnosis: The subfamily Nacobboderinae
can be recognized by the slender young females that trans-
form into posteriorly swollen, fully developed, females
on which the cuticle remains annulated, and by the second-
stage juvenile without longitudinal lines.

Type genus Meloinema Choi & Geraert, 1973. Diagnosis: As this is the only genus in the subfamily the definition of the genus is the same as for the subfamily.
Type species: *Meloinema kerongense* Choi & Geraert, 1973.

References

Andrássy, I. (1976). "Evolution as a basis for the systematization of nematodes", Pitman Publishing Ltd., London.

Chitwood, B.G. (1949). Root-knot nematodes. Part 1. A Revision of the genus *Meloidogyne* Goeldi, 1887. *Proc.helminth.Soc.Wash.* 16: 90-104.

Cobb, N.A. (1924). The amphids of *Caconema* (nom.nov.) and other nemas. *J.Parasit.* 2: 118-120

Cornu, M. (1879). Études sur le *Phylloxera vastatrix*. *Mém.prés.acad. sci.Paris* 26: 164-174.

Goeldi, E.A. (1892). Relatorio sobre a molestia do cafeeiro na provincia do Rio de Janeiro. *Archiv.Mus.Nac.Rio de J.* 8: 7-121.

Golden, A.M. (1971). *In* "Plant Parasitic Nematodes I. Morphology, Anatomy, Taxonomy and Ecology", (Eds B.M. Zuckerman, W.F. Mai and R.A. Rohde), 191-232, Academic Press, New York and London.

Golden, A.M. and Jensen, H.J. (1974). *Nacobbodera chitwoodi* n. gen., n. sp. (Nacobbidae : Nematoda) on douglas fir in Oregon. *J.Nematol.* 6: 30-37.

Goodey, T. (1932). On the nomenclature of the root-gal nematodes. *J.Helminthol.* 10: 21-28.

Greeff, R. (1872). Ueber nematoden in Wurzelanschwellungen (Gallen) verschiedener Pflanzen. *Sber.Ges.Beford.ges.Naturw.Marburg* 11: 172-172.

Husain, S.I. (1976). Phylogeny and inter-relationship of the superfamily Heteroderoidea (Skarbilovich, 1947) Golden, 1971. *Geobios* 3: 9-12.

s'Jacob, J.J. (1959). *Hoplotylus femina* n.g., n.sp. (Pratylenchinae: Tylenchida) associated with ornamental trees. *Nematologica* 4: 317-321.

Khan, A.M. (1972). Studies on plant parasitic nematodes associated with vegetable crops in Uttar Pradesh. Final Technical Report. Aligarh Muslim University, Aligarh, India. 328pp.

Kirjanova, E.S. and Poghossian, E.E. (1975). A redescription of *Meloidoderita kirjanovae* Poghossian, 1966 (Nematoda: Meloidoderitidae, fam. n. *Parasitologia Tom VII, Academia Nauk SSSR*: 280-285.

Marcinowski, K. (1909). Parasitisch und semiparasitisch an Pflanzen lebende Nematoden. *Arbeit.Kaiserl.Biol.Anat.Land u. Forstw.* 7: 147-171.

Meagher, J.W. (1968). *Acontylus vipriensis* n.g., n.sp. (Nematoda: Hoplolaimidae) parasitic on *Eucalyptus* sp. in Australia. *Nematologica* 14: 94-100.

Müller, C. (1884). Mitteilungen über die unseren Kulturpflanzen schädlichen, das Geschlecht Heterodera bildenden Würmer. *Landwirt Jahrb.Schwiez* 13: 1-42

Poghossian, E.S. (1966). A new nematode genus and species of the family Heteroderidae (Nematoda in the Armenian SSR). *Dokl.Akad. Nauk Arm.SSR* **42**, 117-123. (In Russian.)

Poghossian, E.E. (1975). Description of the male of *Meloidoderita kirjanovae* Poghossian, 1966 (Nematode, Meloridoderitidae). *Dokl. Akad.Nauk Arm.* **60**, 252-255.

Raski, D.J. and Siddiqi, I.A. (1975). *Tylenchocriconema alleni* n.g. n.sp. from Guatemala (Tylenchocriconematidae n. fam., Tylenchocriconemadoidea n. superfam.; Nematoda). *J.Nematol.* **7**, (3), 247-251.

Sher, S.A. (1973). The classification of *Tetylenchus* Filipjev, 1936, *Leipotylenchus* n. gen. (Leipotylenchinae n. sub.f.) and *Triversus* n. gen. (Nematoda: Tylenchoidea). *Nematologica* **19**, 318-325.

Siddiqi, M.R. (1971). Structure of the oesophagus in the classification of the superfamily Tylenchoidea (Nematoda). *Indian J. Nematol.* **1**, 25-43.

Siddiqi, M.R. (1976). New plant nematode genera *Plesiodorus* (Dolichodorinae), *Meiodorus* (Meiodorinae subf.n.), *Amplimerlinius* (Merlininae) and *Gracilancea* (Tylodoridae grad.n.). *Nematologica* **22**, 390-416.

Skarbilovich, T.S. (1959). On the structure of systematics of the nematodes order Tylenchida Thorne, 1949. *Acta Parasitol.Pol.VII, Warszawa*, 117-132.

Sledge, E.C. and Golden, A.M. (1964). *Hypsoperine graminis* (Nematoda: Heteroderidae) a new genus and species of plant parasitic nematode. *Proc.helminthol.Soc.Wash.* **31**: 83-88.

Whitehead, A.G. (1958). *Rotylenchoides brevis* n.g., n.sp. (Rotylenchoidenae n. subfam.: Tylenchida). *Nematologica* **3**, 327-331.

Wouts, W.M. (1972). A revision of the family Heteroderidae (Nematoda: Tylenchoidea), I. The family Heteroderidae and its subfamilies. *Nematologica* **18**, 439-446. (Published 1973).

Discussion

Coomans questioned whether in *Meloidogyne* the presence of two testes, rather than one, could always be attributed to sex reversal. Triantaphyllou said that this was a developmental peculiarity and genetically the genus had one testis, the development of a second one depending on when the hormonal changes causing sex reversal occurred. Coomans said that this could lead to confusion and suggested that the diagnosis of a species should indicate that two testes may be present in the male if there is no sex reversal but Triantaphyllou responded that the problem may be more complicated than presently indicated. Sasser pointed out that species with a raised vulval area also have the neck misplaced from the longitudinal axis of the body and asked if there was any taxonomic significance in this. Wouts replied that asymmetry could be found in all species.

Triantaphyllou questioned the need to have three sub-families in the family Meloidogynidae, each with a single genus, but Wouts replied that the genera differed so much from each other that it seemed to be necessary in order to emphasise this. Coomans stated his preference for two sub-families, *Meloidoderella* being closer to *Meloidogyne* than to *Meloinema*, and said he would place the first two genera in Meloidogyninae but would split the sub-family into two tribes.

3 TAXONOMY OF THE GENUS *MELOIDOGYNE*

Mary T. Franklin

*Commonwealth Institute of Helminthology,
St Albans, England*

The history of the taxonomy of *Meloidogyne* can be divided
into the period before Chitwood's revision of the genus
in 1949 and the period from 1949 onwards.

Pre-Chitwood Period

The earliest known record of root galls containing
what now seem without doubt to have been various stages
of *Meloidogyne* is that by Berkeley, in 1855, who observed
the galls on cucumber in an English glasshouse. In 1875
Licopoli described galls containing small worms on the
roots of *Sempervivum tectorum* L. in Italy and in 1878
Jobert described a disease of coffee in Brazil in which
the roots bore galls containing "cysts" and nematode
eggs in all stages of development. The latter was the
first record of disease in a field crop due to root-knot
nematodes. No scientific name was given to the organisms
by any of these authors.

Meanwhile, in France, Cornu was investigating the
Phylloxera disease, then devastating the grapevines, and
in the course of his examination of root swellings on
various plants he found nematodes in the roots of sainfoin, *Onobrychis sativa* Lam. In 1879 he published a
description of the galls and the nematodes that he had
first observed in January 1874 on the roots of sainfoin

growing in sandy soil in the Loire valley. He compared
the nematodes with *Anguillula* (now *Anguina*) *radicicola*
of Greeff and *Heterodera schachtii* Schmidt. Deciding
that they resembled *Heterodera* the more closely but not
wishing to mis-identify them he made a new species,
A. marioni, which was later transferred to the genus
Heterodera by Müller (1884).

Müller was studying pests of cultivated plants in
Germany and described and illustrated root-knot nematodes
on the roots of *Musa rosacea* Jacq. Comparing his specimens
with *H. schachtii* he observed that the male of the latter
develops within a cuticle with no tail whereas there was
a sharply-pointed tail on the cuticle of the immature
male of his species. In addition, the swollen female of
H. schachtii lacked transverse striae on the body, but
had a "subcrystalline membrane", and developed on the
surface of the host root. In contrast the species on
M. rosacea, which he described as "pear shaped", had
transverse striae on the integument and he noted that
the nematodes were endoparasitic, living in galls on
the roots. An illustration is given of the striae on the
terminal area of a female showing what is now called the
perineal pattern, but it cannot be identified as to
species. Müller called his nematodes *H. radicicola* and
this name was generally applied to root-knot nematodes
until 1932.

Soon after Müller's publication Treub (1885) briefly
described *H. javanica* from sugar cane in Java, distinguishing it from Müller's species by measurements which
were obviously erroneous. In the United States, Neal
(1889) gave a detailed description of root-knot nematodes
which he named *Anguillula arenaria*, found damaging
various crops in Florida, and later in the same year
Atkinson (1889) gave an account of the life-history of
root-knot nematodes that he identified as *Heterodera
radicicola* and studied on various plants at Auburn,

Alabama. Neither Neal nor Atkinson mentions a paper by
Goeldi, following up the observations of Jobert in Brazil,
published first in 1887 and reprinted in 1892, in which
he gave a detailed description of root-knot nematodes
from coffee in Brazil and gave them the name *Meloidogyne
exigua*. The new generic name was derived from Greek words
meaning "apple-shaped" and "female" and the specific name
referred to their small size. This work was generally
overlooked or ignored by other nematologists and the name
Heterodera radicicola was used in spite of a proposal by
Cobb in 1924 that the root-knot nematodes should be placed
in a new genus or sub-genus, *Caconema*, differing from
Heterodera in being truly endoparasitic, in being less
specialized in their mode of parasitism and in the males
having two testes and amphids protected by "cheeks". The
whole question of nomenclature was thoroughly reviewed
by Goodey in 1932. He showed that Müller was wrong to
apply the specific name *radicicola* to root-knot nematodes,
having confused them with what is now known as *Subanguina
radicicola*, and Goodey considered that the correct name
should be *Heterodera marioni* Cornu, 1879. In the following year, in his book "Plant Parasitic Nematodes and the
Diseases they cause", Goodey (1933) listed all the names
given earlier to root-knot nematodes as synonyms of *H.
marioni*. The name *Heterodera marioni* for root-galling
nematodes lasted for 17 years until Chitwood's revision
of the genus published in 1949, and even today is occasionally found in the literature.

Chitwood And After

Increasing interest in the USA in diseases associated
with root-knot nematodes and, in particular, the work of
Christie and his collaborators in the 1940s on the host-
parasite relationships of a number of root-knot populations from different areas and different hosts, showed
that plants might be heavily galled by one population or

race of *H. marioni* while showing no evidence of attack by others. This led Chitwood (1949) to study the morphology of the various races and he found sufficient differences to enable him to designate five species and one variety and finally to separate the root-knot nematodes from the cyst-forming nematodes.

Chitwood re-defined the genus *Meloidogyne*, synonymized Cobb's genus *Caconema* with it and differentiated it from *Heterodera*. He was unable to obtain specimens from type collections of any of the species previously described, probably because no specimens had been preserved, and he based his descriptions on material from the best source available to him. He redescribed *Meloidogyne exigua*, the type species, from specimens collected on coffee in the New York Botanical Garden, remarking that Goeldi's description was erroneous but nevertheless adequate to identify the genus and that coffee may possibly be attacked by more than one species of root-knot nematode. Topotypes from sugarcane in Java were used for his description of *M. javanica*, again with the remark that this host may harbour more than one species of root-knot nematode and that Treub, in describing this nematode, may possibly have examined more than one species. For his description of *M. incognita* Chitwood used a population on carrot from Texas, presumed to be the type host and locality of the species designated by Kofoid & White (1919) as *Oxyuris incognito*. The original description was based on eggs from faecal samples from soldiers at a military camp in Texas and the eggs in Chitwood's population differed in no way from those described by Kofoid & White. Neal (1889), when describing root-knot disease, mentioned a number of plants severely affected by the nematode that he named *Anguillula arenaria*. He was probably dealing with several species but as Chitwood had to decide on one for his description he chose as *M. arenaria* the species commonly causing severe galling on peanut, *Arachis hypogaea* L. in Florida. Chitwood pointed out that none of the original descriptions was adequate to identify beyond the genus, but he retained the specific names for nematodes from hosts or localities as nearly equivalent to the original as he could. The errors of Chitwood's somewhat arbitrary action have been pointed out both by Gillard (1961) and by Whitehead (1968) but no one has seriously challenged it and it is to be hoped that no one will attempt to re-name his species without very serious study and consideration of the confusion that would result. Chitwood did not give diagnoses for the species he described but provided a key in which he uses, as one of the main distinguishing characters, the pattern of striations or annulations on the posterior region of the mature females, now usually called the perineal pattern. In addition he used the shape of the stylet knobs and the stylet length in males and females. By modern standards Chitwood's descriptions could be criticized for giving insufficient information on, for example, the number of specimens measured and the variability in the perineal patterns, but many descriptions published since 1949 have similar shortcomings. His work provided the stimulus to many other workers to look more closely at all stages of root-knot nematodes, with the result that today, 90 years after the description of *M. exigua* and 29 years after Chitwood's paper, about 36 species have been named. In spite of the large number of species now recognised there have been very few studies of the genus as a whole, that of Whitehead (1968) being the only comprehensive one, while a compendium was compiled by Esser *et al.* in 1976.

Means of Identification

The 36 species listed (Table I) (*M. spartinae* has now been omitted because its position in the genus is uncertain) have been described with varying degrees of detail. It has not proved possible to make a workable key

TABLE I

Meloidogyne species arranged in
order of approximate larval length

1. *M. kikuyensis* de Grisse, 1960
2. *M. exigua* Goeldi, 1887 (type species)
3. *M. artiellia* Franklin, 1961
4. *M. hapla* Chitwood, 1949
5. *M. ovalis* Riffle, 1963
6. *M. incognita* (Kofoid & White, 1919) Chitwood 1949
7. *M. lordelloi* Ponte, 1969
8. *M. oteifae* Elmiligy, 1968
9. *M. deconincki* Elmiligy, 1968
10. *M. microtyla* Mulvey, Townshend & Potter, 1975
11. *M. coffeicola* Lordello & Zamith, 1960
12. *M. inornata* Lordello, 1956
13. *M. propora* Spaull, 1977
14. *M. kirjanovae* Terentyeva, 1965
15. *M. litoralis* Elmiligy, 1968
16. *M. tadshikistanica* Kirjanova & Ivanova, 1965
17. *M. acronea* Coetzee, 1956
18. *M. megriensis* (Pogosyan, 1971) Esser, Perry & Taylor, 1976
19. *M. ethiopica* Whitehead, 1968
20. *M. indica* Whitehead, 1968
21. *M. ardenensis* Santos, 1968
22. *M. javanica* (Treub, 1885) Chitwood, 1949
23. *M. mali* Ito, Ohshima & Ichinohe, 1969
24. *M. africana* Whitehead, 1968
25. *M. thamesi* Chitwood in Chitw., Specht & Havis, 1952
26. *M. naasi* Franklin, 1965
27. *M. graminis* (Sledge & Golden, 1964) Whitehead, 1968
28. *M. graminicola* Golden & Birchfield, 1965
29. *M. megadora* Whitehead, 1969
30. *M. ottersoni* (Thorne, 1969) Franklin, 1971
31. *M. arenaria* (Neal, 1889) Chitwood, 1949
32. *M. lucknowica* Singh, 1969
33. *M. brevicauda* Loos, 1953
34. *M. decalineata* Whitehead, 1968
35. *M. carolinensis* Fox, 1967
36. *M. megatyla* Baldwin and Sasser, 1978

without relying heavily either on measurements, as in
Whitehead's key for larvae, or on somewhat subjective
opinions of the shape and pattern of the striae in the
vulval region of the female. As is well known, measurements
can vary between populations and perineal patterns often
vary greatly within populations. A compendium of diagnostic characters such as that published by Esser *et al.* is

a very useful source of information and an invaluable
tool for identifications. As the authors point out, it
can be used to eliminate most of the species in a search
for the indentification of an unknown population provided
that enough carefully observed information is available.
By reducing the possibilities to a small number of species
the searcher can go to the original descriptions for the
final identification. At this stage the inadequacies in
some original descriptions become obvious, e.g. the
numbers measured may be small, or not stated, or too few
perineal patterns may be illustrated.

Female Infestations of *Meloidogyne* are usually first
found by examination of galled roots from which mature
or developing females are dissected. The female characters
used to identify the species are, in order of importance,
the perineal pattern, stylet length, position of the
excretory pore and the host plant and locality. Of these,
the perineal pattern is by far the most important. The
stylet length is of value only if the stylet is exceptionally long or short, e.g. in *M. ovalis* and *M. brevicauda* it is long (17-24μm and 17-25μm respectively) compared with a range from about 11-16μm to about 14-20μm in
most of the other known species. The stylet in *M. ottersoni* is one of the shortest known, being 10-11μ long, and
in *M. hapla* it is 10-13μm (mean 11). The position of the
excretory pore from the anterior end is related to stylet
length by Esser *et al.* while Whitehead (1968) relates it
to the number of annules from the anterior end. The first
authors found this to be a good corroborating character
but Whitehead, who looked at 14 species, found his method
of measurement to be too variable within a species and
with too great a range and too much overlap between species. The host plant and the geographical location of the
infestation can provide broad hints as to which species
are probably not present and so restrict the choice. For

instance, it is unlikely (but not impossible) that galling of tomato roots in West Africa would be caused by *M. naasi*. This leaves the perineal pattern as the prime means of identification of females, in spite of the serious disadvantages due to variability within species and the difficulty of describing the pattern in clear objective and unambiguous terms.

Larva Larvae can be of great help in identification and in one or two species are diagnostic at the present time, though the discovery of new species may change this. Body lengths (Fig. 1) of most species vary from mean values of about 320-543μm. *M. spartinae*, which is omitted from the graph, has unusual features including larval length 612-921μm that may eventually lead to its removal from the genus. Obviously, it is possible to place an unknown species within a group of known species and so to eliminate from consideration those species whose length is well outside the range of the unknown. The second most useful morphometric character is the tail length which varies from 17μm in *M. indica* (c = 25) and 18μm in *M. propora* (c = 22) to 71μ in *M. graminis* (c = 6) (Fig. 2). These extremes are easily observed at first glance but other species can be grouped according to relative tail length (e.g. c = <10, or c = 10-15) and the unidentified species can be compared with the species most resembling it (Fig. 3). Stylet length can sometimes be helpful in identification but the anterior end is difficult to determine accurately and comparisons of measurements by different workers could be misleading unless the suggestion made by Esser *et al.* for a standard measurement taken from the telorhabdion base to the anterior extremity of the head is adopted. Whitehead gives his own measurements of the larval stylets of 14 species; the mean for 25 larvae varies from 9.7μm in *M. hapla* to 14.8μm in *M. africana*. However, Whitehead observed

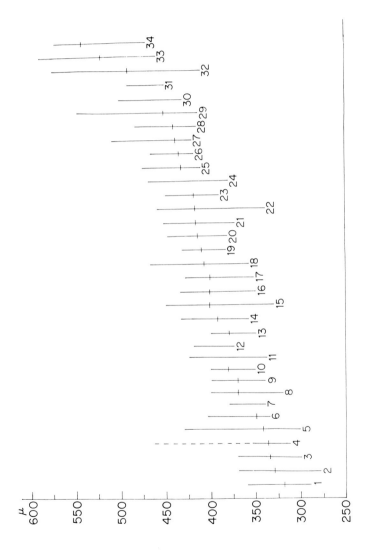

Fig. 1. *Meloidogyne* larvae. Body lengths, ranges and means (μm). Species numbers correspond to those in the list of species; *M. carolinensis* and *M. megatyla* are not included).

significant differences in stylet length as well as total
length between different populations of the same species.
This illustrates the importance of treating measurements
with great caution. More reliable are differences such as
the position of the hemizonid; this is usually in front
of the opening of the excretory pore but is behind it in

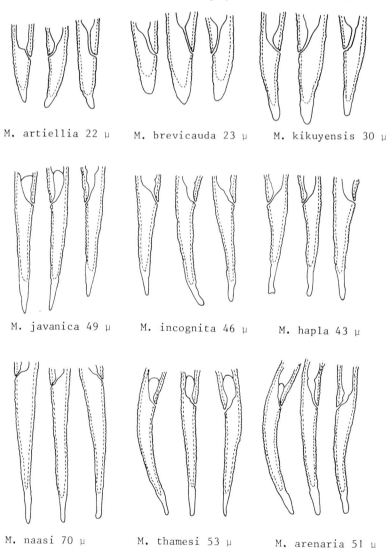

M. artiellia 22 μ   M. brevicauda 23 μ   M. kikuyensis 30 μ

M. javanica 49 μ   M. incognita 46 μ   M. hapla 43 μ

M. naasi 70 μ   M. thamesi 53 μ   M. arenaria 51 μ

Fig. 2. Tails of larvae of selected *Meloidogyne* species.

M. ardenensis, M..graminis, M. thamesi and *M. spartinae*
(Fig. 4). In a few species the rectum of the larva is
dilated (Fig. 4). This is probably a useful character but
may be hard to see in fixed specimens and needs further
investigation. Even more difficult to distinguish are the
annulations on the head, but Whitehead observed specific
differences both in head shape and in number of annules.
The use of modern techniques in microscopy might yield
valuable information on the surface markings of the head

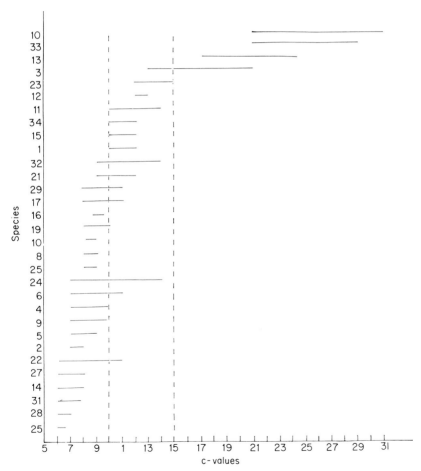

Fig. 3. *Meloidogyne* larvae. c-values. (Species numbers correspond to
those in the list of species).

as has been the case in other nematode genera, e.g. *Heterodera*.

Male The males are less useful for species determination than females and larvae; they are less numerous and not always easily associated with their females. They vary greatly in size due to conditions during development: measurements of total length are useless both because of variability and because of the small numbers of individuals available. One or two species have outstanding features by which they can be identified, e.g. *M. megadora* has unusually large stylet knobs, elongated

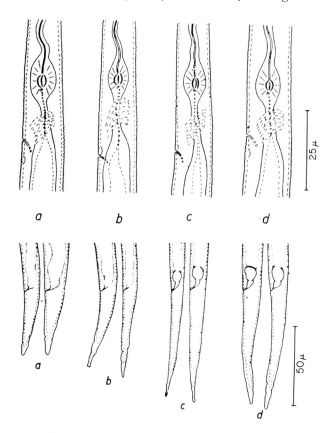

Fig. 4. Oesophagi and tails of larvae of a) *M. ardenensis*, b) *M. hapla*, c) *M. naasi*, d) *M. graminis*. Note swollen rectum in c and d and vesicles in oesophagus of c. (After Sturhan, 1976)

longitudinally: *M. decalineata* has ten longitudinal inci-
sures on the lateral field. The head shape and the number
of post-labial annules can sometimes be of use in identification but in some species, especially in *M. incognita*,
there is marked variation both within and between populations. Terentyeva (1967) uses head measurements of males
of 2 populations of *M. incognita* and one of *M. incognita*
acrita to show statistically significant differences
which she uses as a strong reason for rejecting the
synonymy of *M. i. acrita* with *M. incognita*. But it should
be pointed out that the numbers of individuals measured
are small (6 for *acrita*, 8 and 11 for *incognita*) and only
one population of *acrita* and two of *incognita* were measured. The work of Whitehead (1968), who examined males of
several populations of *M. incognita*, has shown that there
is often considerable between-population variability in
some morphometric characters. The occurrence of male
intersexes in a population suggests that it is *M. javanica*
since they are commoner in this than in other species.

Eggs The eggs of root-knot nematodes have been examined
critically in only a few species. Whitehead measured 20
fixed eggs of eight species and concluded that the differences were not sufficient to be of diagnostic value.
However, Lordello (1956a, 1956b) used egg size as one of
several characters to separate *M. javanica bauruensis*
from *M. javanica*, and *M. inornata* and *M. incognita*.

Other means of identification

Methods of identification examined include analyses of
proteins (Dickson *et al*., 1970, 1971) and chromosome counts
(Triantaphyllou and Hussey, 1973). So far, biochemical
analyses have been made in only a few species and,
although some interspecific differences have been demonstrated, many more species need to be examined before
this technique can be used in species identification.

Chromosome counts have been made for about 10 *Meloidogyne* species, partly with a view to using them for identification. In a field population on grapevine in California Hackney (1977) found 6 chromosomal populations representing four species. Counts were 43 and 47 for two populations of *M. javanica*, 42 for *M. incognita*, 37 and 52 for *M. arenaria* (2n and 3n form respectively) and 17 for *M. hapla* race A (with amphimixis and meiotic parthenogenesis). He suggests that chromosome number may be used in association with other morphological characters in identifying species. Species identification is of considerable help, though not a cast iron indication of use in formulating crop rotations to avoid infestations.

Differences in biology of the different species of root-knot nematodes that may be of use in identification include not only their host preferences but also the type of galling produced. It is well known that galls produced in response to feeding by *M. hapla* are small and often accompanied by an unusual number of lateral roots. *M. microtyla* also produces small galls while *M. naasi* causes spindle-shaped and club-shaped terminal galls on barley roots and *M. incognita* galls are often exceptionally large and confluent. Although host preferences can be good indicators of *Meloidogyne* species, the commoner species have wide and overlapping host ranges and some species comprise several pathotypes with different abilities to reproduce on the same host plants and with inconsistent behaviour at different temperatures. The use of host range tests for identification needs great care in monitoring and in the choice of test plant cultivars.

Conclusions

Since Chitwood's revision of the genus *Meloidogyne* great emphasis has quite rightly been placed on perineal patterns for specific identification but more recently morphological characters have been found in larvae that

are sometimes more easily recognized and less variable
than perineal patterns. Treated with caution, larval body
length together with tail length can provide good clues
to identification which, together with the form of the
rectum and any unusual characters such as the relative
position of the excretory pore and the hemizonid and the
structure of the median oesophageal bulb (e.g. in
M. naasi), can all give strong support to an identification that is often somewhat hesitantly based on perineal
patterns. It should be emphasized that when using morphological characters an adequate number of specimens must
be examined and that as many characters as possible
should be considered.

References

Atkinson, G.F. (1889). A preliminary report upon the life history
    and metamorphoses of a root-gall nematode, *Heterodera radicicola*
    (Greeff) Mull., and the injuries caused by it upon the roots of
    various plants. *Sci.Contr.Agric.Exp.Stn.Alabama Polyt.Inst.*, 1,
    177-226.
Berkeley, M.J. (1855). Vibrio forming cysts on the roots of cucumbers.
    Gdnrs' Chron., 220.
Chitwood, B.G. (1949). Root-knot nematodes. *Proc.helminth.Soc.Wash.*,
    16, 90-104.
Cobb, N.A. (1924). The amphids of *Caconema* (*nom. nov.*) and other
    nemas. *J.Parasit.*, 11, 118-120.
Cornu, M. (1879). Études sur le *Phylloxera vastatrix*. *Mém.Acad.Sci.*,
    Paris, 26, 163-175, 328, 339-341.
Dickson, D.W., Sasser, J.N. and Huisingh, D. (1970). Comparative disc-
    electrophoretic protein analyses of selected *Meloidogyne*, *Ditylen-
    chus*, *Heterodera*, and *Aphelenchus* spp. *J.Nematol.*, 2, 286-293.
Dickson, D.W., Huisingh, D. and Sasser, J.N. (1971). Dehydrogenases,
    acid and alkaline phosphatases, and esterases for chemotaxonomy of
    selected *Meloidogyne*, *Ditylenchus*, *Heterodera* and *Aphelenchus* spp.
    J.Nematol., 3, 1-16.
Esser, R.P., Perry, V.G. and Taylor, A.L. (1976). A diagnostic com-
    pendium of the genus *Meloidogyne* (Nematoda : Heteroderidae). *Proc.
    helminth.Soc.Wash.* 43, 138-150.
Gillard, A. (1961). Onderzoekingen omtrent de biologie, der verspreid-
    ing en de bestryding van wortel knobbelaaltjes (*Meloidogyne* spp.)
    Meded.LandbHoogesch.Gent., 26, 515-646.
Goeldi, E.A. (1887). Relatoria sôbre a molestia do cafeiro na prov-
    incia da Rio de Janeiro. *Archos Mus.nac.*, *Rio de J.* 8, 7-112 (1892).
Goodey, T. (1932). On the nomenclature of the root-gall nematodes.
    J. Helminth. 10, 21-28.
Goodey, T. (1933). *In* "Plant parasitic nematodes and the diseases

they cause." 306, Methuen, London.

Hackney, R.W. (1977). Identification of field populations of *Meloidogyne* spp. by chromosome number. *J.Nematol.*, 9, 248-249.

Jobert, C. (1978). Sur une maladie du cafeier observée au Bresil. *C.r.hebd.Séanc.Acad.Sci. Paris*, 87, 941-943.

Kofoid, C.A. and White, W.A. (1919). A new nematode infection in man. *J.Amer.med.Assoc.*, 72, 567-569.

Licopoli, G. (1875). Sopra alcuni tubercoli radicellari contenenti anguillole. *Rc.Accad.sci.fis.Napoli.*, 14, 41-42.

Lordello, L.G.E. (1956a). Nematóides que parasitam a soja na região de Bauru. *Brabantia*, 15, 55-64.

Lordello, L.G.E. (1956b). *Meloidogyne inornata* sp.n. a serious pest of soybean in the State of São Paulo, Brazil (Nematoda, Heteroderidae). *Revta bras.Biol.* 18, 375-379.

Müller, C. (1884). Mittheilungen über die unseren kulturpflanzen schädlichen, das Geschlecht *Heterodera* bildenden Wurmer. *Landw.Jbr.* 13, 1-42.

Neal, J.C. (1889). The root-knot disease of the peach, orange and other plants in Florida, due to the work of the *Anguillula*. *Bull. U.S.Bur.Ent.*, 20, 1-31.

Sturhan, D. (1976). Freilandvorkommen von *Meloidogyne*-Arten in der Bundes republik Deutschland. *NachrBl.dt.PflSchutzdienst (Braunschweig)*, 28, 113-117.

Terentyeva, T.G. (1967). *In* "Nematode diseases of crops" (Ed. N.M. Sveshnikova) 227, Moscow, Izdatelstvo "Kolos".

Trueb, M. (1885). Onderzoekingen over Sereh-Ziek Suikkeriet gedaan in s'Lands Plantentuin te Buitenzorg. *Medad.PlTuin.,Batavia*, 1885, 1-39.

Triantaphyllou, A.C. and Hussey, R.S. (1973). Modern approaches in study of relationships in the genus *Meloidogyne*. *Bull.OEPP*. 9, 61-66.

Whitehead, A.G. (1968). Taxonomy of *Meloidogyne* (Nematodea : Heteroderidae) with descriptions of four new species. *Trans.zool.Soc. Lond.*, 31, 263-401.

Discussion

In opening the discussion Sasser said that he agreed that *incognita* and *acrita* are one and the same species but he expressed concern about the status of *thamesi*. With populations established at Raleigh, North Carolina, the progeny of what appeared to be a typical *thamesi* female turned out to have perineal patterns of the *arenaria* type. He asked whether *thamesi* could be regarded as a good species or alternatively suggested that perhaps it does not exist in cultures at Raleigh. Mary Franklin replied that she had no experience with material other than that

used by Chitwood for his description of the species. Netscher remarked that variability is common among the progeny of a single female, and gave the example of *javanica*
where specimens exist with or without "wings". When separate cultures were made of both these forms and continued for four generations the population of each then
displayed the same variation as present in the original
population.

Because populations in the field may be variable and
indeed may consist of mixtures of different species Sikora asked whether it would be better to work with single
egg masses instead of field populations for experimental
studies. Mary Franklin replied that if it is the intent-
ion to describe a species the progeny of a single egg
mass is not sufficient since a species must represent
the whole population. Wouts remarked that whilst this
reply was applicable to description of a species, in
his opinion single egg masses are preferable for use in
experiments and if specimens are preserved for later
identification a check is always possible on the material
used for experiments. If a field population is used for
experiments the result may be invalidated when it later
proves to be a complex of species; he gave an example of
experiments with *Heterodera trifolii* which in many cases
had been carried out with mixed populations. Except for
the experiments of Norton (D.C. Norton, Phytopathology
57, 1305 (1967)), who used a single egg mass to start
with, it is now virtually impossible to know whether
many of the experiments with *H. trifolii* involved one or
more species. Netscher said that since many mixed populations of *Meloidogyne* exist in the field one should
first try to find out whether they represented one species or several and that this was particularly important
in breeding for resistance where tests should be made
with various field populations and not with a single

egg mass.

Questioned about the use of larval length as taxonomic character, Mary Franklin said that this varied a lot and that its use was limited except where species differed considerably (Fig. 1.). It was suggested that comparisons between species would be more effective if they were all grown on the same host but Coomans said that this would then make them valid only for those species on that particular host, which may be a good host for some species but not for others; and hence the latter would be produced under less favourable environmental circumstances and would not develop as well as on a good host, thus falsifying the comparison. It was generally agreed that not too much weight should be given to measurements and that any close comparisons based on such measurements are reliable only if carried out by the same person. It was also generally agreed that there should be uniformity in the methods used for preparing specimens for taxonomic purposes and it was also pleaded by Bird that attention should be paid to the physiological circumstances which may affect measurements, e.g. usually only one particular type of plant is used and this may give only a very partial picture of a very complex situation.

4 OBSERVATIONS ON THE VARIABILITY OF BIOMETRICAL CHARACTERS IN THE PERINEAL REGION OF *MELOIDOGYNE INCOGNITA*

S. Grimaldi De Zio, L. Padula, F. Lamberti,
M.R. De Lucia Morone and M. Gallo D'Addabbo

*Istituto di Zoologia ed Anatomia Comparata dell'Università
and Laboratorio di Nematologia Agraria, C.N.R., Bari, Italy*

The possibility of using some measurements or ratios to characterize quantitatively the perineal pattern of *Meloidogyne incognita* (Kofoid and White) Chitw. was studied by examining populations originating from Bulgaria, Italy, Rumania, U.S.A., and Venezuela, reared on different hosts. A total of 176 specimens was examined.

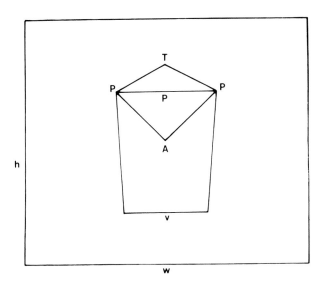

Fig. 1. Schematized perineal pattern of *Meloidogyne incognita*: h, maximum height; w, maximum width; v-A, shortest distance vulva-anus; v-p, shortest distance vulva-level of phasmids; v-T, shortest distance vulva-tail (A = anus; P = phasmids; T = tail; v = vulva).

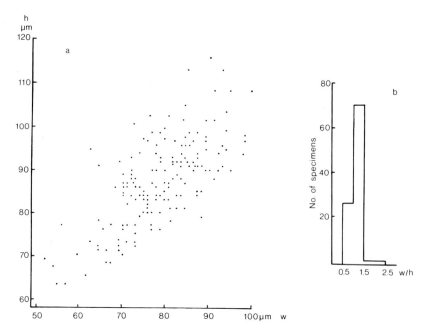

Fig. 2. a) relation between the maximum width (w) and maximum height (h) of the perineal region of *M. incognita*; b) frequency of the ratio w/h.

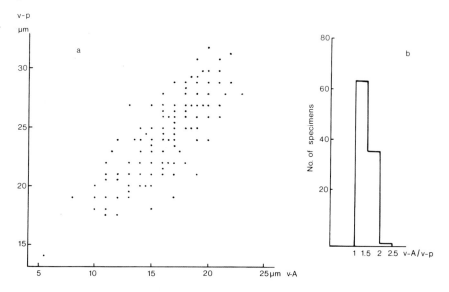

Fig. 3. a) relation between the shortest distance vulva-anus (v-A) and the shortest distance vulva-level of the phasmids (v-p) in the perineal region of *M. incognita*; b) frequency of the ratio v-A/v-p.

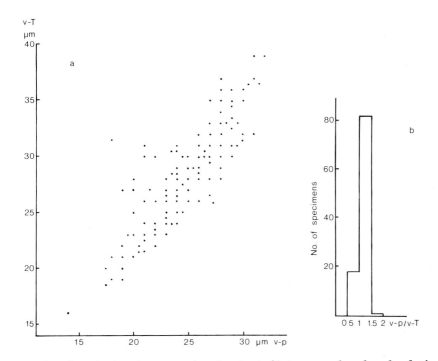

Fig. 4. a) relation between the shortest distance vulva-level of the phasmids (v-p) and the shortest distance vulva-tail (v-T) in the perineal region of *M. incognita*; b) frequency of the ratio v-p/v-T.

The perineal patterns was schematized (Fig. 1) by drawing imaginary lines to define dimensions and distances between organs.

The analysis of measurements of the defined parameters confirmed the wide variability of the perineal pattern. However, three ratios appeared to be consistently constant within and between populations:

i) maximum width (w)/maximum height (h), w/h, (Fig. 2) of the region as delimited by the outermost well defined striae;

ii) shortest distance vulva-anus (v-A)/shortest distance vulva-level of phasmids (v-p), v-A/v-p (Fig. 3);

iii) shortest distance vulva-level of phasmids (v-p)/shortest distance vulva-tail (v-T), v-p/v-T (Fig. 4).

If studies of other species of root-knot nematodes

confirm the stability of these ratios around values
typical for each species, such measurements could be useful in diagnostic work.

5 MORPHOLOGY AND ULTRASTRUCTURE

A.F. Bird

*CSIRO, Division of Horticultural Research,
Adelaide, South Australia*

Introduction

Differences in size and morphology between species or
populations of *Meloidogyne* and between the different
developmental stages and the sexes of the genus have been
described in these Proceedings by Mary Franklin and differences that occur in the morphology of these nematodes
during different stages of their life cycle by De Guiran
and Ritter. This chapter is confined to descriptions of
the morphology and ultrastructure that is characteristic
of the genus and no attempt is made to describe any differences that may exist between the species.

The first work on the fine structure of *Meloidogyne*
using the transmission electron microscope (TEM) was in
1965 (Bird and Rogers, 1965a; 1965b), so that information on
the ultrastructure of this genus is comparatively recent.
However, a number of excellent papers on this topic have
been published more recently (Baldwin and Hirschmann, 1973;
1975; 1976; Endo and Wergin, 1977; Wergin and Endo, 1976)
and some aspects of the fine structure of *Meloidogyne*
have also been described in books and reviews on nematode
ultrastructure (Bird, 1971; 1976; McLaren, 1976a; 1976b).

The four stages to be considered are the second stage
larva (L2), the male, the female, and the egg. The approximate dimensions of the various stages of *Meloidogyne*

to be described are as follows. The L_2 are about 400 µm long with a width of about 15 µm and have a dry weight of 21 ng. Males are about 1400 µm long by 30 µm wide and females approximately 650 µm long and about 400 µm wide at their greatest width. The eggs are about 90 µm long and about 40 µm wide.

Terminology

In the Nematoda there is some confusion about the nomenclature of certain anatomical structures which are known by a variety of synonyms. I propose to use terms which have become established in the literature and are used for all forms whether they be free living or parasites of animals or plants. These terms are commonly used in several books that have been published on the anatomy of nematodes (Bird, 1971; Chitwood and Chitwood, 1950; Crofton, 1966; De Coninck, 1965; Hyman, 1951). Some of the synonyms used here may become generally used in the years ahead, but it must be emphasised that unless very sound reasons are advanced for the use of a new term its introduction into the literature should be avoided as this only leads to greater confusion.

Cuticle

Most modern authors agree that the nematode cuticle appears to consist of, or to have evolved from, a three-layered or three-zoned structure (Bird, 1971; Dick and Wright, 1973; Johnson *et al.*, 1970; Lee and Atkinson, 1976) and *Meloidogyne* fits into this pattern. These cortical, median and basal layers represent the outermost, middle and inner layers of the cuticle. Synonyms recently used by De Grisse (1977) for these layers are epicuticle, exocuticle, mesocuticle and endocuticle. The term epicuticle is also used by Lee and Atkinson, 1976. It corresponds with the external cortical layer and is the triple layered surface structure common to all nematodes, being about 35 nm thick in the L_2 of *Meloidogyne*.

I prefer to use the term external cortical layer for this structure (Bird, 1976), although in the large ascarids the dimensions of the external cortical layer are similar to those of the insect epicuticle which is about 1 μm thick (Locke, 1966; Neville, 1975). The cuticulin layer of insect cuticle on the other hand has similar dimensions (about 20 nm) to most nematode external cortical layers (25-40 nm) and may be a more appropriate term for this structure (Bird, 1971) which is also known as the outer epicuticle (Neville, 1975). The point to remember is that the term epicuticle in current literature (De Grisse, 1977; Lee and Atkinson, 1976) is synonymous with the term external cortical layer (Bird, 1976).

It has recently been suggested (Johnson and Graham, 1976) that what I have referred to in *Meloidogyne* as the inner cortical layer should more properly be called the outer median layer and the cortex is equivalent to the epicuticle. I would accept this although there may be problems adapting it to cuticles of other species, particularly those parasitic in animals. For *Meloidogyne* it has the advantage of retaining in the adult female cuticle the basic three-zoned structure (common to all nematodes) of cortex, median and basal zones because the internal cortical layer that is retained in the adult female nematode is equivalent to the outer median layer. In describing the *Meloidogyne* cuticle (Fig. 1.) I shall include alternative nomenclatures.

On hatching, the preparasitic L_2 has a cuticle that is approximately 0.3 μm thick (Bird, 1968; Johnson and Graham, 1976). It consists of an outermost triple-layered external cortical layer or epicuticle [cortical layer (Johnson and Graham, 1966)] that is about 35 nm thick and which can range from this thickness to about 100 nm if the L_2 develops into a male and to over 100 nm if the L_2 develops into a female (Baldwin and Hirschmann, 1975; Bird, 1971; Bird and Rogers, 1965a).

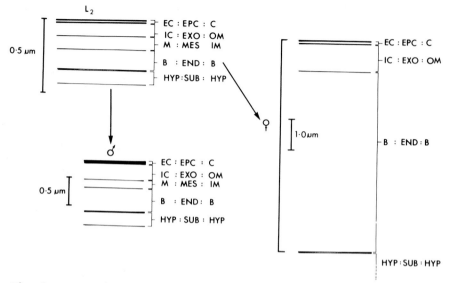

Fig. 1. Scale diagram showing the sizes of the cuticles and their layers or zones in the preparasitic second stage larva (L₂), male (♂) and female (♀) *Meloidogyne*. EC = external cortical layer, EPC = epicuticle, C = cortical layer, IC = internal cortical layer, EXO = exocuticle, OM = outer median layer, M = median layer, MES = meso-cuticle, IM = inner median layer, B = basal layer, END = endocuticle, HYP = hypodermis, SUB = subcuticle.

The surface of the external cortical layer is indented by transverse striations or annulations. These are 0.8-1.0 μm apart in the L₂ (Bird, 1971; Johnson and Graham, 1976), about 2.0 μm apart in the males (Baldwin and Hirschmann, 1975) and varying distances apart in the female, the variations being caused by swelling.

The internal cortical layer (exocuticle, outer median layer), which appears fibrillar in structure in transmission electron microscopy ranges from 50-100 nm in thickness in the L₂, depending on whether measurements are taken between the body annulations or under them (Johnson and Graham, 1976). It becomes about 300 nm thick in males and 500 nm thick, or more, in females (Bird, 1971)

The median layer (mesocuticle, inner median layer) is thought to be fluid filled in the living nematode and contains globular electron-dense bodies in the L₂ and

male. In the L_2 it ranges in thickness from 100 nm thick
between cuticular annulations and apparently is absent
beneath these annulations. In males it appears to be
about 200 nm thick and to exist beneath the annulations
as well as between them, particularly in the anterior
region of the nematode (Baldwin and Hirschmann, 1975).
In the females, of course, as mentioned above, this
layer ceases to exist.

The basal layer of the cuticle (endocuticle) is
characterized by parallel striations which are about 20
nm apart in transverse sections and 18 nm apart in longitudinal sections. In the L_2 this zone is about 125 nm
thick and in males it is 300-400 nm thick. These striations do not occur under the lateral fields in either
L_2s or males. They fork at the edge of the lateral field
and are replaced by layers of obliquely orientated fibres.
It is interesting to note that several days after entry
of the larva into its host, the striations of the basal
layer atrophy and are not found again in the course of
development unless the L_2 develops into a male.

The regular spacings of the basal layer indicate that
there are close linkages between the molecules. It appears to be a natural form of fibrous collagen, being a
tough almost crystalline protein structure, and probably
acts as the main skeletal layer in the nematode's cuticle.
It is absent in the sedentary and relatively protected
female where it becomes fibrous and much thicker, being
6 μm or more in the posterior region.

The cuticle is invaginated at the mouth, rectum,
cloaca, vagina and excretory pore and at the paired
openings of the amphids anteriorly and the phasmids
posteriorly.

Hypodermis

The hypodermis, or subcuticle, is a plasma membrane-
bound layer that exists between the cuticle and somatic

muscles. It is about 100 nm thick in the L_2s, 10 nm thick
in the male and of variable thickness in the female. It
projects into the pseudocoelomic cavity in the lateral,
ventral and dorsal regions to form four hypodermal chords.
The interchordal hypodermis shows little variation in
thickness throughout the individual (Baldwin and Hirschmann, 1975).

Anteriorly and posteriorly the somatic hypodermis connects respectively with the stomatal and rectal hypodermis. Thus radii of hypodermal tissue extend between
the stylet protractor muscles and connect the somatic
hypodermis with the hypodermis surrounding the stoma
(Baldwin and Hirschmann, 1975).

The hypodermis in *Meloidogyne* is syncytial with most
of the nuclei occurring in the lateral chords and, to a
lesser extent, in the ventral chord and the dorsal chord
in the anterior region of the nematode. The hypodermis
also contains nerves, nerve cells, glands and the duct
of the excretory gland. Hypodermal chords usually contain large numbers of various organelles and inclusions
such as large lipid globules, mitochondria, rough endo-
phasmic reticulum (RER), ribosomes, β-glycogen and Golgi
bodies.

Interchordal hypodermis is usually free of organelles
but may contain peripheral nerve processes and always
contains hemidesmosomes in the preparasitic L_2s and males.
The hemidesmosomes are modifications of the hypodermal
plasma membrane and act as points of attachment of cuticle and somatic muscle.

Musculature

The muscle cells of nematodes are spindle-shaped and
consist of a contractile portion, made up of obliquely
striated muscle fibres with filaments of myosin and
actin, and a noncontractile portion that contains the
mucleus, many mitochondria, glycogen particles and

lipid globules.

The musculature of *Meloidogyne* can be considered as somatic or specialized. Both have the same origin but specialized muscles have a specific purpose and are limited to a particular part of the body.

Somatic muscle

The somatic muscle of *Meloidogyne* is classified according to its shape and arrangements as platymyarian-meromyarian, which means that the striated portion lies next to the hypodermis and that there are two to five rows of muscle cells between the chords. The number of somatic muscle cells can vary in different parts of the body. Thus in *Meloidogyne* preparastiic L_2s there are eight in cross section in a small part of the anterior region and four in cross section further posteriorly and throughout most of the length of the nematode (Fig. 2.).

In males the number of muscle cells between the chords varies from two in the cephalic region to four or five throughout most of the oesophageal region (Baldwin and Hirschmann, 1975). The thick filaments (myosin) are 22-24 nm diameter and thin filaments (actin) 6 nm in diameter (Baldwin and Hirschmann, 1975; Bird, 1971). During contraction it is thought that the actin filaments slide past the myosin filaments and the energy necessary for this process is thought to be derived from high energy phosphate bonds split by the myosin.

In cross section the bands of muscle filaments in *Meloidogyne* are seen to follow the same repeated sequence. A band (I) with only fine filaments, a band (A) containing thick filaments surrounded by a hexagonal pattern of 10 - 15 fine filaments, a narrow band (H) with thick filaments only and then bands A and I again. Approximately five to six such cycles occur in the broadest portion of a muscle. The tips of the muscle cells contain only fine filaments.

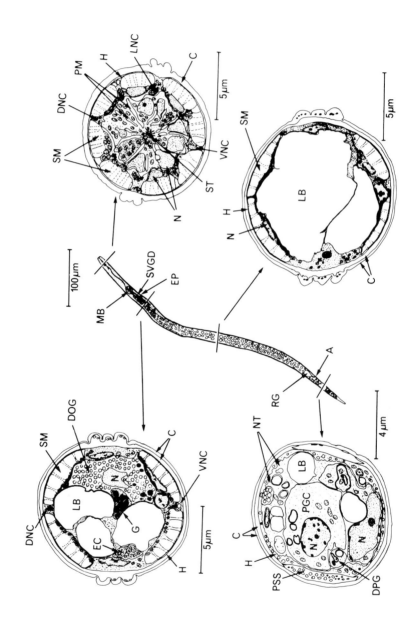

The number of somatic muscle cells in *Meloidogyne* has not been counted but nematodes with a meromyarian type of musculature generally have very few cells, probably 50-100. Sixty-five have been counted in *Oxyuris equi*, 87 in *Strongylus* sp. (Bird, 1971) and 95 in *Caenorhabditis elegans* (White *et al.*, 1976). These muscles are attached to the hypodermis through hemidesmosomes in the region of the I bands. The hemidesmosomes bind the sarcolemma of the muscle to the basal lamina and this in turn is connected to hemidesmosomes under the cuticle by fine fibrils that extend across the hypodermis.

As mentioned earlier, the adult female *Meloidogyne* loses much of its somatic muscle. This atrophy can be detected within a day of the L2 penetrating a root by comparing the times taken for L2s dissected from roots to move through a column of sand with those of preparasitic L2s of similar age (Bird, 1967).

Specialized muscle

The cephalo-oesophageal muscles in plant parasitic nematodes have become modified into stylet protractor muscles. In the males of *Meloidogyne* there are three stylet protractor muscle cells (Baldwin and Hirschmann, 1976). The contractile portion of each of these muscles is attached posteriorly to a stylet knob. Anteriorly the protractors branch into ten protractor elements which attach to the basal lamina surrounding the body wall and

---

Fig. 2. (opposite) Scale diagram of larva of *Meloidogyne* as seen with the light microscope with sections cut through four different regions as seen with the electron microscope. DNC = dorsal nerve chord, SM = somatic muscle, DOG = dorsal oesophageal gland, C = cuticle, H = hypodermis, VNC = ventral nerve chord, MB = median bulb, SVGD = subventral oesophageal gland duct, EP = excretory pore, A = anus, RG = rectal gland, NT = nerve tissue, PSS = phasmidial sensory structures, N = nucleus, DPG = duct of phasmidial gland, LB = lipid body, PGC = phasmidial gland cell, LNC = lateral nerve chord, PM = stylet protractor muscles, ST = stylet, EC = excretory canal, DNC = dorsal nerve chord, G = granules (from DOG)

vestibule extension near the cephalic framework. In the
L2 the protractors appear to branch into twelve protractor elements that have similar attachments (Fig. 2).

The non-contractile portion of each muscle cell is
located in the oesophagus and contains numerous mitochondria, Golgi bodies, β-glycogen and a large nucleus.

Muscles associated with the digestive and reproductive
systems are described later. The only other set of
specialized muscles that have been examined in *Meloidogyne*
are the anal muscles. The most commonly described of
these is the H-shaped *depressor ani*, which I have observed in transverse sections of the preparasitic L2. It
consists of two vertical groups of fibres connecting the
dorsal surface of the rectum or cloaca to the dorsal
body wall. These fibres are connected by a horizontal
band of sacroplasm which contains the nucleus. In the
female of *Meloidogyne*, in which there does not appear
to be any connection between the digestive tract and the
anus, it functions as a pump to extrude the gelatinous
matrix from the rectal glands. During this process the
depressor ani muscle contracted rhythmically every 10
seconds over a period of 4 hr during which it was being
observed (Maggenti and Allen, 1960).

Nervous System

At the moment there is no precise information on the
numbers of nerve cells in the various stages of *Meloidogyne*, but it is roughly estimated that there may be
around three hundred in the preparasitic L2. This number
has been estimated to occur in the small free-living
nematode *Caenorhabditis elegans* (Ward *et al.*, 1975). The
nuclei of these nerve cells occur in clusters in the
oesophagus on either side of the nerve ring and as a cluster in the tail area. The nerve ring itself, at least in
the L2 stage, does not appear to contain nuclei (Bird,
1971; Bird and Saurer, 1967). The nerve ring is thought

to be made up of nerve processes and to be an important co-ordinating centre (Kanagasuntheram *et al.*, 1974). It is thought that it is here that impulses coming from sensory cells are passed through synapses to motor nerves and at least two different synapses have been described in this region (Kanagasuntheram *et al.*, 1974).

In addition to the circumpharyngeal or circumoesophageal nerve ring there is a posterior nerve ring or rectal commissure which connects the ventral nerve and the dorsal rectal ganglion. There are also connections in this region between the lumber ganglia and the dorsal ventral nerves. These ganglia and the posterior nerve ring have not, so far as I am aware, been described in L_2s or males of *Meloidogyne*.

A number of longitudinal nerves run backwards from the circumoesophageal nerve ring. Usually there are at least eight, consisting of a ventral, a dorsal, two laterals, two subventrals and two subdorsals. The first four are easily recognized in transverse sections of L_2s (Fig. 2) and these and the subdorsals and subventrals have been identified and illustrated in transverse sections of males of *Meloidogyne* (Baldwin and Hirschmann, 1975).

As these nerves run posteriorly they are all connected by transverse commissures, but these are difficult to detect in sections. The large longitudinal nerves run obliquely outwards from the oesophagus and become embedded in the hypodermal chords. It is thought (Crabb and Matthews, 1977) that the spontaneous contractions of the somatic muscles are co-ordinated and modulated by these ganglionated dorsal and ventral nerves. The ventral nerve in particular, because of its size, is thought to play a major role in the co-ordination of movement in nematodes. It seems that the dorsal and ventral nerves may be motor in activity whereas the lateral nerves may be sensory.

A great deal of work has been done recently on nematode cephalic sense organs (Ware *et al.*, 1975) including

some excellent studies specifically on these structures
in *Meloidogyne* (Baldwin and Hirschmann, 1973; 1976; Endo
and Wergin, 1977; Wergin and Endo, 1976), two recent reviews (McLaren, 1976a; 1976b) and a comprehensive study
of the nerves in the cephalic regions of some nineteen
genera of plant parasitic nematodes (De Grisse, 1977).

Six nerve bundles run forward from the circumoesophageal nerve ring. These are two subventrals, two subdorsals and two lateral nerves, the latter containing lateral papillary and amphidial dendrites (Baldwin and Hirschmann, 1975; Ware *et al*., 1975). The cell bodies of all
these neurons lie near the nerve ring and their axons
either project into the ring or the ventral ganglia.

Most tylenchid nematodes have cephalic sense organs
that fit the *en face* plan first put forward by De Coninck (1942). This comprises sixteen sensory structures
and a pair of amphids. The sensory structures are arranged in three circles and transverse sections cut

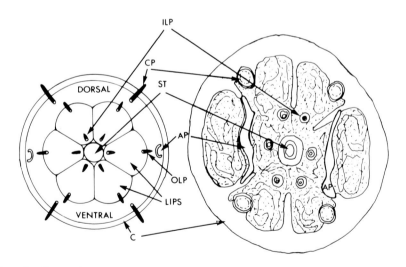

Fig. 3. Comparison of sensory structures in the head region of L2 of
Meloidogyne [after Wergin and Endo, 1976] with *en face* view of a
typical nematode [after De Coninck, 1965]. ILP = inner labial papillae, CP = cephalic papillae, ST = stylet, AP = amphidial pits,
OLP = outer labial papillae, C = cuticle.

through the head should show the innervations of these
structures in concentric circles: a circle of six inner
labial nerves, a circle of six outer labial nerves, a
circle of four cephalic nerves, all of which lead to papillae or setae. The laterally placed amphids (Fig. 3a)
are also visible in transverse sections.

The anatomical pattern for cephalic receptors in various tylenchid nematodes is strikingly similar (De Grisse,
1975). The six inner labial nerves lead to one or two
dendritic processes. The six outer labial nerves bear
only one dendritic process and may be reduced and more
difficult to see in some transverse sections. The four
cephalic nerves lead to one or two dendritic processes,
the amphidial nerve to seven and the amphidial gland
bears three or four dendritic processes, so that about
fifty nerve receptors may be involved in the cephalic
region of an average tylenchid (De Grisse, 1975).

In the L$_2$ and male of *Meloidogyne* (Baldwin and Hirschmann, 1973; Endo and Wergin, 1977) the six outer labial
receptors that have been reported for other tylenchids,
albeit somewhat reduced (De Grisse, 1975) have not been
detected. Otherwise, both males and L$_2$s are similar (Fig.
3b) and have six inner labial organs (each consisting of
two modified cilia) that can be seen in sections as pores
surrounding the stoma, four cephalic receptors innervated
by a single axon or dendrite that terminates beneath the
cuticle and seven cilia in each amphid, together with
three to five "accessory" cilia that remain external to
the amphidial canal and terminate beneath the cuticle in
the cephalic region. Thus, approximately forty neurons
in the head region of *Meloidogyne* have been observed so
far.

In *Caenorhabditis elegans* it has been shown (Ward *et
al.*, 1975) that one of the ciliated sensory neurons in
each of the six inner labial sense organs makes direct
chemical synapses on to a muscle so that these are

sensory-motor neurons.

The amphidial gland is a large irregular cell that en-
circles the cilia and nerve processes. It contains numerous mitochondria but no nucleus or other cell organelles
have been observed.

It seems possible that those structures that open to
the outer surface of the nematode through cuticular pores
such as amphids and labial organs may have a chemorecep-
tive or secretory function whereas those that lack a cuticular opening, such as the cephalic and accessory receptors, may be tactoreceptive.

The only other sensory structures so far described in
Meloidogyne, other than specialized sexual structures
such as spicules, are a number of cuticular sensory
structures in which nerve endings are found within the
cuticle (Bird, 1971). Perhaps the most pronounced of
these are associated with the phasmids in the tail region
(Fig. 2) where in L_2s the fibrous basal layer adjacent
to the lateral line region becomes permeated with nerve
endings and neuron tissue. The phasmids open to the exterior through a pair of ventro-lateral ducts (Fig. 2),
into each of which a single cilium apparently projects.

The Pseudocoelom

It has been pointed out by Maggenti (1976) that this
term is "only a designation for a morphologically unde-
fined body cavity". This is because there is no general
agreement regarding its embryonic origin. Without considering the various arguments for or against its origin
the nematode pseudocoelom is usually depicted as a cavity
whose outermost walls, consisting of muscle and hypodermal chords, are of mesodermal origin and its inner walls,
consisting of gut and reproductive system, are of endo-
dermal origin. This is in contrast to the true coelom
which is lined by membranes of mesodermal origin. It can
be stated simply that the body cavity of nematodes is

lined by a membrane and that this is supported by electron microscopy (Zacheo *et al.*, 1975). I shall refer to
this cavity as a pseudocoelom only because it has been
called this in the past. In *Meloidogyne* this cavity is
much reduced, particularly in the posterior region. Information on its fine structure, its continuity throughout the organism or on the presence of coelomocytes has
not, as far as I know, been recorded.

Excretory System

As with pseudocoelom, the term excretory system could
well be a misnomer and future experiments may show that
it has a secretory role that is more important than any
excretory one. However, until results are published which
show conclusively that these structures have no excretory
function in a wide range of nematode families, and this
seems unlikely, the term excretory system should be retained because of its familiar usage.

The excretory pore is located near the basal region
of the median bulb in the L_2 of *M. javanica*, although
there is some variation between different species. In
female nematodes of this species the excretory pore is
located about 25 μm in front of the anterior portion of
the median bulb (Bird, 1959), so that its position varies
considerably within the same species at different stages
of development. The excretory system in female members
of the genus is seen most clearly in living specimens
just after the completion of the final moult.

The terminal duct that opens through a pore on the
ventral surface is the one structure common to all nematode excretory systems. This duct is thought to arise
as an invagination of the hypodermis which meets and
fuses with the excretory sinus (Bird, 1971). The cuticular part of the duct is shed during moulting along with
other cuticular invaginations. The excretory duct of the
L_2 is smaller than that of the female, having a diameter

of 0.3 μm and a lumen of little less than 0.2 μm compared
with a diameter of 1.0 μm and a lumen of about 0.6 μm.
The duct is surrounded by an array of vesicles which project from its surface and whose membranes appear to connect with those on the surface of the duct (Bird, 1971).
In living specimens of female *Meloidogyne* the excretory
duct can be traced back from the pore for a distance of
about 70 μm during which it runs back across the median
bulb to the surface of the oesophageal gland and then
across to the hypodermis as it moves from the neck region
into the main body of the nematode.

In view of the virtual atrophy of the intestine in
this nematode, with subsequent loss of connection between
mouth and anus, it would seem that the excretory system
may have an important role in the elimination of unwanted
metabolites. It has been shown that antigenic substances,
which are presumed to have a relatively high molecular
weight, are liberated from the excretory pores of both
L_2s and females of *Meloidogyne* (Bird, 1964), and it is
not uncommon to find fixed material in sections of ducts
viewed with the TEM.

Digestive System

The digestive system in most nematodes is simply a
tube into which a number of glands open. It commences at
the mouth as a cuticular invagination known as the stomodeum and terminates at the anus in another cuticular invagination known as the proctodeum. The cuticular linings
of the stomodeum and proctodeum are shed during moulting.

In the females of *Meloidogyne* there is no direct connection between the mouth and anus and the intestine as
such becomes vestigial. It retains a somewhat cellular
structure with cell borders visible in the TEM (Dropkin
and Acedo, 1974) but the characteristic intestinal lumen
with microvilli is missing, or very difficult to resolve,
and the whole structure is more of a storage organ than

a passageway where food is broken down and absorbed, or transported as a waste to the anus.

The intestinal epithelial cells in the L_2 of *Meloidogyne* are not easy to resolve and have not been delineated clearly in any published work. However, they have been illustrated in the closely related L_2s of *Globodera* (*Heterodera*) *rostochiensis* (Wisse and Daems, 1968) where it has been shown that in cross section there are two cells visible. Each contain large lipid droplets, glycogen particles, a few mitochondria and sometimes a nucleus. The lumen enclosed by the cells is small with few microvilli. The cell membranes of the two intestinal epithelial cells are joined to each other near the lumen by desmosome-like junctions. A similar pattern seems to occur for the L_2 of *Meloidogyne* but these cells are harder to resolve and the lipid globules, particularly in the posterior region, tend to displace evidence for an intestinal lumen (Fig. 2).

Recently, Baldwin and Hirschmann (1976) have described the fine structure of the stomatal region of males of *M. incognita* in detail. The overall mean length of the stomatostylet is 23 μm, the diameter of the lumen in the knob and shaft region being about 0.3 μm whereas in the area where the cone narrows, the lumen changes in position and shape from central and circular to ventral and oblong (0.14 x 0.23 μm). The lumen opens on the ventral side so that the outermost part of the stylet is solid. The conical part of the stylet is located in the stomatal cavity which extends to the base of the stylet cone. The vestibule extension (Fig. 4) is a three-layered structure which extends from the vestibule of the cephalic framework to the posterior end of the cone; its innermost layer is continuous with the external cortical layer of the cuticle. The basal ring of the cephalic framework is continuous with the basal layer of the cuticle. There are six cephalic framework blades (Fig. 4) - one dorsal

A.F. BIRD

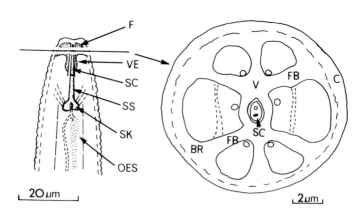

Fig. 4. Head region of male *Meloidogyne* based on the work of Baldwin and Hirschmann (1973). F = framework, VE = vestibule extension, ST = stylet cone, SS = stylet shaft, SK = stylet knob, OES = oesophagus, FB = framework blades, BR = basal ring.

two subdorsal, one ventral and two sub-ventral. Because of the size of the lateral sections this framework is bilaterally rather than radially symmetrical. When the stylet is retracted the stomatal opening becomes a slit that is a little over 1 μm long.

The pharynx or oesophagus of nematodes is morphologically and functionally very diverse. In *Meloidogyne* the oesophagus is made up a procorpus which runs from the base of the stylet to the median bulb which is known as the metacorpus. The lumen of the procorpus is cylindrical with a cuticular lining. This duct, sometimes known as the salivary duct, runs posteriorly towards the median bulb where its lumen changes from cylindrical in the anterior part of the bulb to triradiate. The muscle cells of the metacorpus in *Meloidogyne* are well supplied with mitochondria. The myofilaments of the muscles fan out radially to the periphery of the bulb where they are attached to the basement membrane.

Nine single celled glands are known to discharge their
products into the digestive tract of the female *Meloidogyne*. All are highly developed and are related to the
parasitic feeding habit and reproductive capacity of the
nematode. The first three, the dorsal and two subventral
oesophageal glands, have already been considered in some
detail in the paper on histopathology and physiology of
syncytia (see Bird in these Proceedings). The remaining
six are the rectal gland cells (RGCs) which are responsible for the production of the relatively huge amount
of gelatinous matrix that is exuded around the eggs. The
RGCs are thought to be developed from the rectal gland
or dilated rectum that can be seen in the preparasitic
L_2, at least of some species, particularly if they are
treated with Periodic Acid/Schiff reagents for carbohydrate (Bird and Saurer, 1967). The function and ultrastructure of the rectal gland in the L_2 has not been
described. Transverse sections cut through this region
of the L_2 and viewed by TEM show that it is filled with
a fine particulate material that is probably exuded on
contraction of the *depressor ani* muscle. The term rectal
gland is more appropriate here because it is not just a
cuticular dilation and it contains a chemically and morphologically distinct substance.

During the early stage of development the RGCs are
inconspicuous but they grow rapidly after moulting has
been completed and their nuclei and nucleoli increase
in size to relatively giant proportions.

Studies on the histochemistry and ultrastructure of
the RGCs and of the gelatinous matrix that they produce
(Bird and Rogers, 1965b; Bird and Soeffky, 1972) have
shown that the matrix is an irregular meshwork, containing protein, carbohydrate and certain enzymes, which is
transformed into a uniform granular mass of much greater
density when dehydrated. The RGCs have large nuclei and
a dense cytoplasm containing many Golgi bodies and sinus

canals which carry the matrix to the exterior where it is
secreted through the anus. Secretion precedes and accom-
panies egg laying so that the eggs become surrounded and
encased in a gelatinous mass.

Reproductive System

The reproductive system is essentially similar in both
sexes in that it consists of one or two tubular gonads;
one or two of them in males and two in the female. The
male reproductive system is unusual in this respect
because in the Tylenchida males normally only have a
single testis. Males with two testes in *Meloidogyne* are
thought to have developed from female larvae following
sex reversal induced by adverse environmental conditions
(Hirschmann, 1971). Thus the gonads in *Meloidogyne* may
be either monorchic or diorchic and these are directed
anteriad and reflexed.

Sperm can be seen in the vas deferens, particularly
if either phase contrast or Nomarski interference cont-
trast optical systems are used. They are approximately
5 μm in diameter and irregular in outline, being amoeboid
and aflagellate. *Meloidogyne* sperm resemble those described from some cyst nematodes (Shepherd *et al.*, 1973)
in possessing a layer of cortical microtubules which
cover the entire surface of the sperm. The surface membrane of the sperm in *Meloidogyne* is a triple-layered
structure with a total thickness of about 25 nm. The
microtubules which emerge from it have a diameter of
60-70 nm. No discrete nuclei have been observed, although
irregular masses of what appears to be chromatin can be
seen in the cytoplasm.

The microtubules which, at the moment, appear to be
peculiar to the Heteroderidae are in fact structures
commonly found surrounding spermatozoa from a very wide
range of animals and even some plant spermatozoids
(Reger and Fain-Maurel, 1973). Their function is unknown

but it has been suggested that in *Heterodera* their function may be skeletal rather than contractile (Shepherd
et al., 1973).

In nematodes, sperm are transferred to females with
the aid of copulatory spicules. These are typically
paired cuticularized structures that are lodged in invaginations of the cloaca. They show considerable diversity
of size and shape in different species and this has
proved to be useful for taxonomic purposes. Normally
spicules consist of three parts - head, shaft and blade.
Although there is no published information on the fine
structure of spicules in *Meloidogyne* these organs have
been studied in eleven species of *Heterodera* and *Globodera* (Clark *et al.*, 1973). In these nematodes each spicule has an inner swollen cylindrical shaft extending
for about a quarter of its length, and then a flattened
blade with incurved edges. The gubernaculum, an unpaired
structure about a third of the length of a spicule, is
also present dorsal to the spicules. The cuticular blades
of these genera have incurved interlocking wings which
form an enclosed tubular structure through which sperm
can be transferred when the spicules are inserted into
the female reproductive tract. Specific differences were
observed only in the shape of the tip. In general appearance and size, the spicules of *Meloidogyne* resemble
those of *Heterodera*, generally ranging from 25-30 μm in
length (Whitehead, 1968).

Spicules are thought to have an important sensory
function because in a wide range of different species
examined all spicules contain one to several nerve axons
running throughout their length. In *Globodera rostochiensis* (Clark *et al.*, 1973) the spicular nerve separates
into a pair of dendritic elements which communicate with
the exterior through distinct pores at the tip of each
spicule.

Reproduction in *M. javanica* is parthenogenetic and

males occur in significant numbers only during periods of environmental stress. It seems reasonable to assume that fertilized eggs may be more resistant than unfertilized ones, but this is very difficult to prove.

Sperm has been found in the uterus of M. *javanica* near its junction with the spermatotheca and sections through these sperm demonstrated a very close association with the uterine walls (McClure and Bird, 1976).

The following comments on the female reproductive tract of *Meloidogyne* and the development of the egg within it are based on the works of three authors (Bird and McClure, 1976; McClure and Bird, 1976; Triantaphyllou, 1962). At the tip of the ovary the developing eggs or oogonia are radially arranged around a central rachis to which they are attached by cytoplasmic bridges. This is the germinal zone of the ovary and is followed by a longer part of the ovary known as the growth zone. As the oocytes mature the rachis disappears and they pass into the oviduct in tandem. Some distance along the oviduct the oocyte, which is packed with lipid globules, refringent protein bodies and glycogen, enters a structure known as the spermatotheca-oviduct valve (SOV). The spermatotheca follows immediately after this and the rest of the reproductive tract as far as the vagina is known as the uterus. The SOV consists of 2 rows of four staggered cells containing large irregular nuclei and a cytoplasm rich in various organelles, highly suggestive of great metabolic activity. The lumen in the valve is narrow and is stretched by the oocytes which assume the ovoid shape that is retained in the mature egg.

Egg shell formation begins in the spermatotheca where the oolemma is modified to form the vitelline layer. The egg shell in *Meloidogyne*, and in most other nematodes, consists of an outermost vitelline layer, a chitinous layer and a glycolipid layer (Bird and McClure, 1976; Tarr and Fairbairn, 1973). In *Meloidogyne* the vitelline

layer is about 30 nm thick, the chitin layer about 400
nm, and the glycolipid layer is variable in thickness,
being much thicker at the poles. This layer also contains
membranes.

The cells lining the proximal portion of the uterus
contain large intracytoplasmic spaces bordered by endoplasmic reticulum whereas the cells lining the distal
part of the uterus have dense cytoplasm with large areas
of compact endoplasmic reticulum. The chitin layer begins
to be laid down in the portion of the uterus closest to
the spermatotheca and the glycolipid layer is formed in
the mid region of the uterus. After this egg shell formation is complete and its structure and general imperm-
eability resembles that of eggs that have been laid. The
glycolipid layer is responsible for the remarkable chemical resistance of *Meloidogyne* eggs.

References

Baldwin, J.G. and Hirschmann, H. (1973). Fine structure of cephalic
    sense organs in *Meloidogyne incognita* males. *J.Nematol.*, 5, 285-
    302.
Baldwin, J.G. and Hirschmann, H. (1975). Body wall fine structure of
    the anterior region of *Meloidogyne incognita* and *Heterodera gly-
    cines* males. *J.Nematol.*, 7, 175-193.
Baldwin, J.G. and Hirschmann, H. (1976). Comparative fine structure
    of the stomatal region of males of *Meloidogyne incognita* and
    Heterodera glycines. *J.Nematol.*, 8, 1-17.
Bird, A.F. (1959). Development of the root-knot nematodes *Meloidogyne
    javanica* (Treub) and *Meloidogyne hapla* Chitwood in the tomato.
    Nematologica 4, 31-42.
Bird, A.F. (1964). Serological studies on the plant parasitic nema-
    tode, *Meloidogyne javanica*. *Explt.Parasitol.* 15, 350-360
Bird, A.F. (1967). Changes associated with parasitism in nematodes.
    J.Parasitol., 53, 768-776.
Bird, A.F. (1968). Changes associated with parasitism in nematodes.
    J.Parasitol., 54, 475-489.
Bird, A.F. (1971). The Structure of Nematodes. 318, Academic
    Press, New York.
Bird, A.F. (1976). *In* "The Organization of Nematodes." (Ed. N.A.
    Croll), 107-137, Academic Press, New York.
Bird, A.F. and McClure, M.A. (1976). The tylenchid (Nematoda) egg
    shell : structure, composition and permeability. *Parasitology*, 72,
    19-28.
Bird, A.F. and Rogers, G.E. (1965a). Ultrastructure of the cuticle

and its formation in *Meloidogyne javanica*. *Nematologica* 11, 224-230.

Bird, A.F. and Rogers, G.E. (1965b). Ultrastructural and histochemical studies of the cells producing the gelatinous matrix in *Meloidogyne*. *Nematologica* 11, 231-238.

Bird, A.F. and Saurer, W. (1967). Changes associated with parasitism in nematodes. *J.Parasitol.*, 53, 1262-1269.

Bird, A.F. and Soeffky, A. (1972). Changes in the ultrastructure of the gelatinous matrix of *Meloidogyne javanica* during dehydration. *J.Nematol.*, 4, 166-169.

Chitwood, B.G. and Chitwood, M.B. (Eds.) (1950). *In* "An introduction to Nematology." 213, Monumental Printing Co., Baltimore, Maryland.

Clark, S.A., Shepherd, A.M. and Kempton, A. (1973). Spicule structure in some *Heterodera* spp. *Nematologica* 19, 242-247.

Crofton, H.D. (1966). *In* "Nematodes." 160, Hutchinson, London.

Croll, N.A. and Matthews, B.E. (1977). *In* "Biology of Nematodes." 201, Blackie, Glasgow.

De Coninck, L. (1942). De symmetrie-verhoudingen aan het vooreinde der (vrijlevende) Nematoden. *Naturw.Tijdschr.*, 24, 29-68.

De Coninck, L. (1965). *In* "Traite de Zoologie : Nemathelminthes." (Ed. P. Grassé), 4, 3-600, Masson, Paris.

De Grisse, A.T. (1975). The ultrastructure of some ciliary receptors in the cephalic region of tylenchids (Nematoda). *Med.Fac.Landbouw. Rijks.Univ.Gent.*, 40, 473-487.

De Grisse, A.T. (1977). *In* "De ultrastruktur van het zenuwstelsel in de kop van 22 soorten plantemparasitaire nematoden, behorende tot 19 genera (Nematoda : Tylenchida). Rijksuniversiteit Gent.

Dick, T.A. and Wright, K.A. (1973). The ultrastructure of the cuticle of the nematode *Syphacia obvelata*. *Can.J.Zool.*, 51, 187-196.

Dropkin, V.H. and Acedo, J. (1974). An electron microscope study of glycogen and lipid in female *Meloidogyne incognita* (root-knot nematode). *J.Parasitol.*, 60, 1013-1021.

Endo, B.Y. and Wergin, W.P. (1977). Ultrastructure of anterior sensory organs of the root-knot nematode, *Meloidogyne incognita*. *J.Ultrastruct.Res.*, 59, 231-249.

Hirschmann, J. (1971). *In* "Plant Parasitic Nematodes." (Eds. B.M. Zuckerman, W.F. Mai and R.A. Rohde), 1, 11-63. Academic Press, New York.

Hyman, L.H. (1951). The Invertebrates. 572, McGraw-Hill, New York.

Johnson, P.W., Van Gundy, S.D. and Thomson, W.W. (1970). Cuticle formation in *Hemicycliophora arenaria*, *Aphelenchus avenae* and *Hirschmanniella gracilis*. *J.Nematol.*, 2, 59-79.

Johnson, P.W. and Graham, W.D. (1976). Ultrastructural studies on the cuticle of the second-stage larvae of four root-knot nematodes: *Meloidogyne hapla*, *M. javanica*, *M. incognita* and *M. arenaria*. *Can. J.Zool.*, 54, 96-100.

Kanagasuntheram, R., Singh, M., Ho, B-C., Yap, E-H., and Chan, H-L., (1974). Some ultrastructural observations on the microfilaria of *Breinlia sergenti* nervous system. *Int.J.Parasitol.*, 4, 489-495.

Lee, D.L. and Atkinson, H.J. (1976). *In* "Physiology of Nematodes." 215. MacMillan, London.

Locke, M. (1966). The structure and formation of the cuticulin layer in the epicuticle of an insect, *Calpodes ethlius* (Lepidoptera, Hesperiidae). *J.Morph.*, **118**, 461-494.

McLaren, D.J. (1976a). Nematode sense organs. *Advances in Parasitology* **14**, 195-265.

McLaren, D.J. (1976b). *In* "The Organization of Nematodes." (Ed. N.A. Croll). 139-161, Academic Press, New York.

McClure, M.A. and Bird, A.F. (1976). The tylenchid (Nematoda) egg shell : formation of the egg shell in *Meloidogyne javanica*. *Parasitology* **72**, 29-39.

Maggenti, A.R. (1976). *In* "The Organization of Nematodes." (Ed. N.A. Croll). 1-10. Academic Press, New York.

Maggenti, A.R. and Allen, M.W. (1960). The origin of the gelatinous matrix in *Meloidogyne*. *Proc.Helm.Soc.Wash.*, **27**, 4-10.

Neville, A.C. (1975). Biology of the arthropod cuticle. Springer-Verlag, Berlin.

Reger, J.F. and Fain-Maurel, M.A. (1973). A comparative study of the origin, distribution, and fine structure of extracellular tubules in the male reproductive system of species of Isopods, Amphipods, Schizopods, Copepods and Cumacea. *J.Ultrastruct.Res.*, 44, 235-252.

Shepherd, A.M., Clark, S.A. and Kempton, A. (1973). Spermatogenesis and sperm ultrastructure in some cyst nematodes, *Heterodera* spp. *Nematologica* **19**, 551-560.

Tarr, G.E. and Fairbairn, D. (1973). Ascarosides of the ovaries and eggs of *Ascaris lumbricoides* (Nematoda). *Lipids* **8**, 7-16.

Triantaphyllou, A.C. (1962). Oögenesis in the root-knot nematode *Meloidogyne javanica*. *Nematologica* **7**, 105-113.

Ward, S., Thomson, N., White, J.G. and Brenner, S. (1975). Electron microscopical reconstruction of the anterior sensory anatomy of the nematode *Caenorhabditis elegans*. *J.Comp.Neur.*, **160**, 313-338.

Ware, R.W., Clark, D., Crossland, K. and Russell, R.L. (1975). The nerve ring of the nematode *Caenorhabditis elegans* : Sensory input and motor output. *J.Comp.Neur.*, **162**, 71-110.

Wergin, W.P. and Endo, B.Y. (1976). Ultrastructure of a neurosensory organ in a root-knot nematode. *J.Ultrastruct.Res.*, **56**, 258-276.

White, J.G., Southgate, E., Thomson, J.B. and Brenner, S. (1976). The structure of the ventral cord of *Caenorhabditis elegans*. *Phil.Trans.R.Soc.Lond.*, **275**, 299-325.

Whitehead, A.G. (1968). Taxonomy of *Meloidogyne* (Nematodea : Heteroderidae) with descriptions of four new species. *Trans.Zool.Soc. Lond.*, **31**, 263-401.

Wisse, E. and Daems, W.Th. (1968). Electron microscopic observations on second-stage larvae of the potato root eelworm *Heterodera rostochiensis*. *J.Ultrastruct.Res.*, **24**, 210-231.

Zacheo, T.B., Grimaldi, S., Lamberti, F. and De Lucia, M.R.M. (1975). Nematodes do have a coelcematic cavity. *Nematol.medit.*, **3**, 109-112.

Discussion

Asked whether spicules functioned solely as copulatory organs, Bird replied that Samoiloff *et al* (1973)[1] in

Canada had irradiated spicules with a laser beam and destroyed the ability of the males to locate females. However, Triantapjyllou queried whether the reaction observed was due to destruction of the spicules or to general
damage to the tail region.

Referring to the nature of the gelatinous matrix, Bird
explained that it is a glycoprotein of complex structure
which is extremely hard and difficult to freeze fracture.
He considered that it probably has great survival value
because of its size, the nature of the large cells that
synthesize it, and the amount of energy that must be
involved in the whole process; however, there is no experimental evidence to indicate the function of the gelatinous matrix.

Coomans observed that the pesudocoelom has many different definitions, and evidence from the literature
suggests that the true pseudocoelom is difficult to define. In nematodes, the cavity is difficult to observe
but membranes have been seen (Zacheo *et al.*, 1975, in
refs). Some evidence points to nematodes having a true
coelom that has become much reduced during evolution.

---

[1]Samoiloff, M.R., McNicholl, P., Cheng, R. and Balakanichi, S. (1973).
Regulation of nematode behaviour by physical means. *Experimental
Parasitology* **33**, 253-262.

6 CYTOGENETICS OF ROOT-KNOT NEMATODES

A.C. Triantaphyllou

*Department of Genetics, North Carolina State University
Raleigh, North Carolina, USA*

Introduction

Intensive cytogenic studies during the past 15 years probably have contributed more to our understanding of the
biology and interrelationships of root-knot nematodes
than any other single line of investigation. Diverse and
often opposing opinions expressed by various investigators
in the past, and during these proceedings, about the taxonomic treatment of the group are at least partially due
to the cytogenetic complexity of root-knot nematodes. In
my opinion, if one becomes thoroughly familiar with the
cytogenetic characteristics of the group, one can no
longer be dogmatic about its taxonomic treatment. Arguments e.g. about the taxonomic distinction of *Meloidogyne
acrita* from *M. incognita*, the recognition of *M. arenaria
thamesi* as a species distinct from *M. arenaria*, etc.,
have no scientific basis and at times may suggest lack
of understanding of the actual problem. In order to have
a better grasp of the biological complexity of root-knot
nematodes, one should consider that the group has a long
history of evolution toward parthenogenesis and, in this
process, it has undergone extensive cytological modifications involving among others polyploidy, aneuploidy and
increases or decreases in the amount of genetic material
(DNA). All these changes have contributed to the

diversification of the group from a genetic point of view
and consequently, from a taxonomic and behavioural point
of view.

It should be recognized that very little is known about
the inheritance and evolution of mitotically parthenogenetic (apomictic) organisms in general, and there has
been no consensus among taxonomists about the taxonomic
treatment of such organisms. Normally, reproductive females of apomictic organisms give rise to clones of genetically almost identical progeny, whose evolution is
relatively independent of that of other clones. For this
reason, the biological species concept is not applicable
in apomictic organisms and any taxonomic system will
have to be subjective.

Historically, research on the cytogenetics of root-
knot nematodes is recent. However, awareness of the reproductive and developmental peculiarities of these nematodes was indicated in early reports (Tyler, 1933). Reproduction in the absence of males (parthenogenesis or
hermaphroditism) was known to occur and the role of the
environment in causing fluctuations in the sex ratio of
these nematodes had been noticed and actually demonstrated. Further progress was made after the recognition that
root-knot nematodes constituted a complex that could be
subdivided into discrete forms (species), each one with
specific behavioural and morphological characteristics
(Christie and Albin, 1944; Chitwood, 1949). The effect
of the environment during postembryogenesis on sex differentiation of the larvae and the development of males,
male intersexes and female intersexes was investigated.
Also, cytological methods were developed to study the
process of gametogenesis, mode of reproduction and the
chromosomal complement of many root-knot nematodes.
There are extensive reviews on this work (Triantaphyllou
and Hirschmann, 1964; Triantaphyllou, 1971, 1973). The
present report summarizes earlier work, incorporates new

or unpublished information and attempts to relate the
available cytogenetic information to the evolution, taxonomy, genetics and behaviour of root-knot nematodes.

Mode of reproduction

Root-knot nematodes are known to reproduce by one of
the following methods:
a) cross fertilization (amphimixis),
b) facultative meiotic (automictic) parthenogenesis and
c) obligatory mitotic (apomictic) parthenogensis (Triantaphyllou, 1971). Of the 12 species of root-knot nematodes and an additional two undescribed forms that have
been studied cytogenetically, only three, namely *M. carolinensis*, *M. microtyla* and *M. megatyla*, the recently described "Pine Root-Knot nematode" (Baldwin and Sasser,
1978), appear to reproduce exclusively by cross-fertilization (Table I). Cytological evidence indicates that
another 6 species can reproduce by both cross-fertilization and meiotic parthenogenesis. When males are not
present in a culture, females of these species produce
eggs which, during maturation, undergo meiosis that results in the reduction of the chromosome number from the
somatic to the haploid number. Re-establishment of the
somatic chromosome number takes place later through
fusion of the second polar nucleus with the egg pronucleus (automixis). When males are present in a culture
they mate freely with females and reproduction is by
cross-fertilization. Sperm is seen in the spermatotheca
and inside the oocytes of inseminated females and finally
fusion of sperm and egg pronuclei takes place as evidenced from numerous cytological observations. There is
no genetic evidence that amphimixis actually takes place
in these nematodes. However, preliminary crosses have
been attempted between females from a population of
M. hapla with 15 chromosomes, not capable of reproducing
on strawberry, and males from a population with 17

TABLE I

Cytogenetic characteristics of root-knot nematodes

Species	Chromosone No.	Mode of Reproduction
M. microtyla	n=19	
M. megatyla	n=18	Amphimixis
M. carolinensis	n=18	
M. exigua		
M. graminicola		Facultative
M. naasi	n=18	Meiotic
M. graminis		Parthenogenesis
M. ottersoni		
M. hapla – race A	n=17, 16, 15	
M. hapla – race B	3n=45, 48	
M. arenaria – race A	3n=50-56	
M. arenaria – race B	2n=36	Obligatory
M. incognita – race A	3n=40-46	Mitotic
M. incognita – race B	2n=32-36	Parthenogenesis
M. javanica	3n=43-48	
Rice root-knot nematode	2n=36	
Oak root-knot nematode	2n=30-32	

chromosomes capable of reproducing on strawberry. A small percent of the progeny of such crosses had the capacity to reproduce on strawberry, indicating that cross-fertilization had occured. There was also cytological evidence that female progeny recovered from strawberry were truly hybrids in having 15 bivalent and two univalent chromosomes during metaphase I in maturing oocytes (unpublished data).

A third group comprises obligatorily mitotic parthenogenetic species and includes three of the agronomically most important species of root-knot nematodes, namely M. incognita, M. javanica and M. arenaria (Table I). Females of this group produce oocytes which do not

undergo meiosis. At the end of the maturation period the
eggs have the somatic chromosome number and can proceed
with cleavage without the need of being activated or
fertilized. When males are present in a population, they
usually inseminate the females and sperm are found in
the spermatotheca and inside the maturing oocytes. However, in a few cases, where the fate of the sperm nucleus
inside an oocyte could be followed, it was observed that
the sperm nucleus did not fuse with the egg pronucleus
(Triantaphyllou, 1962). Instead, the sperm nucleus de-
generated in the cytoplasm of the oocyte and, therefore,
did not participate in actual fertilization. Evidence
that fertilization does not take place in mitotically
parthenogenetic forms is also provided by the absence of
individuals of higher degree of ploidy in a population.
If true fertilization were taking place and the sperm
nucleus were contributing its chromosomes in the formation of a zygote nucleus, the progeny of inseminated
females would be expected to have the somatic chromosome
number of the females, plus the somatic or the reduced
chromosome number of the males, i.e. they would have one
and a half or twice the chromosome number of parthenogenetic progeny. This evidently does not happen, since
the chromosome number of a given population is quite
stable. Slight variation in a population usually does
not exceed ± 1 or 2 chromosomes and can be attributed to
errors in counting or to aneuploidy.

The chromosomal complement

The chromosomal complement of root-knot nematodes has
been studied more extensively than that of any group of
plant-parasitic nematodes.

Some observations have been made on chromosomal figures of oogonial divisions taking place in the apical
germinal zone of the ovary of egg-laying females. Prophase figures of such divisions have distinct, rod-shaped

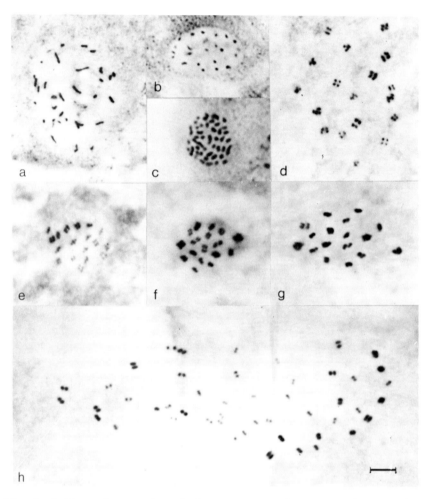

Fig. 1a-h Photomicrographs of chromosomes during oogenesis of *Meloidogyne* spp. a,b. Prophase chromosomes of oogonial divisions in *M. incognita* and *M. exigua*, respectively; c. Polar view of metaphase figure of an oogonial division of *M. incognita*; d,e. 18 bivalent chromosomes (tetrads) at prometaphase of the first maturation division of oocytes of *M. graminis* and *M. exigua*, respectively; f,g. 17 bivalent chromosomes at prometaphase I of two populations of *M. hapla*; 48 univalent chromosomes (dyads) at prometaphase of the first and only maturation division of *M. javanica*. Bar in Fig 1h equals 3 μm. Scale in Figs. a-f as in Fig. 1h.

chromosomes of 0.5 - 2.5 μm in length and 0.4 - 0.5 μm in diameter (Fig. 1a,b.). Usually the chromosomes are spread all over the nucleus and can be counted accurately, but

counting is tedious and often very difficult. At metaphase
when the chromosomes are further condensed and arranged on
one plane, counting is easier, especially in perfect polar
views (Fig. 1c). Difficulties are encountered when some
chromosomes of a metaphase plate are not discrete because
they are in contact with others.

Most of the observations regarding the chromosomal complement of root-knot nematodes have been made on chromosomal figures during prometaphase and metaphase of the first
maturation division of oocytes of egg-laying females. In
amphimictic and facultatively parthenogenetic species, the
chromosomes at this stage are bivalent and in the reduced
(haploid) number. Each bivalent consists of two univalents
and each univalent consists of two distinct chromatids that
lie parallel to each other. Therefore, each bivalent is
tetrapartite (tetrad) consisting of four chromatids, arranged in a characteristic configuration (Fig. 1d,e,f). In
mitotically parthenogenetic species (apomictic), the prometaphase and metaphase chromosomes are univalent and in
the unreduced (somatic) number. Each univalent is bipartite
(dyad) consisting of two chromatids that lie parallel to
each other (Figs 1h, 2a,b). In general, prometaphase I figures are the most favourable for studying and counting the
chromosomes of these nematodes. Often, however, anaphase I
and telophase I figures, in polar view, are also useful for
counting the chromosomes and determining their relative
length (Fig. 2c). Amphimictic and facultatively parthenogenetic species have chromosomes of relatively uniform size
(Fig. 1d-g). Mitotically parthenogenetic species show more
variation in chromosomes size, but the variation is continuous, so that it is very difficult to identify individual
chromosomes, or classify them into groups. Usually, variation in the degree of condensation of the chromosomes during maturation, even among oocytes of the same female,
makes such identification of individual chromosomes almost
impossible.

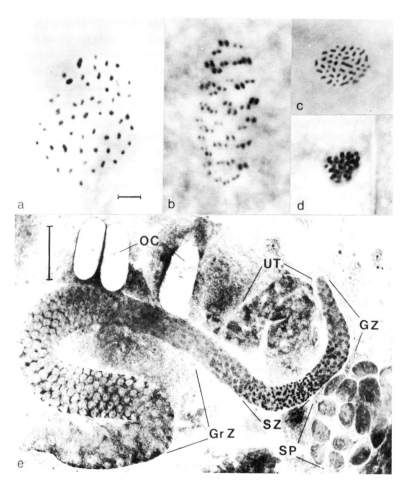

Fig. 2a-e Oogenesis in *Meloidogyne* spp. a,b. Univalent chromosomes (dyads) at prometaphase of the first maturation division of oocytes of *M. arenaria* (Race A) and *M. javanica*, respectively; c. Polar view of telophase I figure of an oocyte of *M. javanica*; d. Prophase I of an oocyte of *M. incognita*. Most oocytes present in the uteri of a female of *M. incognita*, are at this stage, with the prophase chromosomes characteristically packed together in a small, more or less, spherical area; e. General view, in small magnification (x10 objective), of a cytological preparation (smear), showing part of the female reproductive system of *M. arenaria*, stained with propionic orcein. Included are: the apical, germinal zone of the ovary (GZ) where multiplication of oogonia takes place; the synapsis zone (SZ) and part of the growth zone (GrZ) filled with small oocytes; the spermatotheca (SP), the beginning of the uterus (UT) and three oocytes (OC) at metaphase I (side view). Bar in Fig. 2a equals 3 μm. Scale in Figs b-d as in Fig. 2a. Bar in Fig. 2e equals 100 μm.

The karyotypes of these nematodes would probably be better characterized from mitotic divisions of somatic cells. However, such divisions in nematodes occur only during embryogenesis when their study is extremely difficult because of the reduced permeability of egg membranes and the small size of the nuclei of blastomeres. No cultures of somatic cells of nematodes have been established, thus far. Such cell cultures would be very helpful for cytological, genetic and biochemical studies in the future.

In spite of the above difficulties, extensive variation has been detected among populations within species, especially within *M. incognita*. Thus, some populations of *M. incognita* have one very long (3 μm) chromosome, whereas others have two to five long chromosomes (2.0 - 2.5 μm) that can be consistently recognized from the rest of the complement. Still other populations have no such distinct chromosomes.

The most striking feature of the chromosomal complement of root-knot nematodes is the extensive variation in chromosome numbers between and within species, especially among the agriculturally important species (Table I). Because the amphimictic and the facultatively parthenogenetic species of *Meloidogyne* - with the exception of *M. hapla* and *M. microtyla* - have a haploid chromosome number of n=18, this number has been assumed to be the basic number for the genus. Populations with a somatic chromosome number of 36 are regarded as diploids and those with 54 chromosomes are considered to be triploids. Populations with 45 to 48 chromosomes are suspected to represent "hypotriploids", derived from diploid ancestors with reduced chromosome numbers, as will be explained in the following subdivision of this chapter. Deviations from the euploid, diploid or triploid forms are attributed to aneuploidy, or to various types of chromosomal rearrangements.

Cytogenetic relationships of species and races

It is evident that evolution of the numerous forms of root-knot nematodes has been influenced significantly by the establishment of parthenogenesis, polyploidy and aneuploidy. The ancestral stock from which the group has evolved must have been a diploid (n=18), amphimictic form, probably living in temperate climates and having a limited host range. Descendents of this ancestral stock may be the present diploid, amphimictic species which have undergone extensive genetic evolution, but have maintained the amphimictic mode of reproduction and the original karyotype probably with some changes not involving polyploidy or aneuploidy. Such species are *M. carolinensis*, *M. megatyla*, and probably many of the other rare *Meloidogyne* species that have been described from specific hosts, in isolated regions of the world, and which have not yet been studied cytogenetically. *M. microtyla* which is amphimictic and has n=19 chromosomes (unpublished data) probably has evolved secondarily from a progenitor with n=18 chromosomes through a cytological mechanism that led to chromosome number increase, such as centric fission or transverse fragmentation (John and Lewis, 1968). It represents the only known case involving an increase of the basic chromosome number in the genus.

The first, very important step in the cytogenetic evolution of the root-knot nematodes is represented by such diploid species as *M. graminis*, *M. graminicola*, *M. naasi*, *M. ottersoni*, and *M. exigua* which, although basically amphimictic, have acquired the capacity to reproduce also by meiotic parthenogenesis.

The agriculturally most important species of root-knot nematodes, *i.e.* *M. hapla*, *M. arenaria*, *M. javanica* and *M. incognita*, represent the most advanced steps in the evolution of the genus. The cytogenetic pathway of their evolution, as visualized at this time, can be presented as follows: *M. hapla* populations with n=17, 16 or 15 chromosomes are thought to have been evolved from a progenitor with 18 chromosomes through a mechanism of chromosome number reduction, such as centric fusion. Such a reduction in chromosome number, which usually results in reduced rate of genetic recombination, may have contributed to the successful adaptation of *M. hapla* populations which reproduce by mitotic parthenogenesis (Race B) and have 45 or 48 chromosomes (several populations from Chile, S.A. have 48 chromosomes - unpublished data) may have been evolved from meiotic parthenogenetic populations of *M. hapla*, probably through hybridization. Thus, *M. hapla* with 45 chromosomes may have been derived through fertilization of an unreduced egg containing 30 chromosomes with a sperm having the haploid chromosome number of 15. *M. hapla* with 48 chromosomes may have been evolved in a similar manner from a facultatively parthenogenetic *M. hapla* with n=16 chromosomes. The establishment of mitotic parthenogenesis following such an event apparently stabilized the triploid progeny reproductively and guaranteed their success and establishment as distinct biological entities. Morphological and behavioural similarity between the meiotic and mitotic forms attest to their close genetic relationship and, for the same reason, their close phylogenetic relationship.

Evolution of *M. arenaria* can be explained in a similar manner as in *M. hapla*. Most *M. arenaria* populations are mitotically parthenogenetic and have approximately 54 chromosomes. This form of *M. arenaria* (triploid) may have been evolved through hybridization from a facultatively parthenogenetic form (meiotic) with n=18 chromosomes. The process requires fertilization of an unreduced egg with 36 chromosomes with a normal sperm with 18 chromosomes and the subsequent establishment of apomictic reproduction among the progeny. Some *M. arenaria* populations which have 36 chromosomes (diploid form) may have been evolved from a similar facultatively parthenogenetic stock following

the abolishment of meiosis and the establishment of apomictic reproduction.

Evolution of *M. javanica* is more difficult to explain
because of the extensive and continuous variation in chromosome numbers existing within this species. All *M. javanica* populations studied thus far (about 75) are apomictic, but chromosome numbers vary from 43 to 48. Presumably
they have been evolved from a facultatively parthenogenetic stock or stocks of the meiotic type with reduced basic
chromosome number (less than 18). The process may have
been similar to that described for the apomictic form of
M. hapla. Some additional variation in chromosome numbers
may have occurred subsequently to account for the aneu-
ploid forms *i.e.* forms with chromosome numbers other than
45 and 48. All *M. javanica* forms may thus be regarded as
triploid, but with a reduced chromosome number (hypotrip-
loid).

M. incognita, probably the most important species agronomically among the root-knot nematodes, reproduces exclusively by mitotic parthenogenesis and appears to have two
chromosomal forms. The diploid form has 32 - 36 chromosomes and the triploid, which is the predominant, has 40 -
46 chromosomes. There is a cytological feature that uni-
fies all *M. incognita* (*sensu lato*) populations and differ-
entiates them from all other *Meloidogyne* species studied
thus far. It concerns the behaviour of the chromosomes
during prophase of the first and only maturation division
of the oocytes. In other species, the maturing oocytes
advance to metaphase and the chromosomes are distinct and
well separated from each other as soon as the oocytes pass
through the spermatotheca into the uterus. In *M. incognita*, oocytes passing into the uterus remain in prophase
with the chromosomes grouped together and do not advance
to metaphase until much later, when the oocytes approach
the posterior part of the uterus. This characteristic
feature of *M. incognita* may suggest a pathway of evolution

different from that of other species of *Meloidogyne*. Also
it may suggest a monophyletic evolution of *M. incognita*,
in spite of the presence of two degrees of ploidy in the
species. The predominant form of *M. incognita* with 40 - 46
chromosomes which may be considered to be a triploid may
have evolved from a facultatively parthenogenetic ancestor
with a reduced chromosome number, as in the case of triploid forms of *M. hapla* and *M. javanica*. The diploid form
of *M. incognita* with 32 - 36 chromosomes may represent an
intermediate evolutionary step in which apomixis was
established without polyploidization of the reduced or
the normal chromosomal complement of the same hypothetical
ancestor.

An alternative explanation of the derivation of the
various cytological forms of the species with 40 - 48
chromosomes is possible. It can be assumed that chromosomal fragmentation has occurred and has resulted in an increase of the diploid chromosomal complement from 2n=36
to 40 - 46 in *M. incognita* (race A), 43-48 in *M. javanica*
and 45 or 48 in *M. hapla* (race B). The suspected diffuse
type of kinetochore activity of chromosomes of these nematodes would allow chromosomal fragmentation without loss
of the chromosomal fragments which, in subsequent division, would behave like independent chromosomes. However,
studies of the DNA content of nuclei of these nematodes do
not support this explanation. Thus, Lapp and Triantaphyllou (1972) have found that forms with higher chromosome
numbers have proportionally higher DNA content, indicating
that such forms are truly polyploid, or derivatives of
polyploids.

Genetic and evolutionary implication of polyploidy and parthenogenesis

The unprecedented success of root-knot nematodes as
expressed by their world-wide distribution and very
extensive host range, undoubtedly is related to their

polyploid, parthenogenetic nature. Amphimixis, the most
common method of reproduction in higher animals, is not
as common among the prevalent root-knot nematodes. The
major advantageous feature of amphimixis is its potential
for increased genetic variability in freely interbreeding
populations (panmictic populations). Root-knot nematode
populations, because of their limited capacity for migration in the soil and the sedentary life style of the
females, are not panmictic and tend to undergo a high
degree of inbreeding which normally leads to homozygosity.
Thus, root-knot nematodes could not take full advantage
of amphimixis and probably, for this reason, amphimixis
is not common among the successful root-knot nematodes.
The few species which appear to reproduce exclusively
by amphimixis (*M. carolinensis*, *M. microtyla* and *M. megatyla*) are very limited in distribution and host range.

   Obligatory, mitotic parthenogenesis (apomixis) ex-
cludes genetic recombination, but it has reproductive
advantages since the entire population consisting of
females (thelytoky) can give progeny. In root-knot nematodes, however, some sexual waste occurs at adverse environmental conditions when a high percentage of the
progeny may develop into males or male intersexes
through various developmental processes including sex
reversal (Triantaphyllou, 1973). Probably the major advantage of apomixis is that it supports extensive heterozygosity, but this is likely to happen only when apomixis is combined with polyploidy. Any gene mutation and
most chromosomal mutations that may occur are less likely
to be eliminated and tend to accumulate in the genome.
This makes individual nematodes highly heterozygous for
many loci and possible capable of adaptation to a greater
variety of environments. The extensive geographical distribution and host range of apomictic species like *M. incognita*, *M. javanica*, *M. arenaria* and part of *M. hapla*
(race B) may be explained or attributed to their

apomictic mode of reproduction and also to their polyploid
nature.

Facultative parthenogenesis, which involves meiotic
type of parthenogenesis combined with amphimixis, occurs
in many species of root-knot nematodes (Table I). This
type of reproduction combines the advantages, but also
the disadvantages of amphimixis and parthenogenesis. The
overall influence of facultative parthenogenesis upon
the success of a species is not predictable and appears
to depend on various genetic and environmental factors.
Most often, parthenogenesis in root-knot nematodes becomes seasonal. In the spring and early summer, when
nematode populations are of low density, males are rare
and reproduction is primarily parthenogenetic, facili-
tating a fast build up of the population. At the end of
the growing season, when populations are dense and the
food conditions become adverse, many males are usually
produced and reproduction is primarily amphimictic.
Amphimixis promotes genetic recombination for better
adaptation and probably better larval overwintering.
Theoretically, facultative parthenogenesis may be the
most advantageous method of reproduction for root-knot
nematodes living in temperate or colder climates, with
a definite winter period. This may explain why practically all successful facultatively parthenogenetic species
(*M. hapla* race A; *M. naasi*, *M. graminis*, etc.) are distributed primarily in temperate regions of the world and
none in low altitudes in tropical areas.

In general, it can be stated that obligatorily amphimictic species are the least successful biologically and
the least important agriculturally forms of root-knot
nematodes. Facultatively parthenogenetic species are
relatively more successful than amphimictic species,
especially with regard to their geographical distribution and frequency of occurrence. They are also more important as plant parasites and most of them are adapted

to temperate regions of the world. Among them *M. hapla* (race A) which is by far the most successfully adapted as a plant parasite in temperate climates has undergone, in addition, some chromosomal evolution expressed as a reduction of the chromomosal complement from n=18, the basic number of the genus, to n=17, 16 and 15.

Mitotically parthenogenetic, polyploid forms like *M. javanica, M. incognita* (race A), *M. arenaria* (race A), and *M. hapla* (race B) are the most important root-knot nematodes. They have an extensive host range and are widely distributed geographically, especially in the tropical and subtropical regions of the world. Apparently they are best adapted to warmer climates and have been favoured by the special environment of crop lands. The present agricultural systems prevailing in the world have assisted these nematodes to become major pests and have contributed to their biological success. These nematodes are not adapted to non-cultivated lands. Mitotically parthenogenetic, diploid forms, like *M. arenaria* (race B), *M. incognita* (race B), the rice and the oak root-knot nematodes are not very successful forms as judged by their limited distribution, infrequent occurrence and, for some of them, their limited host range. It is obvious that mitotic parthenogenesis benefited root-knot nematodes only in cases where it is combined with polyploidy.

Cytotaxonomy

Cytological information can assist in identification of the most common nominal species of root-knot nematodes. Especially when identification of a population on a morphological and/or host specificity characteristics is only tentative, cytological information can be very useful, either in confirming or disputing the previous identification. The cytological information concerning the major species has quite a broad basis since it has

resulted from an extensive study of more than 300 populations from about 50 countries of five continents. And
although some additional variation may be revealed in
future studies, I feel confident that a good deal of the
existing variation is represented in the observations
that have already been made.

Contrary to the prevailing opinion, cytological preparations for identification purposes are easy to make as
described in the next subdivision of this chapter.
Oocytes that have just passed through the spermatotheca
into the uterus of your females are at prometaphase or
metaphase of the first maturation division and provide
the most useful and easy to observe cytological characteristics. Table II is a guide that can assist in identification of the most common species.

Many more diploid apomictic forms, like the rice and
the oak root-knot nematodes (Table I), will undoubtedly
be found as more cytological surveys are conducted. However it is very likely that such forms will be limited
in their distribution, host range and their importance
as plant parasites.

Staining of chromosomes

Two methods have been successfully employed in staining the chromosomes of root-knot nematodes. The "*Feulgen
method*", based on the use of the "Schiff's aldehyde reagent", is a classical cytological method, specific for
staining nucleic acids (Sharma & Sharma, 1972). Only the
chromosomes and no other part of the oocytes are stained,
therefore results are very reliable. Nevertheless, this
method is not the best for routine counts of *Meloidogyne*
chromosomes, because it involves a rather complicated
procedure. Also, chromosomes stained by this method are
small in size, faintly stained and are distributed in a
small area. Thus, precise chromosome counts and detailed
observations of chromosome morphology and behaviour are

TABLE II

Prophase chromosomes nondiscrete, grouped
together in a small area (Fig. 2d).
Most oocytes of a female are at prophase
I, and only a few may have advanced to
metaphase I .... *M. incognita*

Prophase chromosomes well separated from
each other (Figs. 1d-h, 2a,b) Most oocytes
of a female have advanced to anaphase and
telophase I 1

1. Prometaphase and metaphase chromosomes
are bivalents (tetrads - Fig. 1d-g)
and in small numbers (15-19) 2

   Prometaphase and metaphase chromosomes
   are univalents (dyads - Figs. 1h, 2a,b)
   and in large numbers (30-56) 3

2. Number of bivalent chromosomes n=19 .... *M. microtyla*
   Number of bivalent chromosomes n=18
   (Fig. 1d,c) .... Several species[1]
   Number of bivalent chromosomes n=17,
   16 or 15 (Fig. 1f,g) .... *M. hapla* (race-A)

3. Number of univalent chromosomes 2n=36 ...... *M. arenaria* (race-B)
   Number of univalent chromosomes 3n=
   43-48 (Fig. 2b) .... *M. javanica*[2]
   Number of univalent chromosomes 3n=
   45 or 48 .... *M. hapla* (race-B)[2]
   Number of univalent chromosomes 3n=
   50-56 (Fig. 2a) .... *M. arenaria* (race-A)

---

[1]In this group there can be any one of the following species: *M. carolinensis, M. megatyla, M. exigua, M. graminicola, M. naasi, M. graminis* or *M. ottersoni*. Also many other rare, described species, with limited host range and geographical distribution, are expected to belong to this category.
[2]Although there is an overlap in chromosome numbers between *M. javanica* and *M. hapla* (race-B) the chances of error in identification are not great if morphological or host specificity characteristics are considered at the same time. Furthermore, the known geographical distribution is instructive. *M. javanica* is widely distributed in tropical, subtropical and some temperate regions. *M. hapla* (race-B) with 45 chromosomes is very rare and is known only from a few populations in the USA and one greenhouse population from England. *M. hapla* (race-B) with 48 chromosomes is known to occur only in Chile, S.A.

difficult.

A "modified propionic orcein method", described in detail below, has been developed specifically for the study of the extremely small chromosomes of root-knot nematodes. By this method the chromosomes appear swollen and therefore, much larger, they stain heavily and they are spread over a larger area, compared to chromosomes stained by the Feulgen method. Furthermore, the cytoplasm is not hardened and remains semifluid for several days after staining, so that further spreading of the chromosomes is possible by applying pressure on the coverslip.

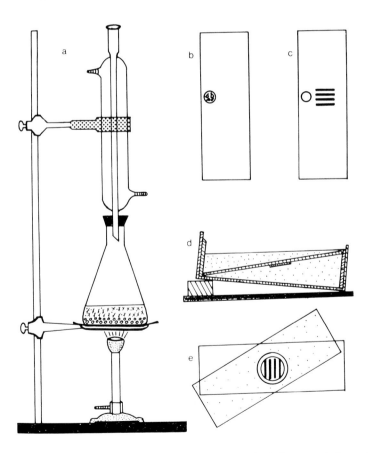

Fig. 3. Apparatus for the modified orcein method for staining chromosomes.

Preparation of stain Combine 2 g of orcein stain
(reagent, certified for cytology) and 100 ml of 45% propionic acid in a 500 ml pyrex flask. Adapt a condenser over
the flask and circulate cold water (Fig. 3A). Heat the
contents of the flask to boiling and continue boiling
gently for 1/2 to 1 hr. The vapours condense on the cold
walls of the condenser and drip back into the flask, thus
maintaining the concentration of the acid approximately
constant. *Caution:* slight overheating may cause violent
boiling and often dangerous explosions. Addition of some
glass beads in the flask usually helps prevent violent
boiling.

Nematode material For best results, obtain galled roots
from a 40-50 day-old culture of the nematode maintained
on a suitable host plant, growing under the best conditions and with light infection. Young females with medium-
size, white egg masses are dissected with forceps and a
needle from the roots in 0.9% NaCl solution under a
stereoscope.

Preparing smears Transfer four females to a drop of water
on a microscope slide (Fig. 3B). Using fine forceps, remove one female from the drop, and place it in the centre
of the same slide with as little water as possible. Before
the female dries out, make a cut at the neck region with
a sharp eyeknife. Holding the female body with the forceps, draw it along the slide applying slight pressure, so
that the body contents are smeared uniformly along a strip
of about 1 cm long. Discard the cuticle. Smear the other
three females in a similar manner, in parallel strips
(Fig. 3C).

Hydrolysis (5 to 10 min at room temperature.) Immerse
the slide with the smears in 1N HCl (about 10 ml of 37.8%
reagent HCl in 100 ml distilled water). To avoid loss of
material during immersion, invert the slide, and holding
it exactly horizontally, lower it into a staining dish

filled with the HCl solution, so that the entire slide
will touch the surface of the solution at the same time
(Fig. 3D).

Fixation   (20 to 40 min). Remove slide from the HCl solution and wipe it dry with tissue paper leaving wet only
the smeared material and a small area surrounding it.
Immerse slide in the fixative which consists of three
parts of absolute ethyl alcohol and one part glacial
acetic acid.

Staining   (20 to 30 min.) Remove slide from the fixative
and wipe it dry with tissue paper. Place slide on a perfectly level surface and apply one or two drops of orcein
stain on the material. Cover the drop of stain with a
deep well (cavity) slide to prevent evaporation (Fig. 3E).

Mounting   Remove cavity slide and drain stain to the side
of the slide on tissue paper. Some stain will adhere to
the smears. To remove this stain, immerse the slide for
3-5 seconds in 45% propionic acid. Place the wet slide on
absorbent paper on a level surface and apply a No. 1,
22mm, square coverslip over the smears. Absorb the excess
of propionic acid with tissue paper, without touching the
coverslip or forcing it sideways. Seal the coverslip by
applying sealing medium with a hot needle made of thick
wire. To prepare sealing medium, add equal weights of
paraffin and lanoline in a beaker and place in a hot bath
until the paraffin melts. Pour the mixture in small flat
containers (e.g. slide boxes) and allow to cool and soli-
dify.

Examination of slides   Use a research microscope with a
x10 objective to survey the smears. Eggs stained red or
nontransparent eggs, filled with fat globules, are not
good for cytological observations. Eggs slightly tinted
with stain and appearing semitransparent are the best
for further observation (Fig. 2E). They are located in

the uterus close to the spermatotheca. Observe such eggs using a x90 or x100 objective, and a No. 58 (546 nm) green filter (commonly used in photography) to increase the contrast. Use a camera lucida to make sketches for reliable and unbiased counts of the chromosomes.

Acknowledgement

This is paper number 5608 of the Journal Series of the North Carolina Agricultural Experiment Station, Raleigh, North Carolina. This study was supported in part by National Science Foundation Grant No. DEB. 76-20968 A01.

References

Baldwin, J.G. and Sasser, J.N. (1978). *Meloidogyne megatyla* n. sp., (Heteroderidae). A root-knot nematode from loblolly pine. *J.Nematol.* 10, (In press).

Chitwood, B.G. (1949). Root-knot nematodes. *Proc.Helminth.Soc.Wash.* 16, 90-104.

Christie, J.R. and Albin, F.E. (1944). Host-parasite relationships of the root-knot nematode, *Heterodera marioni*. *Proc.Helminth.Soc. Wash.* 11, 31-37.

John, B. and Lewis, K.R. (1968). The chromosomes complement. *Protoplasmatologia* 6A, 206.

Lapp, N.A. and Triantaphyllou, A.A. (1972). Relative DNA content and chromosomal relationships of some *Meloidogyne, Heterodera,* and *Meloidodera* spp. (Nematoda : Heteroderidae). *J.Nematol.* 4, 287-291.

Sharma, A.K. and Sharma, A. (1972). *In* "Chromosome techniques, theory and practice." 575, Butterworths London, University Park Press, Baltimore.

Triantaphyllou, A.C. (1962). Oogenesis in the root-knot nematode *Meloidogyne javanica*. *Nematologica* 7, 105-113.

Triantaphyllou, A.C. (1971). *In* "Plant parasitic nematodes" (Eds. B.M. Zuckerman, W.F. Mai and R.A. Rohde) 2, 1-32, Academic Press, New York.

Triantaphyllou, A.C. (1973). Environmental sex differentiation of nematodes in relation to pest management. *Ann.Rev.Phytopath.* 11, 441-462.

Triantaphyllou, A.C. and Hirschmann, H. (1964). Reproduction in plant and soil nematodes. *Ann.Rev.Phytopath.* 2, 57-80.

Tyler, J. (1933). Development of the root-knot nematodes as affected by temperature. *Hilgardia* 7, 391-415.

Discussion

Dalmasso said that his biochemical investigations were

complementary to cytogenetic studies of Triantaphyllou. He agreed that species with mitotic parthenogenesis are more heterozygous than those with meiotic parthenogenesis, the former species showing more variation in their proteins. Dalmasso also agreed that *M. incognita* is different from other *Meloidogyne* species because of the more clustered chromosome pattern, though he used Schiff's reagent instead of acetic orcein. Asked if something like a world map had been prepared to show geographical differences in the chromsome patterns between species, Triantaphyllou replied that although such a map is under preparation it is not very informative because populations with the same number can be found in different places throughout the world so that there is no real pattern in distribution. Netscher asked why reproduction by meiotic and mitotic parthenogenesis is so successful in *Meloidogyne*, whereas *Heterodera* produces amphimictically. Triantaphyllou replied that every genus is different with regard to evolution of reproduction and that parthenogenesis has proven to be the most advantageous method of reproduction for *Meloidogyne*.

In response to an enquiry about the fate of the polar bodies Triantaphyllou explained that in amphimixis the polar bodies are extruded; in meiotic parthenogenesis the first polar body is also extruded but the second one is used to re-establish the diploid condition; in mitotic parthenogenesis the single polar body is not extruded but disintegrates in the cytoplasm. In parthenogenetic forms a spermatozoon can enter an oocyte but later it disintegrates. It is not known what happens to the DNA of the sperm and it cannot be ruled out that part of it is not utilized by the egg, with some possible genetic effects. Bird queried whether there is any influence on the survival of the eggs from the same nematode when it can be amphimicitic as well as parthenogenetic but Triantaphyllou said that both types of eggs appear to develop normally.

According to Coomans the environmental conditions can
determine whether a species reproduces amphimictically
or parthenogenetically when both types of reproduction
are possible; this has been established in other invertebrates as well as for example in *Cladocera* and *Rotifera*.
Triantaphyllou said that the same holds true for some
root-knot nematodes, like *M. hapla*, but others are obligatorily parthenogenetic. Coomans then went on to state
that Triantaphyllou's conclusion that the parthenogenetic
reproduction is the most advantageous type for *Meloidogyne*
may well depend on the climatic conditions and the geographical areas where the species have so far been investigated. He suggested that in those areas where at some
time in the life cycle adverse conditions may occur, the
combination of amphimixis and parthenogenesis may be best
for the species. Asked whether there is likely to be
more variability in species with mitotic parthenogenesis
than in those with amphimictic reproduction, Triantaphyllou replied that individuals of mitotically parthenogenetic population of small size may be genetically similar to each other, but most likely are heterozygous for
many loci. In other words, they do not show much genetic
variation, but have the genetic potential for variation.
In amphimictic populations individuals are genetically
different from each other, because of the effect of gene
recombination, but they may be homozygous for many loci.
These two genetic situations interact differently with
various environments and this interaction is expressed
as a short or long-term adaptation. Certain environments
favour apomictic reproduction and others favour amphimictic reproduction or a combination of amphimixis and
parthenogenesis.

Wouts said that the amount of DNA is considered as a
good character for distinguishing between different
Heterodera species and asked whether it could also be
used for distinguishing *Meloidogyne* species.

Triantaphyllou replied that the DNA content of nuclei of *Meloidogyne* is about half that of *Heterodera*, and consequently determination of the amount of DNA in *Meloidogyne* is much more difficult. The DNA content of closely related species of *Meloidogyne* is proportional to their chromosome numbers and this indicates that species with high chromosome numbers have been derived through polyploidy or aneuploidy. Thus DNA content provides some information about the relationship of *Meloidogyne* species, but cannot be considered as an easy and reliable taxonomic character. Triantaphyllou referred to the micro-spectrophotometrical method he has been using recently in determining the amount of DNA in the sperm nucleus of *Heterodera* species and which can be used also for *Meloidogyne*.

Replying to queries about the use of the generic name *Hypsoperine*, Triantaphyllou said that he considered this as a valid genus because of the small amount of DNA and because n=7 in *H. spartinae*. Mary Franklin also agreed that *H. spartinae* is very different from other species.

Asked whether a chromosome number could be used in practice as a taxonomic character, Triantaphyllou replied that the technique could be learned quickly enough though some skill is necessary to apply it. Netscher said that he had used the technique for taxonomic purposes but had found if difficult to establish good counts of chromosomes, though this was with *M. incognita* which seems to be the most difficult of the species. Triantaphyllou said that type of maturation of oocytes and chromosome counts had been successfully employed for identification of root-knot nematodes and could give useful information such as indicating mixed populations even if exact chromosome numbers could not be established. When this information is combined with morphological characters such as perineal pattern, then a reasonable identification of species is possible.

A. Dalmasso and J.B. Bergé

*I.N.R.A. - Station de Recherches sur les
Nématodes, Antibes, France*

The polyacrylamide gel electrophoresis technique can be
used to identify proteins in single females of *Meloidogyne*
species (2-10 µg of proteins equivalent BSA). Variability
in the migration of molecules in the gel depends on their
net surface charge and size and reflects part of the DNA
changes, through the messenger RNA, that might be indicative of the phylogenetic evolution of an organism (Lewontin, 1974). A study of enzymatic polymorphism is thus
possible as an aid to clarification of the systematic
position of individual organisms.

The technique has been used to identify eleven loci
in 74 populations and strains (22,000 females analysed)
of *M. hapla*, *M. naasi*, *M. arenaria*, *M. javanica*, *M. incognita* and *Meloidogyne* sp. with the following observations and conclusions:-

i) Two non-specific esterase loci, β and b, show inter-
and intraspecific variations in *M. hapla* (Bergé and Dalmasso, 1976). Esterase b gives a good segregation for
the different *Meloidogyne* spp. (Table I).

ii) *M. arenaria* and *M. javanica* are closely related;
M. incognita is phylogenetically more distant from them.
A Spanish population of *Meloidogyne* sp. from Seville with
36 chromosomes is related to *M. incognita*, but differs
from it in the coding for one of its two esterase β

A. DALMASSO AND J.B. BERGÉ

TABLE I

Isozyme characters in *M. arenaria*, *M. javanica*, & *M. incognita* complex

Species	Chromosome numbers	Esterase β	Esterase b	MDH β	MDH b	α GPDH	Tetrazolium oxidase
M. arenaria	50–56	0.19 + 0.24 +	0.30 + 0.36 +++ 0.38 +++	0.19 + 0.24 +	0.30 +++ 0.35 ++ (0.40) +	(0.35) 0.38 0.40	Two strong and scattered bands at same levels as in *M. javanica*
M. javanica	43–47	0.19 + 0.24 +	0.30 +++ 0.36 ++ 0.38 ++	0.19 + 0.24 +	0.30 +++ 0.35	0.38 0.40	Two strong and scattered bands at same levels as in *M. arenaria*
M. incognita	40–44	0.19 + 0.24	0.30 +++		0.30 +++ 0.35 ++	0.35 +++	Two strong and close bands at same levels as in *Meloidogyne* sp.
Meloidogyne sp.	36	0.24 + 0.34 +	0.30 +++		0.30 +++ 0.35 ++	0.35 +++	Two strong and close bands at same levels as in *M. incognita*
Electrophoresis systems		1		3		2	2

+ slight band   ++ medium band   +++ strong band

Electrophoresis systems:
1 – Gel 7% acrylamide; pH 8.9; bromophenol is used as migration reference 5 cm from tops of microhematocrite tubes; 60 V. 2 – Same as 1 but electrophoresis at 120 V and 7 cm for bromophenol migration. 3 – Gel 4% acrylamide; pH 8.9; 60 V; migration 5 cm.

( ) Only for Grau-du-Roi population

alleles. In fact both show permanent heterozygosity for
the esterase β locus; alleles 0.18/0.24 for *M. incognita*
and 0.24/0.34 for *Meloidogyne* sp.

iii) Three other investigated loci - tetrazolium oxidase,
α glycerophosphate dehydrogenase (GPDH) and malate dehydrogenase (MDH) - confirm this phyletic theory. Seven remaining loci, in which no variation has been found, may
be considered as characteristic of the genus *Meloidogyne*
in comparison with other related genera.

There is a good agreement between karyotypes observed,
the host range of each population and the biochemical
data.

M. arenaria and *M. javanica* seem to have a higher degree of permanent heterozygosity. In these polyploid forms
occurrence of mutation must be increased and new alleles
are protected by the non-reducing parthenogenesis. This
may be an ecological advantage and can explain their
wide world distribution. However, there is an exception
- an apparent polyploid population of *M. hapla* (Saulcy,
45 chromosomes) has been found homozygous for all the
investigated loci. A recent origin of the population
would explain this.

Data also show intraspecific polymorphism e.g. a population of *M. arenaria* from Grau-du-Roi possesses a
supernumerary fast band of MDH (three instead of two).
In this species α GPDH activity is represented by one to
three more or less coloured bands according to the population. Such intraspecific genetic variation is not
surprising in relation to the numerous ecological differences reported in several *Meloidogyne* species.

Polyacrylamide gel electrophoresis is a useful tool
for the investigation of *Meloidogyne* taxonomy and similar
investigations of other nematode genera could be profitable.

A. DALMASSO AND J.B. BERGÉ

References

Bergé, J.B. and Dalmasso, A. (1976). Variations génétiques associées
à un double mode de reproduction parthénogénétique et amphimitique
chez le nématode *Meloidogyne hapla*. *C.r.Acad.Sci.*, Paris, série D,
282, 2087-2090.
Lewontin, R.C. (1974). *In* "The Genetic Basis of Evolutionary Change"
346, Columbia University Press.

D.R. Viglierchio

Division of Nematology, University of California,
Davis, California, USA

Introduction

Physiology deals with the life processes of an organism;
composition and function of tissues, organs and morphological structures; modes and mechanisms of internal reactions and communications; as well as those means by
which the organism perceives and copes with the environment. There exists among nematologists the temptation to
supplement the few factual observations with an abundance
of speculation to create an illusion of understanding.
This practice has merit as a developmental tool but risks
the confusion of fantasy with fact. In the interest of
perspective, this review attempts to summarize factual
information and point out complexities within selected
areas of the physiology of root-knot nematodes, dealing
in part with internal activities as well as those processes involved in the interface with the environment. The
important interactions of root-knot nematodes with the
host have been treated by others in these Proceedings.

Composition of tissues and fluids is a basic initial
bit in the summation of information leading to an improved
understanding of vital processes. Of the some 40 species
identified as root-knot nematodes only four have been investigated to some extent. Except for *Meloidogyne incognita* female cuticle and egg shell, the data from the investigations refer to the composition of the entire nematode. Whole animal homogenates are of doubtful value for

other than comparative purposes. It is evident that in the
gross listings of root-knot nematode constituents (Table
I), lipid material is in greatest abundance and therefore
consistent with analysis for other nematodes (Krusberg,
1971). In the one case, M. javanica L_2, the duplicate lipid content reports are not in agreement; there is no way
of assessing whether this is due to nematode biotype or
analytical method. To understand a nematode's organ and
tissue function, more specificity in the compositional
analysis is essential.

Compositional analysis

Amino acid analysis serve as helpful bases for the formulation of hypothetical physiological processes taking
place in the living nematode. For illustration, amino acid
analyses of several fractions of root-knot nematodes are
given in Table II. Comparison of the amino acid analysis o
M. incognita egg shell shows a mole percent amino acid
difference between ruptured and hatched egg shells. The Δ
mole percent column shows a decrease in the hatched egg
shell mole percent amino acid in 14 of 18 amino acids measured. This may suggest that in the hatching process a
tissue layer consisting of these 14 amino acids was degraded from the original egg shell. The increase in the proportion of four amino acids could be relative, merely reflecting unchanged amounts in other strata of the egg
shell, increasing in proportion of the total as the others
decreased. Analysis of the exudate of M. incognita larvae
shows a wide array of amino acids leaving the body of the
nematode. It appears that the physiology of the larvae
does not require an amino acid-sparing function. Furthermore the presence of arginine, asparagine, citrulline,
glutamine, and ornithine may in part possibly serve a
nitrogen-excreting function in M. incognita larvae. It is
striking that methionine was not found in any of the fractions examined and that cystine and cysteic acid were

TABLE I Various constituents of *Meloidogyne* spp.[a]

	Glycogen	Percent glycosides	Percent lipids	Percent protein	Percent chitin	Percent amino acids (dry wt)	ATP
M. arenaria							
adult			40				
L2			48				
egg			67				
M. incognita							
adult			46				
L2	+[b]		+			32	
♂	+		+			22	
♀			+			36	
♀ cuticle						91	
egg	+	+	64			26	80 pg/egg
egg shell				50	30		
M. javanica							
L2			30				
	6.9		40.4	40.9			

[a] Date from Krusberg, *et al.*, 1973; Van Gundy, *et al.*, 1967; Reversat, 1976; Bird and McClure, 1976; Dropkin and Acedo, 1974; Spurr, 1976.
[b] Indicates presence

TABLE II

Amino acid composition of various fractions of
root-knot nematodes[a]

	M. *hapla* and M. *javanica* Egg sac	Female cuticle	M. *incognita* Exudate	Ruptured egg shell mole %	Hatched egg shell mole %	Δ mole %
Alanine	+[b]	+	+	8.50	8.83	+0.33
Arginine	+	++	+	0.85	0.35	-0.50
Aspartic acid	++	+	+	5.58	5.91	+0.33
Glutamic acid	+	+	+	5.19	4.13	-1.06
Glycine	+	+	+	4.03	3.79	-0.24
Histidine	++	+	+	0.29	--	-0.29
Leucine	+	+	+	2.59	2.16	-0.43
Isoleucine				1.44	1.20	-0.24
Lysine	+	+	+	3.36	2.61	-0.75
Hydroxyproline	tr	tr		3.24	2.86	-0.38
Proline	+	++	+	35.42	40.23	+4.81
Serine	+		+	4.81	4.77	-0.04
Cystine		+				
Cysteic acid		+				
Threonine	++	+	+	5.60	5.03	-0.57
Tyrosine	tr	tr	tr	1.30	0.77	-0.53
Valine	+	+	tr	3.81	2.73	-1.08
Phenylalanine			+	4.42	1.71	-2.71
Methionine						
α-aminobutyric			tr	0.17	--	-0.17
Asparagine			+			
Citrulline			tr			
Glutamine			+			
Methionine sulfoxide			+			
Ornithine			+	0.36	0.76	+ .40

[a]Date from Bird, 1958; Myer & Krusberg, 1965; Bird & McClure, 1976

[b]Presence

TABLE III   Enzymes identified in various root-knot nematode fractions as present (+) or absent (-)[a]

Enzyme	Meloidogyne spp. exudate	M. javanica homogenate	M. javanica matrix	M. hapla	M. arenaria exudate	M. incognita exudate ♀	M. incognita L2	M. incognita egg mass	M. mali ♀
Amylase	+			-	-	-			
Protease	+							+	
Cytochrome oxidase		-	-				+		
Peroxidase		-	-	+	+	+			
Polyphenol oxidase	+	-,+	+						+
Invertase				+	+				
Cellulase				+	+				
Pectinase				+	+				
Alkaline phosphatase		+		+	+	+			
Acid phosphatase		+	+	+	+	+			
Leucine amino peptidase		-	-						
Esterases		+		+	+	+			
Lactic acid dehydrogenase		+		+	+	+			
Malate dehydrogenase		+		+	+	+			
α-glycerophosphate dehydrogenase		+		+	+	+			
Catalase					+	+			
Glucose 6-phosphate dehydrogenase		+		+	+	+			
6-phosphogluconate dehydrogenase						+			
Cholinesterase activity		+				+			
Trypsin-like enzyme						+	+	+	

[a]Date from Bird, 1958, 1966a, 1966b; Bird & Rogers, 1965; Dasgupta & Ganguly, 1975; Dickenson et al., 1971; Dieter, 1956; Dropkin, 1963; Hussey et al., 1972; Ishibashi, 1970; Mjuge, 1956; Myers, 1963, 1965; Van Gundy et al., 1967; Zinoviev, 1957.

present in the cuticle of the female of *M. hapla* and *M. javanica*. The low level presence of sulphur-containing amino acids in the fractions shown in Table II is somewhat at variance with observations of other nematode cuticles (von Brand, 1966; Krusberg, 1971), and may be a reflection of egg shell or perhaps nematode. The data are much too scant for the evidence to be conclusive.

Enzymes constitute a class of large molecular weight polymeric compounds, catalytically active and absolutely indispensable to physiological processes of life forms as we know them. It has been suggested that the simpler uni-cellular microorganism may contain in the order of 5,000 enzymes whereas the more complex unicellular microorgan-ism may contain in the order of 15,000 to facilitate vital processes. The fact that a compilation (Table III) for the presence of a mixed group of enzymes for various root-knot nematode fractions can be prepared is evidence of our scanty knowledge in this area. Such a compilation tends to conceal as much as it reveals; for example, less than 30% of the observations have been confirmed. Observations have been conducted by different authors with different methods so that with a duplicate observation like that of polyphenol oxidase from *M. javanica*, it is not possible to attribute the discrepancy to differing biotypes, dif-fering stages or differing methods. On the positive side the presence of cellulase and pectinase in an animal is not the usual situation and therefore has some signifi-cance for parasitic nematode relations with the plant host (Dropkin, 1963; Myers, 1965). Without specific aims miscellaneous observations of the presence of enzymes have limited usefulness. When the initial source is an homogenate of a multicellular organism such as a nematode, there is no way to determine what organ system, tissue, particle, or membrane may give rise to a particular enz-yme or whether it was in the proper relation to inter-mediates or other enzymes for a particular degradation,

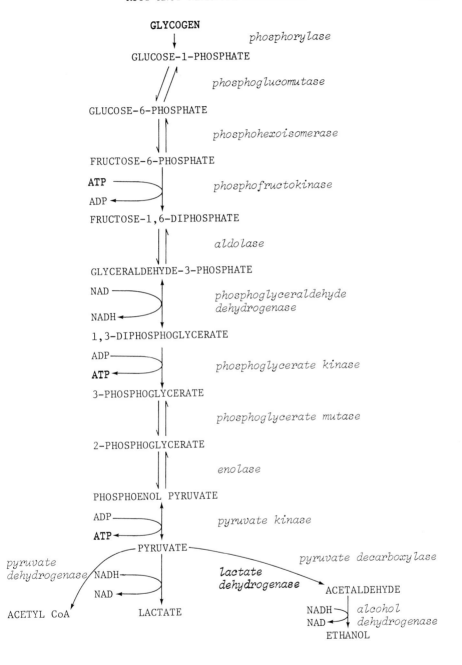

Fig. 1. Embden-Meyerhof pathway (Boldface indicates components found in root-knot nematodes)

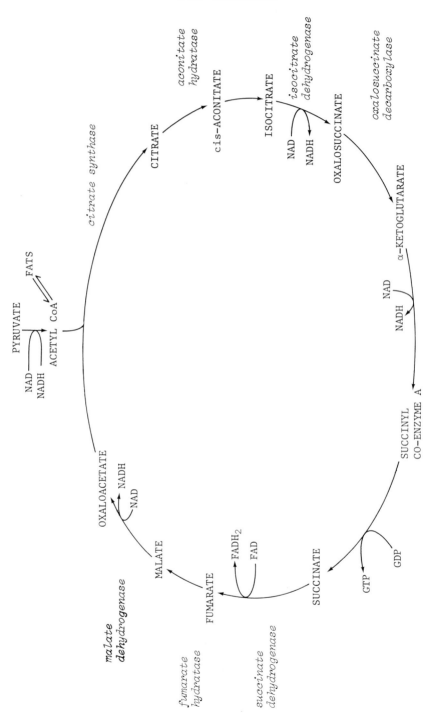

Fig. 2. Tricarboxylic acid cycle (boldface indicates components found in root-knot nematodes).

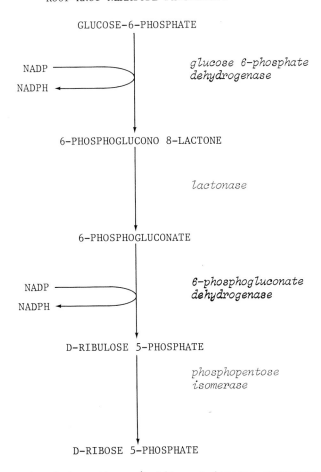

GLUCOSE-6-PHOSPHATE

NADP — glucose 6-phosphate
NADPH ← dehydrogenase

6-PHOSPHOGLUCONO 8-LACTONE

lactonase

6-PHOSPHOGLUCONATE

NADP — 6-phosphogluconate
NADPH ← dehydrogenase

D-RIBULOSE 5-PHOSPHATE

phosphopentose
isomerase

D-RIBOSE 5-PHOSPHATE

Fig. 3. Pentose-phosphate pathway (boldface indicates components found in root-knot nematodes)

transfer or electron transport to take place. If the understanding of a plant-parasitic nematode is to better in terms of physiology, metabolism or its interactions with the environment, a somewhat different orientation and emphasis in experimentation is essential. Our knowledge of root-knot nematodes can be illustrated by utilising several metabolic pathways frequently encountered in biochemical systems. In the Embden-Meyerhof pathway for glucose and glycogen utilization (Fig. 1) three components have been identified in the case of *M. incognita* - glycogen in the female, egg and L_2 stage, ATP also in the egg, and

lactic dehydrogenase in several stages. In the tricarboxylic acid cycle which is the classical system for the oxidation of acetate to produce energy rich phosphate bonds
(Fig. 2) one enzyme, malate dehydrogenase, has been identified from root-knot nematode homogenates. An alternative
route by which oxidation of carbohydrate takes place, the
pentose-phosphate pathway (Fig. 3), is usually found in
the cytoplasm and presumably generates a reducing power in
the cytoplasm as well as a formation of pentoses, particularly D-ribose which is used in the synthesis of nucleic
acids. Two enzymes, glucose-6-phosphate dehydrogenase and
6-phosphogluconate dehydrogenase, have been identified
from root-knot nematode homogenates. In such considerations it is essential to establish the presence of other
components and whether the components are in the proper
orientation for the system to function. There is an added
consideration that though the system may be present and is
able to function it may contribute only in a minor fashion
to the metabolism of the organism.

An alternative pathway of carbohydrate metabolism which
may occur in root-knot nematodes is the fumarate reductase
pathway (Fig. 4). One enzyme of this system has been identified from root-knot nematode homogenates. In this scheme
glucose is degraded in the cytoplasm to malate which then
transfers across the mitochondrial membrane which may continue its degradation via fumarate within the mitochondrion. Of particular interest is the formation of fatty
acids. Despite the scheme there remains some uncertainty
as to whether the fatty acid formation takes place within
the mitochondrion or within the cytoplasm or both (Bryan,
1975). Nematodes tend to have a high proportion of the
body weight in lipids and the body lipids also tend to be
liquids in the natural situation. Furthermore a substantial proportion of the fatty acids identified from animal
parasites as well as plant parasites are uncommon ones;
they vary from short chain to long chain and many are

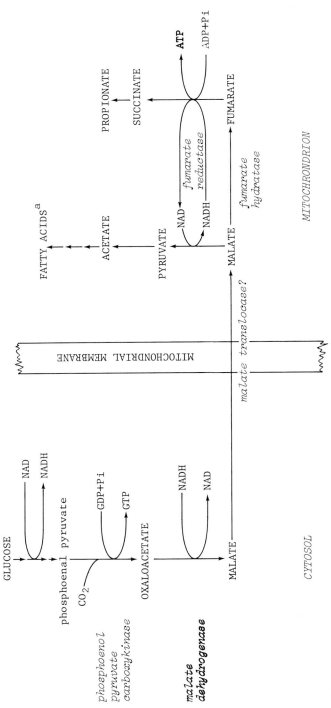

Fig. 4. Fumarate reductase pathway of carbohydrate metabolism occurring in *Ascaris* and some other nematodes (after Saz, 1972) (Boldface indicates components found in root-knot nematodes)

[a]Pyruvate conversion to acetate and to fatty acids may occur outside the mitochondrion (Bryant, 1975).

branched and unsaturated (von Brand, 1966; Krusberg, 1971). They are clearly different from the normal components found in plants or animals. The scheme in Figure 4 is useful in hypothetical explanations of events taking place within a cell, but the bulk of the broader questions concerning lipid metabolism remain unanswered. Root-knot nematode eggs have about two-thirds of their bulk made up by lipids (Table I). Transformations take place as the egg embryonates through the first-stage larva into the second-stage pre-parasitic form to where the bulk of the lipids reappear in the gut lumen. After feeding has begun are lipids synthesized from ingested carbohydrates directly in the gut by a fatty acid synthesizing system or is nutrient absorbed within the body and the fatty acid synthesized within the tissue then exported into the gut lumen? Are glycogen and trehalose involved in this process as they are for some animal parasitic nematodes (von Brand, 1966; Barrett, 1976)? Furthermore, the unusual fatty acids would require a somewhat different fatty acid synthesizing mechanism from the pathways usually observed. In the starving root-knot nematode larvae how does the system reverse itself to supply energy from the gut to the starving animal and subsequently what processes predominate when the intestine essentially transforms into a blind storage organ in the female? Our understanding of other vital processes, *i.e.* electron transport, protein metabolism and trace element nutrition is equally rudimentary and in need of attention.

In addition to the weakness of in-depth knowledge available there are deficiencies in comparative studies with different nematodes by different techniques and different workers that would decrease confidence in the available data. There is substantial fallacy in accepting information from other systems blindly on the assumption that the same events occurring in other nematodes also occur in root-knot nematodes.

Behavioural responses

Root-knot nematodes must express the appropriate behavioural responses if the population is to cope with the
environment and survive. Nematologists many years ago
observed that infective larvae of root-knot nematodes preferred a particular region of a host rootlet though other
portions of the plant are suitable (Linford, 1941). In the
diagrammatic sketch of a rootlet (Fig. 5), the region of
proliferation consisting of a root cap and a meristematic
zone is succeeded by a region of elongation followed by a

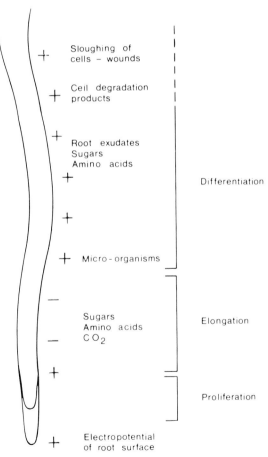

Fig. 5. Schematic diagram of a rootlet indicating characteristics of
various regions (redrawn after Jones, 1960).

TABLE IV

Factors investigated for chemotactic
properties to root-knot nematodes[a]

Attractive	Inactive	Repulsive
Seedling roots	Seedling roots	Seedling roots
Root region of elongation	Host root exudates	Host root exudates
Host root exudates	Non-host root exudates	Non-host root exudates
Non-host root exudates		
Bacteria	pH (?)	
Na hyposulfite	Colchicine	
Cysteine	IAA	
Glutathione	Digitoxin	
Ascorbic acid	Ouabain	
Tyrosine	Tryptophane	
Gibberellic acid	Glucose	
Glutamic acid	Fructose	
	Ribose	
	Xylose	
	Arabinose	
	Galactose	
	Sucrose	
Distilled H_2O		Range of electrolytes and non-electrolytes including Hoagland nutrient solution.
Relative Negative potential		

[a]For references see Croll, 1970.

region of differentiation and then the mature root. The
electrical potential along the root was found to be positive in the region of proliferation, negative in the region of elongation and positive again beyond it (Jones,
1960). Early analyses of emanations from the various

regions of the root were not too helpful in the search for an explanation for the attraction of the zone of elongation to infective root-knot nematode larvae. It was not difficult to show that the root region of a host plant was a preferred place for nematodes even if they could not penetrate the roots or that the roots of some non-host plants would attract while others would repel root-knot nematode larvae (Viglierchio, 1961). Subsequently more sophisticated devices incorporating known gradients and traps were devised in order to determine what substances were truly attractive, inactive, or repulsive. Table IV is a partial compilation of some of these findings. Some organic substances were found attractive; many others were found inactive (Croll, 1970). In a test system with traps utilizing an agar transport medium, distilled water traps invariably collected many more nematodes than a whole range of organic and inorganic substances including Hoagland nutrient solution (Viglierchio, unpublished). Other authors suggested that nematodes were motivated to move towards zones of relatively lower potential. Obviously the situation has yet to be resolved. Attraction appears to exist under certain conditions and it appears to be effective over short distances. Remarkable as it may seem there is yet absolutely no evidence as to which of the sensory organs detect the substances that elicit a positive or negative taxis response by the nematode. Neither is it known whether the physiology of detection involves polarity, conformational fit, redox potential or particular substituent groups.

Temperature is invariably an important consideration for poikilothermic animals subject to a fluctuating environment. For poikilothermic nematodes the range is generally from 0 to 40°C (Croll, 1970); for root-knot nematode species it would probably be more realistic to consider the range for major activity as between 10 and 35°C. Optimum temperatures given for the buildup of any

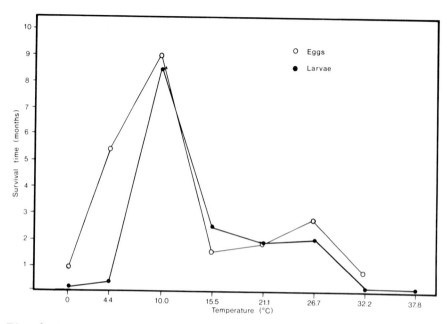

Fig. 6. Time period for survivors of a stored population of *M. incognita acrita* to be reduced to 10% of initial level in relation to storage temperature (data from Bergeson, 1959).

particular species need to be accepted with some reservations inasmuch as any such value is a resultant of an integration of optima of an enormous number of physiological reactions of the nematode as well as the plant, in the case of plant parasites. This notion can be illustrated with several observations, considering mainly the effect of temperature and ignoring side effects (Bergeson, 1959). The curve illustrating survival time of *M. incognita* larvae in soil (Fig. 6) deals primarily with the catabolic processes in starvation. At 0°C, survival time lasts 5-6 days, but with increase in temperature to 10°C, survival time rises dramatically to about one year. Further increases in temperature reduce survival time to about 2½ months; at temperatures greater than 25°C the survival time drops abruptly to the order observed at 0 and 4°C. Apparently at the low temperatures the reaction rates for vital processes are slowed too much to offer survival whereas high temperatures bring about a deactivation of reactions.

In a similar appraisal of survival time of the eggs of
M. incognita in soil (Fig. 6) two maxima in survival time
occur (Bergeson, 1959). The primary optimum as with larvae
occurs again at 10°C but after a sharp drop in survival
time with increasing temperature another peak occurs at
about 25°C. It appears that there may be at least two processes with different temperature maxima involved in egg
survival in the soil.

In a different behavioural response, egg hatching, *M.*
incognita acrita manifests several hatching optima depending upon the frame of reference (Bergeson, 1959) (Fig. 7).
In this situation the optimum hatching rate (greatest
slope, 26.7°C) is different from the optimum total hatch
(21.1°, 26.7°C) which again differs from that of the optimum sustained hatch (longest straight line, 15.5°C);
consequently it is necessary to determine the aspect of
hatching of interest and to select the appropriate optimum for that particular consideration. In a comparison of

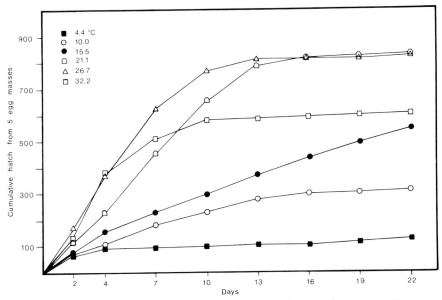

Fig. 7. Cumulative larval hatch of *M. incognita acrita* at different
temperatures (data from Bergeson, 1959).

the hatching at various temperatures of two other root-
knot nematode species for a six-day period (Bird and Wallace, 1965) (Fig. 8), the optimum total hatch occurs at

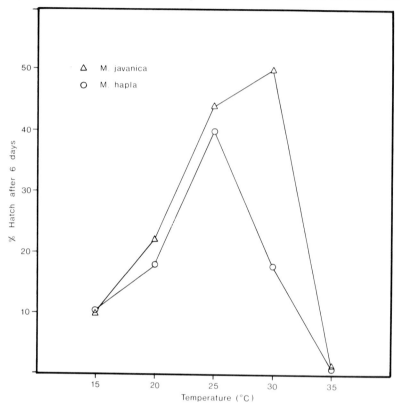

Fig. 8. Influence of temperature on the hatching of *M. javanica* and
M. hapla (redrawn from Bird and Wallace, 1965).

25°C for *M. hapla* and slightly higher at 30°C for *M. java-*
nica. In as much as hatching rate is a function of length
of hatch for *M. incognita* (Fig. 7), a longer hatching
period would have permitted the determination of whether
the results were more a reflection of hatching rate, or
of total hatch, or differential nematode viability.

When mobility (movement through a predetermined column
of sand) as a function of temperature was examined with
the same two species, optimum activity for *M. hapla* was
at 20°C and for *M. javanica* at 25°C (Fig. 9). However, in

penetration of host roots as a function of temperature, it was evident that *M. hapla* penetrated about the same at all temperatures. *M. hapla* penetration correlates well with its mobility as a function of temperature; *M. javanica* penetration is temperature independent which suggests that some process other than mobility is limiting penetration. On the other hand body growth of both nematodes appears to be temperature dependent. If female body size is used as a relative measure of body growth, the rate of increase of *M. hapla* females at 20 - 25°C is essentially the same as the rate of increase of *M. javanica* females at 25 - 30°C. The essential point in these comparison is that the optimal temperature for the buildup of a population of a species may differ from species to species but it is invariably the weighted resultant of many temperature dependent

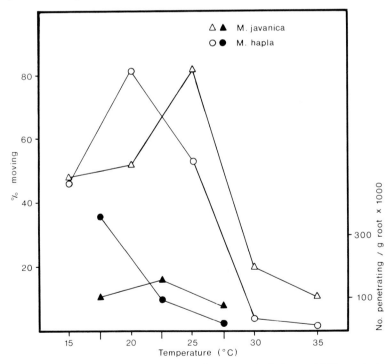

Fig. 9. Determination of mobility (open symbols) at different temperatures and penetration (filled symbols) at different temperature ranges (redrawn from Bird and Wallace, 1965).

and temperature independent physiological reactions govern-
ing not only behavioural responses but also many other processes. Clearly the long-term dynamics of field populations
increase in complexity over those under controlled conditions. Catabolism, anabolism, system destruction, seasonal
and diurnal fluctuations and other factors combine to
yield the resultant temperature function.

One of the few examples of a low temperature controlled
event occurring in Nematoda is found in root-knot nematodes. Of those root-knot nematodes whose biology has been
studied, *Meloidogyne naasi* appears to be the only one with
this particular temperature regime requirement. *M. naasi*,
which reproduces on cereal crops and grasses, has a well
recognized diapausing stage, L_2, in the egg (Evans and
Perry, 1976). The factors which induce diapause are unknown and since this state occurs naturally in the field
or in the laboratory it appears to be obligate. Host root
diffusates and artificial hatch stimulating agents are
ineffective. The only known treatment which breaks diapause and allows the nematode to resume its cycle is winter chilling in the field or a laboratory simulation of
seven weeks at 10°C followed by a return to room temperature for hatching. The physiological modification of the
nature of the process by which the life cycle block is removed is open to question. It may be similar to those processes which allow seeds to germinate after a cold treatment; processes themselves poorly understood.

Solute transport

Nematodes normally live in an aqueous medium and, like
any other organism, must solve the problem of solute transport to cope with their environment. Uptake of metabolic-
ally active substances and elimination of wastes are common vital problems. Any body opening would seem to be a
channel for solute transport; nematodes have several major
body apertures and many more minor ones. The principal

intake aperture is the oral opening through which major
solutes of life support pass. The other body openings, at
least for root-knot nematodes, normally serve as apertures
for outflow; of course in some of the minor apertures,
there may be no net flow as in sensory functions.

Alimentation is one of the major stumbling blocks in
the understanding of biological processes in nematodes
which can be illustrated by the root-knot nematode. Most
plant parasitic nematodes are obligate parasites (Christie,
1959) so that without a suitable piece of host tissue upon
which to feed the life cycle is blocked. Nutrition is the
key which reactivates the cycle and whose quality and
quantity effect a spectrum of consequences (McClure and
Viglierchio, 1966; Trudgill, 1972). It is common knowledge,
for example, that a poor host, high temperatures, or crowding can modify the sex ratios either by redirecting the
apparent sexes or by inhibiting the development of a sex.
Nutrition apparently can change morphology, e.g. in a
Rhodesian *M. javanica* population the females on most hosts
developed a vulva nearly terminal but with the same population on potato the females developed a vulva nearly midbody (Martin, 1960). The importance of nutrition notwith-
standing, the information available from root-knot nematodes can be summarized by stating that the cytoplasmic
contents of plant cells are ingested and processed internally in some fashion to give proteins, carbohydrates and
fats, while the unused remainder is released to the outside. There is little known about the quantity and quality
of what is ingested; what is known about the processes
internally and what is released to the outside have been
discussed earlier. The difficulty of relevant studies not-
withstanding, our knowledge is woefully inadequate.

A different major avenue of transport is the body wall
which for some life-sustaining substances is a principal
pathway of transport. The number of compounds that can
penetrate the body wall is enormous (Marks *et al.*, 1968;

Castro and Thomason, 1973; Husain and Masood, 1974); in
summary it could perhaps be said that some small polar
molecules selectively penetrate readily while other larger
ones do so less well, if at all. The smaller non-polar
molecules appear to penetrate readily, the ease decreasing
with increasing size.

Water is one of the major substances that passes readily through the body wall. This can be appreciated simply
by observing a root-knot larva dry out on a microscope
slide and noting how quickly the water moves through the
body wall into the atmosphere. On the other hand, with a
little care a root-knot larva can be dried out until it is
well wrinkled from loss of water, then, upon addition of
water, it quickly recovers its normal size and activity.
It appears that water can move freely through the body
wall in either direction as has been observed with other
nematodes using tritiated water. Under the usual condition
of rapid drying the nematode appears to die at about the
time the body water vapour pressure equilibrates with that
of the atmosphere. On the other hand, there is a wealth of
practical experience indicating that under different,
slower drying conditions there is a stage of root-knot
nematode which can survive. This suggests that there is
some mechanism by which sufficient water is retained within the nematode to permit it to survive in an anhydrobiotic state. Other invertebrates, including nematodes, have
developed such mechanisms (Crowe, 1971). Recently such a
nematode in the anhydrobiotic state was found to have synthesized trehalose and glycerol, neither of which was present in the active animal (Madin and Crowe, 1975). Perhaps
some similar mechanism occurs with root-knot nematodes
which helps their survival in nature.

The life support of an animal is dependent upon the
energy bound up in the food which it consumes. This energy
is released by the organism in a series of processes of
chemical breakdown which are summarized into one term,

metabolism. As a matter of convenience metabolism has been
divided into aerobic, requiring oxygen, or anaerobic, not
requiring oxygen. Aerobic metabolism can supply far more
energy from food than can anaerobic. Glucose, for example,
is convertible by aerobic metabolism to carbon dioxide and
water to yield 39 energy-rich phosphate bonds available to
power other biological processes. The same quantity of
glucose, however, metabolized anaerobically would be converted to lactic acid to yield only three energy-rich
phosphate bonds. Aerobic metabolism is more efficient and
is the process commonly used by most organisms; consequently animals rely on different systems to keep their
tissues adequately supplied with oxygen. Nematodes lack
specialized respiratory or circulatory organs but some
have developed certain aerating aids (von Brand, 1966).
Nematodes as aquatic animals must derive their oxygen from
that dissolved in the waters of the surrounding medium
and the process by which this enters the tissues is called
diffusion. It is useful to examine diffusion in terms of
how it applies to nematodes and in particular to root-
knot nematodes. Gas diffusion is a function of the partial pressure gradient. The pressure exerted by each gas
in a fixed volume of a gas mixture is equal to the pressure of that gas in the absence of all others in that
fixed volume. Inasmuch as air is one-fifth oxygen by volume, the partial pressure of oxygen in air as well as
water in equilibrium with air is 0.2 atmospheres. Diffusion favours the dispersion of particles from a dense to
a less dense region (Rogers, 1962; Alexander, 1971). Refer to the schematic diffusion diagram (Fig. 10)

R = rate of diffusion cm^3/hr

A = area of surface through which diffusion takes
place

D = diffusion constant

dp = pressure difference between two regions separated
by a short distance

ds = short distance separating pressure regions
Then

Equation (1)   R = -ADdp/ds

The value of the diffusion constant is dependent primarily on the particle diffusing, the medium through which
it diffuses and the temperature at which diffusion takes

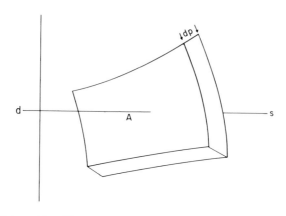

Fig. 10. Schematic diagram illustrating diffusion across an Area, A,
of thickness, ds, with a pressure differential, dp, as explained in
text.

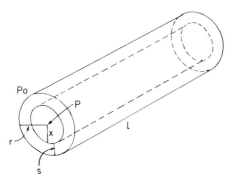

Fig. 11. Schematic diagram illustrating diffusion inwards from the
surface of a cylinder of tissue, as explained in text.

place. The negative sign indicates a decreasing pressure
change.

With the general diffusion equation available it is
now possible to consider it as it applies to a nematode.
As an approximation consider a vermiform nematode without

aerating aids as a cylinder of tissue, Fig. 11. Let the
radius of the outer cylinder be equal to r and the radius
of the inner cylinder be equal to x and the length of the
cylinder be equal to 1. Let the partial pressure of oxygen
at the outer surface be equal to p_0 and the partial pressure along the axis of the cylinder be equal to p.

$$\text{Area of inner cylinder} = 2\pi x 1 \ cm^2$$

$$\text{Volume of inner cylinder} = \pi x^2 1 \ cm^3$$

$$m = \text{oxygen consumption in } cm^3 O_2/cm^3 \text{ tissue hour}$$

$$R = \text{oxygen going through the surface} = \pi m x^2 1$$

$$s = (r-x)$$

Substituting these values in equation (1) and solving for
r_{max}

$$\pi x^2 m 1 = -2\pi x 1 D dp/d(r-x)$$

$$dp = mxdx/2D$$

$$p_0 - p = \frac{m}{2D} \int_0^r xdx$$

$$p_0 - p = mr^2/4D$$

with $\qquad p \geq 0$

then $\qquad p_0 \geq mr^2/4D$

and equation (2)

$$r_{max} \leq 2 \frac{\sqrt{p_0 D}}{m}$$

The maximum radius as expressed in equation (2) can serve
as an approximation for vermiform larvae, adult males,
sausage-shaped juveniles but not adult saccate females.
A different approximation can be derived for adult females assuming a spherical configuration for the adult
female whereby

$$R = 4/3 \ \pi r^3 m$$

$$A = 4 \ \pi \ x^2$$

$$s = (r-x)$$

Then substituting in equation (1)

$$4/3 \ \pi \ r^3 m = -4 \ \pi \ D \ x^2 \ dp/d(r-x)$$

$$dp = \frac{m}{3D} \ xdx$$

$$p_0 - p = m/3D \int_o^r xdx = mr^2/6D$$

so that $\qquad p_0 \geq mr^2/6D$

and Equation (3)

$$r_{max} \leq \frac{\sqrt{6Dp_0}}{m}$$

If $m = 0.1$ cm^3 O_2/cm^3 tissue hour as is the case with many animals including some nematodes and tapeworms and $p_0 = 0.2$ atmospheres as would be the case in a medium in equilibrium with air and $D = 8 \times 10^{-4} cm^2/atm$ h, using the approximation of the value for frog muscle as being equal to nematode tissue, then r_{max} for the vermiform larvae is equal to 800 μm and r_{max} for the spherical shaped female is equal to 980 μm. Despite the numerous crude assumptions made to derive the formulae and the approximations used for the different variables, root-knot nematodes appear to fall well within the size limits predicted by theory beyond which body wall diffusion processes can no longer supply adequate oxygen for tissue respiration. There are a number of limitations to this scheme, however, which it would be best to keep in mind:

i) Root-knot nematodes do not have a truly cylindrical or spherical shape as is well known.

ii) Different layers and tissues of root-knot nematodes may have different diffusion coefficients.

iii) The movement of the body fluids within the root-knot nematode aids in the distribution of oxygen.

iv) The oxygen demands may fluctuate and vary between tissues and layers.

v) Outer stationary layers adjacent to the external surface of the root-knot nematode may modify the supply of oxygen to its surface.

In summary, the large safety factors suggest there is no pressing need to devise more complex processes to explain the transport of an adequate oxygen supply to all nematode body tissues. The difference in size limits between observed and those predicted by theory by a factor of 4 or greater is perhaps best explained by errors in the approximate values used to calculate r_{max} but does not exclude the possible existence of some other limiting process (Baxter and Blake, 1969). It has been established that *M. javanica* on tomatoes goes through its life cycle with oxygen tensions of about 3.5% in the soil (Van Gundy and Stolzy, 1961), but when oxygen fell below this value egg production decreased. Calculations upon insertion of these considerations into equation (3) predict possible difficulties in oxygen supply and therefore are consistent with observation. Similarly determinations of the oxygen requirements of egg hatching in *M. javanica* are not in disagreement with theoretical predictions. Hatching improved continuously as oxygen tensions increased to normal atmospheric levels (Baxter and Blake, 1969). Initially hatched *M. javanica* larvae take up approximately 6.8 μm^3. of oxygen per mg hr whereas by the fifth day a larva consumes approximately one third of this amount (Van Gundy *et al.*, 1967). This indicates a threefold difference in oxygen consumption by one individual and therefore the possibility of a three-fold difference in the value of m as used in diffusion equations. Despite the usefulness of the diffusion equations they must be applied with caution until improved determinations of variables are available.

There is evidence which can be interpreted to mean there are different metabolic rates between root-knot nematode species, as well as within a species, as a function of temperature (Fig. 12). In *M. hapla* the respiration rate, as measured by oxygen uptake, rises logarithmically over the entire range between 5-30°C (Lyons *et al.*, 1975); furthermore it appears that oxygen uptake increases more rapidly and, at higher temperatures, is greater for *M. hapla* than for *M. javanica*. However, the higher consumption may be simply a reflection of nematode sampling rather than true difference between species. *M. javanica* shows a distinct discontinuity in oxygen uptake with increasing temperature; the discontinuity appears between 15 - 20°C.

In an analysis of nematode lipids by electron paramag-
netic resonance (EPR) or electron spin resonance (ESR),
possible phase transition of the lipids were determined
as a function of temperature. The fats of the nematode
were solvent extracted then chromatographically separated
to give a crude polar lipid fraction. A spin label was

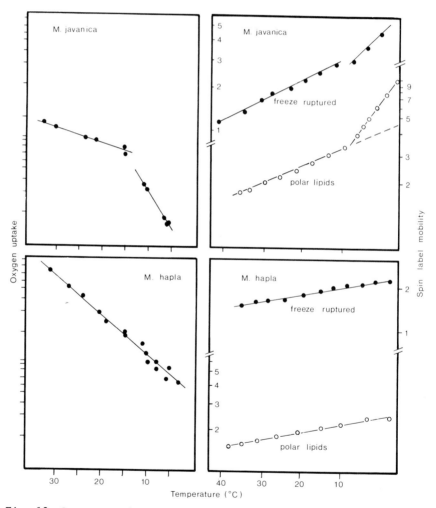

Fig. 12. Oxygen uptake as an estimate of respiration in two species
of root–knot nematode (left) and spin label mobility for extracted
lipids and ruptured nematodes as function of temperature (redrawn
from Lyons *et al.*, 1975).

added to polar lipids or the freeze-ruptured nematodes
and then the energy required for resonance as a function
of temperature was established. For *M. hapla* it is obvious that the curve is continuous over the whole temperature range for both polar lipids and for freeze-ruptured
nematodes (Fig. 12). For *M. javanica* the curve for freeze-
ruptured nematodes may or may not be continuous; the
authors interpreted it as discontinuous. The curve for
polar lipids is unquestionably discontinuous at a point
between 15 - 20°C at which temperature a phase transition
between solid and liquid occurs. The investigators suggested that this discontinuity indicated a phase change
in the membrane lipids of *M. javanica* which in turn exerted control over the respiratory enzymes which in turn
reflected a respiratory discontinuity at that temperature
and was the basis for the chilling injury suffered by this
nematode. This is a very interesting observation but there
are several factors which should be kept in mind. Crude
preparations of polar lipids from higher animals consisting predominantly of phospho-lipids, also contain variable
amounts of fatty acids and steroids. Nematodes with their
very high normal fat content and unusual fatty acid compositions may yield a crude polar lipid fraction different
from that of other animals. Usually for biochemical studies of membranes, phospho-lipids are obtained from tissues relatively low in neutral fats (Maddy, 1966). In
whole nematodes containing about 50% lipids, however, the
phospho-lipids present in the membranes of the order of a
thousand cells is exceedingly small in proportion to the
lipids elsewhere. These ESR curves suggest that the polar
lipids of *M. javanica* and *M. hapla* are substantially different and that perhaps the discontinuity in the respiratory rate of *M. javanica* may be due to modification of
diffusion coefficients with temperature as well as respiratory enzyme function. Clearly there remain many unanswered questions so that, to put respiration and

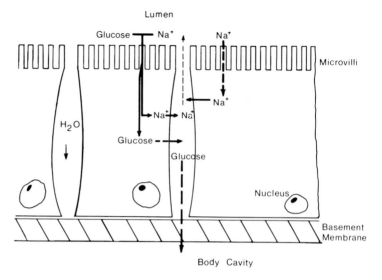

Fig. 13. Schematic representation of electrolyte and glucose transport across the intestinal wall of *Ascaris*. Dashed arrows indicate
diffusion, solid arrows indicate active transport and fused solid
arrows indicate coupled active transport (redrawn from Beames *et al.*,
1977).

metabolic rate in the root-knot nematode into perspective,
there is need for a series of careful determinations of
oxygen consumption of several stages and of several species.

Studies of solute transport through tissues have been
carried out with certain animal parasitic nematodes using
primarily the simpler system of the intestine of *Ascaris
suum*. The system, illustrated schematically in Fig. 13,
shows the intestinal wall comprising essentially three
layers - the microvillae extending into the lumen, the
columnar cells and the basement membrane. An excised section of the intestine immersed in a test bath in such a
fashion as to permit the independent control of solutions
in the lumen and external to the intestine, can be used
to demonstrate water transfer from the lumen side to the
body cavity side of the intestinal wall by way of the
intercellular spaces with the aid of a stimulating drug,

Ouabain (Beames *et al.*, 1977). Glucose and sodium ions are transported across the cell wall against a gradient. If an intestinal section is immersed in an isotonic medium, a potential develops between a negative cavity wall and a positive luminal wall. By applying a potential in the inverse direction from an external source it is possible to reduce the potential difference to zero with passage of a current. Addition of glucose causes an immediate drop in current. Apparently the resultant movement of solutes involves a combination of active and passive transport; the active transport and the maintenance of a potential difference across the intestinal wall is accomplished with the expenditure of glucose. This model, in common with all others, endeavours to depict graphically certain observations but leaves unanswered the fate of other solutes that pass through the gut. Of particular interest here is the nature of the intestine of root-knot nematodes and how the fine structure and transport through the intestinal wall differ from that of *Ascaris*. Basic differences are to be expected inasmuch as the root-knot nematode intestine which accepts and processes incoming nutrients from plant cells also serves as a storage receptacle for reserve lipids. Furthermore, from the literature one is compelled to conclude that the digestive enzymes within the intestinal lumen must originate in part from the intestinal cells (Lee and Atkinson, 1976). The dorsal gland, which vents into the oesophageal lumen just posterior to the base of the stylet and secretes material which has been observed to exude from the stylet to aid in plant cell penetration and extraoral digestion, enlarges as the nematode continues to grow and develop, whereas the subventral oesophageal glands which change in internal composition with penetration of the host and onset of feeding do not (Bird, 1971). As the larva develops into a female, the intestine occludes at the posterior end to become a blind sac open only anteriorly. The digestive

system and associated glands of the root-knot nematode
undergo a series of modifications and activities as it
proceeds through its life cycle. In summary, digestive
mechanisms in root-knot nematodes remain essentially a
matter of speculation based on a few scattered facts.

The body surface of a root-knot nematode constitutes
the interface between the nematode within and the environment without, and the material exchange across the boundary and through the body wall is a matter of survival for

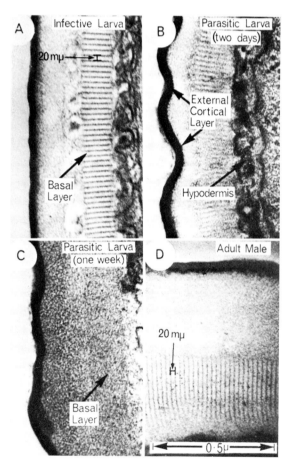

Fig. 14. Electron micrographs illustrating physical changes of the
cuticle associated with the onset of parasitism in *Meloidogyne javanica* and the cuticle of the adult male (reproduced with permission
from Bird, 1971).

the nematode. The body wall external to the hypodermis
consists of a series of different layers and tissues of
different compositions, much more complex than the intestinal wall previously discussed (Bird, 1971). Furthermore
there are changes in the body wall with age, state and
stage. The micrographs of Fig. 14 illustrate the changes
from infective or preparasitic larvae, to parasitic larvae
two days old, to parasitic larvae one week old and finally
the adult male (Bird, 1971). With all these changes

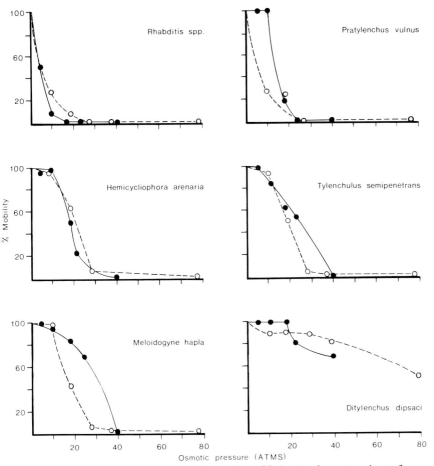

Fig. 15. The mobility of nematodes after 24 hr in hypertonic solutions, followed by transfer to distilled water. Percent mobility
with respect to water controls, solid line indicates urea and dotted
line indicates sodium chloride (Viglierchio *et al.*, 1969).

occurring in the development of *M. javanica*, it would be unrealistic to expect solute transport through the body wall to remain unchanged as the animal goes through all these modifications.

In any event many nongaseous solutes are able to penetrate the nematode body wall with relatively nonpolar organic solutes penetrating most easily. Certain salts and water soluble nonelectrolytes seem to penetrate body walls of plant parasitic nematodes poorly but those of non-plant parasitic forms better (Viglierchio *et al.*, 1969) (Fig. 15). Nematodes vary in these properties; however, the plant parasitic forms appear among the more impermeable ones. Methods generally have been too imprecise to indicate whether small amounts of solute are able to penetrate. Isotopic tracer techniques suggest that certain solutes are able to penetrate slowly (Marks *et al.*, 1968); the times involved, however, leave open to question the relative proportions that enter through the body wall or through other channels, e.g. by oral ingestion.

It has long been believed that the surface of the nematode probably consists of a lipid layer; Chitwood and Thorne discussed this long ago. In ultrastructure studies of root-knot infective larvae the external surface appears as a normal membrane with its typical electron transparent strata sandwiched between two electron-dense lamina (Bird, 1971) (Fig. 16). In serological studies antibodies induced in rabbits by live larvae of *M. javanica* show precipitin reactions principally with emanations from the nematode stylet and the excretory pore but also "some" with body surfaces (Bird, 1964). Unless there is some unique contaminant, the surface is unlikely to be all lipid inasmuch as lipids are not noted for stimulating antibody formation whereas polysaccharides and to a much greater extent proteins are. In a different kind of study with lipase, protease and chitinase hydrolytic enzymes of plant parasitic nematodes other than root-knot, it was

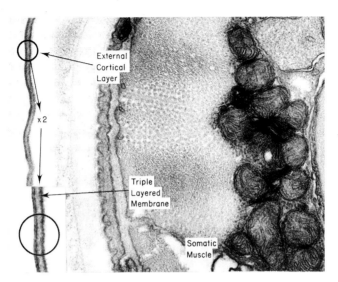

Fig. 16. Electron micrograph of the cuticle of a preparasitic larvae of *Meloidogyne javanica* illustrating the electron dense-transparent-dense membrane-like structure constituting the external layer of the nematode (X40,000) (reproduced with permission from Bird, 1971).

reported that enzyme treatments of the body surface were toxic to the nematodes as a result of a modification (hydrolysis) of the cuticle (Miller and Sands, 1977). In a different series of studies with root-knot nematode species there appeared to be evidence that the lipases and proteases also modify the surface of the nematodes by altering their migration patterns in an electrical field. The evidence suggests that the outer surface of a root-knot nematode (that part that interfaces with the environment) has mixed characteristics - mostly lipids but also some proteins and carbohydrates.

In the current concept a membrane, such as one surrounding a cell, consists basically of a lipid bilayer with some proteins extending across it (Singer and Nicolson, 1972; Brewer and Passwater, 1974). Some of these proteins will have associated carbohydrate on the surfaces away from the cell (Winzler, 1970; Bretscher, 1973). Furthermore, there is additional protein associated with the surface external to the outer lipid portion of the bilayer but much more with the inner cytoplasmic surface of the membrane. Evidence from electron microscopy and biochemical studies indicate that the outermost layer of the nematode body wall manifests a number of properties of the cellular membrane. Membrane components are of cytoplasmic origin and it is difficult to reconcile a vital dependence in a solute transport control activity in the outermost layer of the body wall so remote from the nearest cell layers of the hypodermis. Clearly this question remains in need of resolution.

In summary it appears that there is passive and active transport through the root-knot nematode body wall. Some gases and perhaps certain low molecular weight hydrocarbons permeate the body wall by diffusion. It is also clear that active transport is mediated by one or more semipermeable membranes which are able to select and control solute transport in either direction. The answers to a number of practical questions depend upon an understanding of body wall transport. Solutions to such questions as how root-knot nematodes cope with saline soils, what kind of leakage can be expected from the body wall in relation to the emanations from other orifices, or what kind of formulations and structures predispose fumigant or systemic nematicides to more facile entry into the nematode, lie in large part in the study of body wall transport.

References

Alexander, R.McN. (1971). "Size and Shape". Edward Arnold Ltd.
Barrett, J. (1976). *In* "The Organization of Nematodes" (Ed. N.A. Croll), 11-70. Academic Press, New York.
Baxter, R.I. and Blake, C.D. (1969). Oxygen and the hatch of eggs and migration of larvae of *Meloidogyne javanica*. *Ann.appl.Biol.* 63 191-203.
Beames, C.G., Jr., Mera, J.M. and Donahue, M.J. (1977). *In* "Water Relations in Membrane Transport in Plants and Animals" (Eds. A.M. Jungreis, T.K. Hodges, A. Kleinzeller and S.G. Schultz), 97-109. Academic Press, New York.
Bergeson, G.B. (1959). The influence of temperature on the survival

of some species of the genus *Meloidogyne*, in the absence of a host. *Nematologica* 4, 344-354.

Bird, A.F. (1958). The adult female cuticle and egg sac of the genus *Meloidogyne* Goeldi, 1887. *Nematologica* 3, 205-212.

Bird, A.F. (1964). Serological studies on the plant parasitic nematode *Meloidogyne javanica*. *Exper.Parasitol.* 15, 350-360.

Bird, A.F. (1966a). Esterases in the genus *Meloidogyne*. *Nematologica* 12, 359-361.

Bird, A.F. (1966b). Some observations on exudates from *Meloidogyne* larvae. *Nematologica* 12, 471-482.

Bird, A.F. (1971). "The Structure of Nematodes" 318. Academic Press, New York.

Bird, A.F. and McClure, M.A. (1976). The tylenchid (Nematoda) egg shell: structure, composition and permeability. *Parasitology* 72, 19-28.

Bird, A.F. and Rogers, D.E. (1965). Ultrastructural and histochemical studies of the cells producing the gelatinous matrix on *Meloidogyne*. *Nematologica* 11, 231-238.

Bird, A.F. and Wallace, H.R. (1965). The influence of temperature on *Meloidogyne hapla* and *M. javanica*. *Nematologica* 11, 581-589.

Bretscher, M.S. (1973). Membrane structure: some general principles. *Science* 181, 622-629.

Brewer, A.K. and Passwater, R.A. (1974). Physics of the cell membrane, I. The role of double-bond energy states. *Amer.Laboratory* April, 59-74.

Bryant, C. (1975). Carbon dioxide utilization and the regulation of respiratory metabolic pathways in parasitic helminths. *Adv.in Parasitology* 13, 35-69.

Castro, C.E. and Thomason, I.J. (1973). Permeation dynamics and osmoregulation in *Aphelenchus avenae*. *Nematologica* 19, 100-108.

Christie, J.R. (1959). "Plant Nematodes", 256. H. & W.B. Drew Co., Jacksonville, Florida.

Croll, N.A. (1970). "The Behaviour of Nematodes" 117. St. Martin's Press, Edward Arnold, London.

Crowe, J.H. (1971). Anhydrobiosis: an unresolved problem. *Amer. Naturalist* 105, 563-574.

Dasgupta, D.R. and Ganguly, A.K. (1975). Isolation, purification and characterization of a trypsin-like protease from the root-knot nematode, *Meloidogyne incognita*. *Nematologica* 21, 370-384.

Dickson, D.W., Huisingh, D. and Sasser, J.N. (1971). Dehydrogenases, acid and alkaline phosphatases, and esterases for chemotaxonomy of selected *Meloidogyne*, *Ditylenchus*, *Heterodera* and *Aphelenchus* spp. *J. Nematology* 3, 1-16.

Dieter, A. (1955). Vergleichende experimentelle Untersuchungen an zoophagen und phytophagen Nematoden. *Wissenschaftliche Zeitschrift der Martin Luther Universität Halle-Wittenberg* 5, 157-186.

Dropkin, V.H. (1963). Cellulase in phytoparasitic nematodes. *Nematologica* 9, 444-454.

Dropkin, V.H. and Acedo, J. (1974). An electron microscopic study of glycogen and lipid in the female *Meloidogyne incognita* (root-knot nematode). *J. Parasitology* 60, 1013-1021.

Evans, A.A.F. and Perry, R.N. (1976). *In* "The Organization of

Nematodes" (Ed. N.A. Croll), 383-424. Academic Press, New York.
Husain, S.I. and Masood, A. (1974). Studies on the effect of pipera-
    zine citrate on nematode mortality and larval hatching of *Meloi-
    dogyne incognita*. *Indian J. Nematology* 4, 102-104.
Hussey, R.S., Sasser, J.N. and Huisingh, D. (1972). Disc-electro-
    phoretic studies of soluble proteins and enzymes of *Meloidogyne
    incognita* and *M. arenaria*. *J. Nematology* 4, 183-189.
Ishabashi, N. (1970. Variations of the electrophoretic patterns of
    Heteroderidae (Nematodea: Tylenchida) depending on the develop-
    mental stages of the nematode and on the growing conditions of
    the plants. *Appl.Ent.Zool.* (Tokyo), 5, 23-32.
Jones, F.G.W. (1960). Some observations and reflections on host find-
    ing by plant nematodes. *Meded Landouwhogesch.* Gent, 25, 1009-
    1024.
Krusberg, L.R. (1971). *In* "Plant Parasitic Nematodes" (Eds. B.M.
    Zuckerman, W.F. Mai and R.A. Rhode), II, 213-234. Academic Press,
    New York.
Krusberg, L.R., Hussey, R.S. and Fletcher, C.L. (1973). Lipid and
    fatty acid composition of females and eggs of *Meloidogyne incog-
    nita* and *M. arenaria*. *Comp.Biochem.Physiol.* 45B, 335-341.
Lee, D.L. and Atkinson, H.J. (1976). "Physiology of Nematodes", 215.
    MacMillan Press Ltd, London.
Linford, M.B. (1941). Parasitism of root-knot nematode in leaves and
    stems. *Phytopathology* 31, 634-648.
Lyons, J.M. Keith, A.D. and Thomason, I.J. (1975). Temperature-
    induced phase transitions in nematode lipids and their influence
    on respiration. *J. Nematology* 7, 98-104.
Maddy, A.H. (1966). The chemical organization of the plasma membrane
    of animal cells. *Int.Rev.Cytology* 20, 1-65.
Madin, K.A.C. and Crowe, J.H. (1975). Anhydrobiosis in nematodes:
    carbohydrate and lipid metabolism during dehydration. *J.Exp.Zool.*
    193, 335-342.
Marks, C.F., Thomason, I.J. and Castro, C.E. (1968). Dynamics of the
    permeation of nematodes by water, nematocides and other sub-
    stances. *Exp.Parasitology* 22, 321-337.
Martin, G.C. (1960). Nematological Jottings. Proceedings of the First
    Inter-African Plant Nematology Conference. East African Agricult-
    and Forestry Res. Organ. Kikuyu, Kenya.
McClure, M.A. and Viglierchio, D.R. (1966). The influence of host
    nutrition and intensity of infection on the sex ratio and develop-
    ment of *Meloidogyne incognita* in sterile agar cultures of excised
    cucumber roots. *Nematologica* 12, 248-258.
Miller, P.M. and Sands, D.C. (1977). Effects of hydrolytic enzymes
    on plant parasitic nematodes. *J.Nematology* 9, 192-197.
Mjuge, S.G. (1956). About the physiology of nutrition of the gall
    nematode. *Dokladi Akademii NAUK*, SSSR, 108, 164-165.
Myers, R.F. (1963). Materials discharged by plant parasitic nematodes.
    Phytopathology 53, 884.
Myers, R.F. (1965). Amylase, cellulase, invertase and pectinase in
    several free living, mycophagus and plant parasitic nematodes.
    Nematologica 11, 441-448.
Myers, R.F. and Krusberg, L.R. (1965). Organic substances discharged
    by plant-parasitic nematodes. *Phytopathology* 55, 429-437.

Reversat, G. (1976). Étude de la composition biochimique globale des juvéniles des nematodes *Meloidogyne javanica* et *Heterodera oryzae*. *Cah.ORSTOM.ser Biol.* **XI**, 225-234.

Rogers, W.P. (1962). "The Nature of Parasitism", 287. Academic Press, New York.

Saz, H.J. (1972). *In* "Comparative Biochemistry of Parasites" (Ed. H. van den Bossche), 33-47. Academic Press, New York.

Singer, S.J. and Nicolson, G.L. (1972). The fluid mosaic model of the structure of cell membranes. *Science*, **175**, 720-730.

Spurr, H.W., Jr. (1976). Adenosine triphosphate quantification as related to cryptobiosis, nematode eggs and larvae. *J.Nematology* **8**, 152-158.

Trudgill, D.L. (1972). The influence of feeding duration on moulting and sex determination of *Meloidogyne incognita*. *Nematologica* **18**, 476-481.

Van Gundy, S.D. and Stolzy, L.H. (1961). Influence of soil oxygen concentrations on the development of *Meloidogyne javanica*. *Science* **134**, 665-666.

Van Gundy, S.D., Bird, A.F. and Wallace, H.R. (1967). Aging and starvation of *Meloidogyne javanica* and *Tylenchulus semipenetrans*. *Phytopathology* **57**, 559-571.

Viglierchio, D.R. (1961). Attraction of parasitic nematodes by plant root emanations. *Phytopathology* **51**, 136-142.

Viglierchio, D.R., Cross, N.A. and Gortz, J.H. (1969). The physiological response of nematodes to osmotic stress and osmotic treatment for separating nematodes. *Nematologica* **15**, 15-21.

Von Brand, T. (1966). "Biochemistry of Parasites", 429. Academic Press, New York.

Winzler, R.J. (1970). Carbohydrates in cell surfaces. *Intern.Rev. Cytology* **29**, 77-125.

Zinoviev, V.G. (1957). Enzymatic activity of nematodes parasitizing plants. *Zoologicheski Zhurnal* **36**, 617-619.

Discussion

Replying to a question, Viglierchio said that Lyons *et al.* (1975) made measurements of oxygen consumption and spin-labelling of polar lipids and observed that there is a large step from the results of physiological experiments to the conclusions that can be drawn from them. Wouts queried the emphasis placed on the use of parts of nematodes, rather than the whole animal as used by Dalmasso in his studies. Viglierchio replied that while the whole animal could be used effectively for some types of experimental observations, it nevertheless was just an homogenate in terms of metabolic pathways. Evans asked whether there was any substance in the giant cell which allows a root-knot nematode to become a female, and went on to suggest that the production of males may be triggered by the amount of energy available; glucose was suggested as a possibility. Evans posed the question that if the control of nematode development is regulated by energy availability why are males produced when development could be halted until sufficient food becomes available to continue development to the female state. Triantaphyllou remarked that there have been some attempts to associate sexuality with plant hormones.

Bird noted that the importance had been stressed of membranes at the surface of the egg and the nematode and thought that they are the key to nematode survival. Viglierchio said that the surface indeed is important. There may be difficulties, however, in understanding the permeability of the various states of the membrane as it is a complex system. Possibly it can be studied in *Meloidogyne* females using a hemisphere of cuticle but there are problems in removing the tissue from the inside without causing damage.

Sasser referred to desert lands in the coastal valleys of Peru which had been brought back into cultivation by using water from deep wells, but which suffered root-knot problems in the first year, and queried whether nematodes could survive in a quiescent state for many years. Viglierchio doubted whether they could survive that long; Netscher offered the explanation that nematodes may be able to complete their life cycle on plants that grow quickly after yearly rains; in Mauritania and the Sahara desert he had found *Meloidogyne* at 3 metres depth in plant roots. Orion said that in the Sinai desert a rainfall of 2.5 cm per year is enough to keep *Meloidogyne* nematodes alive. Dalmasso observed that some species of *Xiphinema* and *Tylenchorhynchus* can survive for many years in dry soil. Bird pointed out that organisms do not live for very long without water and 39 years is the maximum time recorded for a nematode in the cryptobiotic state.

A.F. Bird

*CSIRO, Division of Horticultural Research,
Adelaide, South Australia*

Introduction and terminology

This topic has been the subject, at least in part, of a
number of recent reviews to which the reader is referred
(Bird, 1974, 1975; Dropkin, 1976; Endo, 1975). As is well
known, galls induced by root-knot nematodes consist of
hypertrophied cortical cells surrounding the nematode
which has also induced the formation of syncytia or 'giant
cells'. These usually occur in the vascular tissue, are
multinucleate and have a dense cytoplasm and highly invaginated cell walls (Bird, 1961).
    Some confusion has arisen from the use of different
terms. Originally the term giant cell was widely accepted
as a general term for those structures induced by *Heterodera* and *Meloidogyne*, although it was recognised that
this was not a precise term (Bird, 1961). Recently the
term syncytium has been reserved for structures induced
by nematodes such as *Heterodera* (Jones and Northcote,
1972a), *Nacobbus* (Jones and Payne, 1977) and *Rotylenchus*
(Rebois *et al.*, 1975) whilst the term giant cell has been
reserved for the structures induced by *Meloidogyne* (Jones
and Northcote, 1972b).
    If nomenclatorial distinction is to be made between
the structures induced by *Heterodera* and *Meloidogyne*,
assuming that cell wall breakdown occurs in the former

and not in the latter, a point on which there is no general agreement, then those induced by *Heterodera* should
correctly be referred to as syncytia and those induced by
Meloidogyne as coenocytes. A syncytium is defined as a
multinucleate mass of protoplasm formed by fusion of uni-
nucleate cells whereas a coenocyte is defined as a multinucleate mass of protoplasm formed by repeated nuclear
division unaccompanied by cell division. A giant cell, on
the other hand, may simply be an enlarged cell such as the
hypertrophied cortical cells in the gall wall.

It would thus seem that the correct term for the
structures induced by *Meloidogyne* should be coenocyte.
However, a number of workers have reported cell wall
breakdown in the early stages of development of these
structures (Bird, 1972a, 1972b, 1973; Hesse, 1970; Mendes
et al., 1976; Wong and Willets, 1969) and it appears that
some cell wall breakdown and cell fusion occur during
the early and perhaps later development. The term syncytium would thus apply to the structures induced by
Meloidogyne. In fact they seem to be a combination of
syncytia and coenocytes. As the introduction of another
term for these structures would only add to the confusion
in terminology, it is suggested that the structure induced by *Meloidogyne* should be known as a syncytium.

Galling

What exactly are galls? Mani (1964) describes plant
galls as "pathologically developed cells, tissues or
organs of plants that have arisen mostly by hypertrophy
(over-growth) and hyperplasia (cell proliferation) under
the influence of parasitic organisms like bacteria,
fungi, nematodes, mites or insects".

The process of galling by nematodes which is brought
about largely by cell hypertrophy, and that of syncytial
formation, appear to be two quite different responses.
The former starts relatively rapidly, i.e. within a few

hours of infection, and may be induced by larvae feeding at the root surface without actually entering the root (Loewenberg *et al.*, 1960). Galling is not essential for nematode development and growth, and its extent varies in different hosts. It may scarcely occur in some hosts although nematode development may be normal (Bird, 1974), whereas in other instances it can occur in the absence of nematode growth and reproduction (Gaskin, 1959).

It has been suggested that galling may result from the introduction of growth regulators from the subventral glands of the second stage larva (L_2) (Dropkin, 1972). As yet there is no direct evidence to support or refute this statement because sufficiently sensitive histochemical tests for growth regulators are not presently available. Although plant growth promoting substances have been found in *Meloidogyne* (Yu and Viglierchio, 1964) and *Ditylenchus* (Cutler and Krusberg, 1968) it is not known whether they are part of the normal internal physiological processes of the nematode or whether they are exuded during the feeding processes. As in many experiments of this sort, bacterial contamination may be a problem.

Galls induced by *Meloidogyne* appear to contain more growth regulators than adjacent tissues (Balasubramanian and Rangaswami, 1962; Bird, 1962) although it has been stated that on a weight for weight basis the concentration of auxin is the same (Setty and Wheeler, 1968).

It has been suggested that the L_2s of *Meloidogyne* hydrolyse plants proteins to release tryptophane which reacts with endogenous phenolic acids to yield auxins (Setty and Wheeler, 1968). Tryptophane-like substances have been detected histochemically in syncytia and nematodes but not in surrounding plant tissue (Bird, 1962). However, proteolytic enzymes have not yet been detected in stylet exudates (Bird, 1966). The nature of some exudates in relation to syncytial formation is discussed later.

It has been suggested (Wallace, 1973) that galling may be the result of over compensation by the plant, leading to the production of excess auxins, in response to nematode damage. However, work on the responses of different host species to different species of *Meloidogyne* (Viglierchio and Yu, 1968) points to a much more subtle balance. Viglierchio and Yu (1968) state that "plant growth regulators may characteristically be associated with a particular nematode species regardless of whether the healthy host normally contains auxins". They found that different species of *Meloidogyne* induced differences in both quantity and type of auxin in host tissues, suggesting a complexity of biochemical mechanisms. Although the kinds of auxins are characteristic of the species of root-knot nematode they can be moderated by the host, particularly if it is normally high in auxin content (Viglierchio, 1971).

Differences in gall size have been observed to be induced in the same host by different populations of the same species of *Meloidogyne* (Dropkin, 1976) so that this response of the plant to the nematode is variable and sensitive and must reflect, in some way, the changes that take place around the head of the nematode. These changes must ultimately lead to the formation of the syncytia.

Syncytia

Initiation

In order to understand the changes that take place in the cells of the host, it may be profitable to consider first of all some of the changes that occur in the nematode which must have a direct bearing on syncytial initiation.

During the early stages of parasitism, i.e. the first 2-3 days, there is an approximately three-fold increase in the size of the oesophageal glands compared with those

in the preparasitic L_2s (Bird, 1967). The three unicellular glands are the structures most likely to be responsible for the synthesis of materials exuded from the
buccal stylet. Observations of the subventral oesophageal
gland ducts (SVGD) in preparasitic and early parasitic
L_2s, viewed under phase contrast or with Nomarski interference contrast optics, show that pronounced morphological and chemical changes occur in these glands.

In the preparasitic L_2s electron microscopy reveals
that the SVGD contain granules 0.6 - 0.8 μm in diameter,
with a single bounding membrane and a granular matrix of
variable density associated with rough endoplasmic reticulum. These structures have been shown to remain intact
for periods of up to 10 days at 27°C (the maximum time
tested) although food reserves are up. However, within
2-3 days after entry of the L_2 into the plant significant
changes take place. The granules are then much more
difficult to resolve under phase or interference contrast
because they become reduced in size, have an indistinct
membrane which appears to be disrupted and contents with
a speckled appearance (Bird, 1968a). Furthermore, whereas
the SVGD of preparasitic L_2s are negative to Periodic
acid/Schiff (PAS) staining, those of the parasitic L_2s
give a positive response. As well as these marked differences in carbohydrate content, differences in the protein components of the SVGD were observed from spectral
absorption curves obtained from UV microspectrographic
analyses of similarly treated preparasitic L_2s (Bird and
Saurer, 1967).

The granules were first observed in the L_2 in the egg
just before hatching (Bird, 1967) and their presence in
the nematode spans the hatching and root penetration and
parasitic establishment periods. During this period, the
nematode has to penetrate both the egg shell and the host
plant's cell walls. It is very tempting to suggest that
these protein granules are associated with the production

of a variety of enzymes which would assist in the destruction of these barriers. However plausible this may seem, as yet there have been no experiments to show that these structures contain specific enzymes; this is due mainly to lack of sufficiently sensitive histochemical techniques. However, there is evidence which illustrates that the preparasitic L_2s of *M. javanica* can exude cellulase and that this enzyme is produced by the nematodes and not by bacteria in the test media (Bird *et al.*, 1975).

It has been pointed out (Dropkin, 1976) that enzymes exuded from L_2s may come from the anus rather than the buccal stylet and that solutions of carboxymethyl cellulose are not the same as the cellulose found in plant cell walls. This, of course, is true but the same criticism holds for any extracted cellulose material so that the only relevant experiments are those on living plant cell walls. Experiments on just such material (Withers and and Cocking, 1972) have shown that spontaneous fusion of cells can occur when they are treated with cell wall degrading enzymes, leading to the formation of large multinucleate protoplasts. What can be stated is that exudates from L_2s behave in similar manner to commercially purified cellulases when placed in a solution of carboxymethyl cellulose. In nature the mechanical use of the buccal stylet could be expected to play an important role in weakening cell walls and only their partial degradation by enzymes may be necessary. It seems plausible that the early stages of syncytial formation resemble the spontaneous intraspecific fusion of plant cells induced by enzymic degradation described above. However, opinions differ with regard to the origins of these enzymes and it has been suggested that the enzymes responsible for wall degradation do not originate from the nematode but come from the plant (Jones and Dropkin, 1975; Jones and Payne, 1977).

Growth and development

Syncytia undergo a cycle of growth that is directly related to the physiological age of the nematode (Bird, 1972a). The rate at which this development takes place is of course related to the species of host and to the weather, particularly the temperature. However, it can be seen (Fig. 1A,B,C,D) that under optimal conditions the syncytia have a relatively short cycle of activity which, in a good host such as *Vicia faba*, builds up to a peak around about the third or fourth week and then starts to decline. This pattern is much the same whether one measures the area of syncytia per nematode (Fig. 1A), average area of syncytial muclei (Fig. 1B), average mass of DNA per syncytial nucleus (Fig. 1C), or the uptake of ^{14}C per unit area of nematode (Fig. 1D). The peak of the curves corresponds with the onset of egg laying by the nematode.

During the growth and development of syncytia various changes take place in the nematode which are related to the cycle of development mentioned above. These include changes in the cuticle, feeding apparatus, muscles, alimentary tract and in the oesophageal glands which become more compact. In adult females they occupy an area immediately behind the oesophageal bulb and can be shown to contain basic proteins (Bird, 1968b).

The ampulla of the dorsal oesophageal gland (DOG) becomes most discernible in the female nematode just after moulting, which is the time when syncytial growth and nematode development are at their most rapid. In this period the ampulla is full of granules and it seems reasonable to assume that they play an important role in maintaining and controlling syncytial growth. The granules in the DOG are morphologically distinct from those described in the SBGD of the L_2. They do not appear to be bounded by a membrane and have invaginations which give rise to internal irregularities as seen in sections viewed by electron microscopy (Bird, 1969). Also they

stain more intensely and are slightly smaller, having an
average diameter of 0.5-0.6 μm. These granules appear to
break down close to the terminal ducts which connect with
the salivary duct just behind the buccal stylet. The stylet lumen is about 0.5 μm in diameter and the occasional
presence of structures of similar dimensions to the smallest granules has been observed in buccal exudates.

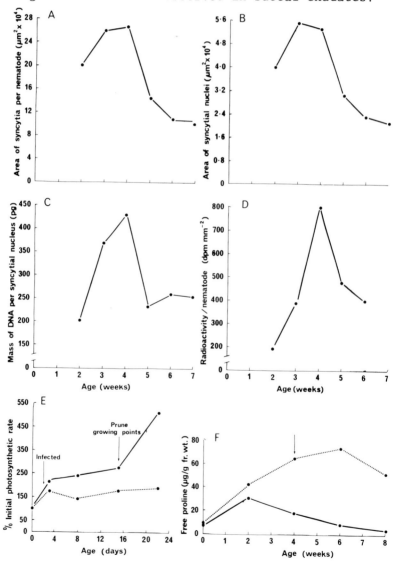

The female nematode is capable of producing an average of approximately 500 μm³ of stylet exudate over a period of several hours at 25°C. Electron micrographs of sections cut through stylet exudates reveal that they consist of a particulate meshwork surrounded by irregular structures which probably arise through precipitation and denaturation at the interface. The particles which make up the meshwork have similar dimensions to ribosomes but are not so discrete. Similarly sized particles have been detected in the cytoplasm of the plant cell adjacent to the stylet tip in *Rotylenchulus reniformis* (Rebois *et al.*, 1975).

Microspectrohotometric analysis of the ampulla of the DOG reveals typical protein-type absorption spectra (Bird, 1968b). No nucleic acids have been detected by this method and the exudate does not stain after 24 hr in gallocyanin. The exudate is stained by bromophenol blue, indicating that it contains protein, and also with fast green FCF, indicating that it is basic protein. This exudate has also been shown to contain a peroxidase (Hussey and Sasser, 1973) and esterases probably exude from the amphidial pouches and mingle with the stylet exudate, these being the type of enzyme that could attack structural lipids vital to the integrity of cell membranes (Owens and Specht, 1964). Thus the exudate is certainly not a simple substance but contains a variety

---

Fig. 1. (opposite) Curves depicting physiological and morphological changes in syncytia (A,B,C,D), whole plants (E) and galls (F) induced by *M. javanica* during their growth and decline. (A) Changes in the size of syncytia in *Vicia faba*; (B) Changes in the size of syncytial nuclei in *Vicia faba*; (C) Changes in the amount of DNA per nucleus in *Vicia faba*; (D) Changes in the amount of ^{14}C incorporated into the female nematode from photosynthates: the plants (*Vicia faba*) were harvested 24 hr after exposure to the label; (E) Changes in relative photosynthetic rates of whole *Lycopersicon esculentum* seedlings induced by infection with *Meloidogyne javanica*, (•——•) control, (•----•) infected; (F) Changes in free proline concentration in *Lycopersicon esculentum* induced by infection with *Meloidogyne javanica*, (•——•) control, (•----•) infected.

of chemical compounds whose possible roles in inducing
and maintaining syncytia can be speculated upon at length.
The nature of the nematode stimulus needs to be investigated using nucleic acid hybridization techniques which
are many times more sensitive than the microspectrophoto-
metric methods used so far and which may not have been
able to detect the minute amounts of either DNA or RNA
potentially involved in changing the expression of the
genes in these plant cells.

The physiological influence of syncytia on the host plant

Many of the measurements of the plant's response to
parasitism by *Meloidogyne* have been made on homogenized,
extracted or dissected material simply because of the
complexity of making reliable physiological measurements
involving whole plants and also because it is necessary
to go beyond this and measure reactions that are occurring
within the syncytia themselves. Various measurements on
whole tomato plants have shown that they respond to
parasitism by *M. javanica* by decreasing their photosynthetic rate (Fig 1E) and accumulating free proline (Fig
1F). The photosynthetic response is quite rapid and can
take place within 1-2 days (Loveys and Bird, 1973). Over
longer periods of time inhibition of photosynthesis
can be detected after inoculation of as few as 2000 L_2s
per plant (Wallace, 1974).

Proline accumulation in the root of infected tomato
plants is not thought to be caused by water stress due
to possible reduction in translocation caused by dis-
ruption of the xylem by syncytia (Meon *et al.*, 1978).
The factors that elicit this response are unknown. The
prevalance of proline in infected roots, particularly in
egg masses and galls as opposed to the tops of these
plants, lends support to the idea of the syncytium functioning as a metabolic sink (Bird and Loveys, 1975; Mc-
Clure, 1977) as considered below.

Attempts to measure physiological processes within
syncytia are fraught with technical problems. However,
these have at least been overcome in two ways, firstly
by the use of fine microelectrodes (tip diameter of 0.5
µm) to measure transmembrane potential differences
between syncytial contents and a reference pipette (Jones
et al., 1974; 1975), and secondly by the use of an ultra-
microanalytical technique (Gommers and Dropkin, 1977).
Syncytia have similar transmembrane potential responses
to normal cells which indicates that their ionic conditions are similar. Ultra-microanalytical analyses of 5-50
ng samples from syncytia shown that the chemical components of these structures vary depending on the host
(Gommers and Dropkin, 1977). Thus when comparisons were
made with actively growing tissues from root tips it was
shown that there is about four times more glucose in the
syncytia of soybean and balsam. However, in balsam the
concentration of free amino acids in syncytia was only
slightly more than in controls, whereas in soybean it
was about six times greater.

The increases in nucleic acid and protein content of
syncytia compared with normal cells have been well docu-
mented by various different workers. The protein content
of tomato syncytia has been shown by cytophotometric
techniques to be about 4-fold in 22 day-old syncytia
(Bird, 1961) and also to range from 3 to 20-fold depending on age (Owens and Specht, 1966) and, using ultra-
microanalytical techniques (Gommers and Dropkin, 1977),
to be about 3-fold in soybean and 10-fold in balsam.

Wall ingrowth formation in syncytia has led to their
being considered as a form of transfer cell (Jones and
Northcote, 1972a; 1972b; Pate and Gumming, 1972). It has
been stated that the main difference between syncytial
transfer cells and normal transfer cells in plants is
that in the former the nutrient sink is the nematode
whilst in the conventional transfer cell the demand for

nutrient is by actively secretory or growing tissue (Jones
and Northcote, 1972a). Evidence is accumulating to support
the hypothesis that the female root-knot nematode function
as a metabolic sink (Bird and Loveys, 1975; McClure, 1977;
Meon *et al.*, 1978). Syncytial transfer cells do not necessarily function in the same manner as other conventional
transfer cells in the same plant. Thus it has been shown
that in the salt resistant soybean cv. Lee transfer cells
in roots are sites of sodium retention under saline conditions (Lauchli, personal communication). However, syncytia
formed in the same variety under similar conditions contai
less sodium and chlorine than adjacent cortical tissues
(Bird, 1977). Thus nematode-induced transfer cells have an
opposite physiological function (Fig. 2) to conventional

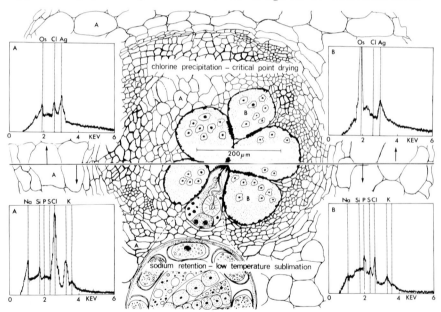

Fig. 2. Scale diagram of a transverse section through a gall in the roc
of *Glycine max* viewed under the scanning electron microscope. The top
part of the diagram illustrates osmium fixed tissues in which chlorine
has been precipitated by silver acetate and then the material has been
dehydrated by means of the critical point technique using CO_2. A and
B indicate areas where EDAX measurements were taken. The bottom part
of the diagram depicts similar root tissues in which the escape of
sodium chloride has been minimized by low temperature sublimation.
Again A and B indicate areas where EDAX measurements were taken.

transfer cells in the same plant cultivar under the same environmental conditions.

Energy dispersive X-ray analysis (EDAX) is a helpful tool in unravelling some of the physiological functions of plant cells. Fig. 2 represents scanning electronmicrographs of thick sections of galls from soybean cv. Lee grown under saline conditions. The top of the figure represents material fixed in 2% osmium tetroxide in cacodylate buffer containing 0.5% silver acetate. Chloride in the tissues reacts with silver acetate to form insoluble silver chloride. The galls were cut and dehydrated using the critical point drying technique. EDAX analysis of this material revealed osmium, chlorine and silver in adjacent vascular tissue and in the surrounding cortical tissue (A) and osmium and silver in the syncytia (B). The silver in the syncytia may be due to its combination with phenolic compounds.

The bottom part of Fig. 2 represents material dropped into liquid nitrogen and then allowed to sublime at -20°C for several days. This prevents leakage of sodium chloride ions. Cortical cells in material treated this way have an EDAX spectrum (A) showing marked sodium, chlorine and potassium peaks whilst syncytia (B) show phosphorus and a little chlorine and potassium.

At present little is known about the physiological functions of transfer cells and it is quite possible that nematode-induced transfer cells may differ very considerably in their function from the much smaller conventional plant transfer cells. An understanding of the biochemical processes associated with syncytial formation remains an important goal.

References

Balasubramanian, M. and Rangaswami, G. (1962). Presence of indole compound in nematode galls. *Nature* **194**, 774-775.

Bird, A.F. (1961). The ultrastructure and histochemistry of a nematode induced giant cell. *J.biophys.biochem.cytol.* **11**, 701-715.

Bird, A.F. (1962). The inducement of giant cells by *Meloidogyne javanica. Nematologica* **8**, 1-10.

Bird, A.F. (1966). Some observations on exudates from *Meloidogyne* larvae. *Nematologica* 12, 471-482.

Bird, A.F. (1967). Changes associated with parasitism in nematodes. *J.Parasitol.* 53, 768-776.

Bird, A.F. (1968a). Changes associated with parasitism in nematodes. *J.Parasitol.* 54, 475-489.

Bird, A.F. (1968b). Changes associated with parasitism in nematodes. *J.Parasitol.* 54, 879-890.

Bird, A.F. (1969). Changes associated with parasitism in nematodes. *J.Parasitol.* 55, 337-345.

Bird, A.F. (1972a). Quantitative studies on the growth of syncytia induced in plants by root-knot nematodes. *Int.J.Parasitol.* 2, 157-170.

Bird, A.F. (1972b). Cell wall breakdown during the formation of syncytia induced in plants by root-knot nematodes. *Int.J.Parasitol.* 2, 431-432.

Bird, A.F. (1973). Observations on chromosomes and nucleoli in syncytia induced by *Meloidogyne javanica*. *Physiol.Plant.Path.* 3, 387-391.

Bird, A.F. (1974). Plant response to root-knot nematode. *Ann.Rev.Phytopath.* 12, 69-85.

Bird, A.F. (1975). Symbiotic relationships between nematodes and plants. *Symp.Soc.Exp.Biol.* 29, 351-371.

Bird, A.F. (1977). The effect of various concentrations of sodium chloride on the host-parasite relationship of the root-knot nematode(*Meloidogyne javanica*) and soybean (*Glycine max* var Lee). *Marcellia* (In press)

Bird, A.F., Downton, W.J.S. and Hawker, J.S. (1975). Cellulase secretion by second stage larvae of the root-knot nematode (*Meloidogyne javanica*). *Marcellia* 38, 165-169.

Bird, A.F. and Lovery, B.R. (1975). The incorporation of photosynthates by *Meloidogyne javanica*. *J.Nematol.* 7, 111-113.

Bird, A.F. and Saurer, W. (1967). Changes associated with parasitism in nematodes. *J.Parasitol.* 53, 1262-1269.

Cutler, H.G. and Krusberg, L.R. (1968). Plant growth regulators in *Ditylenchus dipsaci, Ditylenchus triformis* and host tissues. *Plant & Cell Physiol.* 9, 479-497.

Dropkin, V.H. (1972). Pathology of *Meloidogyne*. Galling, giant cell formation, effects on host physiology. *OEPP/EPPO. Bull.* 6, 23-32.

Dropkin, V.H. (1976). *In* "Encyclopedia of Plant Physiology" (Eds. R. Heitfuss and P.H. Williams). 4, 222-246, Springer-Verlag, New York.

Endo, B.Y. (1975). Pathogenesis of nematode infected plants. *Ann.Rev. Phytopath.* 13, 213-238.

Gaskin, T.A. (1959). Abnormalities of grass roots and their relationship to root-knot nematodes. *Plant.Dis.Reptr.* 43, 25-26.

Gommers, F.J. and Dropkin, V.H. (1977). Quantitative histochemistry of nematode induced transfer cells. *Phytopathology* 67, 869-873.

Hesse, M. (1970). Cytologische untersuchungen an nematodengallen. *Osterr.Bot.Z.* 118, 517-541.

Hussey, R.S. and Sasser, J.N. (1973). Peroxidase from *Meloidogyne incognita*. *Physiol.Plant.Path.* 3, 223-229.

Jones, M.G.K. and Dropkin, V.H. (1975). Scanning electron microscopy of syncytial transfer cells induced in roots by cyst nematodes. *Physiol.Plant.Path.* **7**, 259-263.

Jones, M.G.K. and Northcote, D.H. (1972a). Nematode induced syncytium - A multinucleate transfer cell. *J.Cell.Sci.* **19**, 789-809.

Jones, M.G.K. and Northcote, D.H. (1972b). Multinucleate transfer cells induced in *Coleus* roots by the root-knot nematode, *Meloidogyne arenaria*. *Protoplasma* **75**, 381-395.

Jones, M.G.K., Novacky, A. and Dropkin, V.H. (1974). "Action potentials" in nematode induced plant transfer cells. *Protoplasma* **80**, 401-405.

Jones, M.G.K., Novacky, A. and Dropkin, V.H. (1975). Transmembrane potentials of parenchyma cells and nematode-induced transfer cells. *Protoplasma* **85**, 15-37.

Jones, M.G.K. and Payne, H.L. (1977). Scanning electron microscopy of syncytia induced by *Nacobbus aberrans* in tomato roots. *Nematologica.* **23**, 172-176.

Loewenberg, J.R., Sullivan, T. and Schuster, M.L. (1960). Gall induction by *Meloidogyne incognita* by surfacing feeding and factors affecting the behaviour pattern of the second stage larvae. *Phytopathology* **50**, 322.

Loveys, B.R. and Bird, A.F. (1973). The influence of nematodes on photosynthesis in tomato plants. *Physiol.Plant.Path.* **3**, 525-529.

McClure, M.A. (1977). *Meloidogyne incognita* : A metabolic sink. *J.Nematol.* **9**, 88-90.

Mani, M.S. (1964). *In* "Monographiae Biologicae" (Ed. Dr. W. Junk) **12**, 1-434, The Hague.

Mendes, B.V., Ferraz, S. and Shimoya, C. (1976). Histopatologia de raizes de cafeeiro parasitadas por *Meloidogyne exigua*. *Summa Phytopathologica* **2**, 97-102.

Meon, S., Fisher, J.M. and Wallace, H.R. (1978). Changes in free proline following infection of plants with either *Meloidogyne javanica* or *Agrobacterium tumefaciens*. *Physiol.Plant.Path.* **12**, 251-256.

Owens, R.G. and Specht, H.N. (1964). Root-knot histogenesis. *Contrib. Boyce Thompson Inst.* **22**, 471-490.

Owens, R.G. and Specht, H.N. (1966). Biochemical alterations induced in host tissues by root-knot nematodes. *Contrib. Boyce Thompson Inst.* **23**, 181-198.

Pate, J.S. and Gunning, B.E.S. (1972). Transfer cells. *Ann.Rev.Plant. Physiol.* **23**, 173-196.

Rebois, R.V., Madde, P.A. and Eldridge, B.J. (1975). Some ultrastructural changes induced in resistant and susceptible soybean roots following infection by *Rotylenchulus reniformis*. *J.Nematol.* **7**, 122-139.

Setty, K.G.H. and Wheeler, A.W. (1968). Growth substances in roots of tomato (*Lycopersicon esculentum* Mill.) with root-knot nematodes (*Meloidogyne spp.*). *Ann.Appl.Biol.* **61**, 495-501.

Viglierchio, D.R. (1971). Nematodes and other pathogens in auxin related plant growth disorders. *Bot.Rev.* **37**, 1-21.

Viglierchio, D.R. and Yu, P.K. (1968). Plant growth substances and plant parasitic nematodes. *Exptl.Parasitol.* **23**, 88-95.

Wallace, H.R. (1973). Nematode ecology and plant disease. 228,

Edward Arnold, London.
Wallace, H.R. (1974). The influence of root-knot nematode, *Meloidogyne javanica*, on photosynthesis and nutrient demand by roots of tomato plants. *Nematologica* 20, 27-33.
Withers, L.A. and Cocking, E.C. (1972). Fine structural studies on spontaneous and induced fusion of higher plant protoplasts. *J.Cell. Sci.* 11, 59-75.
Wong, C.L. and Willetts, H.J. (1969). Gall formation in aerial parts of plants inoculated with *Meloidogyne javanica*. *Nematologica* 15, 425-428.
Yu, P.K. and Viglierchio, D.R. (1964). Plant growth substances and parasitic nematodes. *Exptl.Parasitol.* 15, 242-248.

Discussion

The problem of nomenclature for the *Meloidogyne*-induced cells in the plant host was considered generally to relate to their morphological description and recognition of the way in which they were formed. Gommers suggested that it was still unclear whether they were formed by cell wall dissolution or by division of the nucleus without the formation of cell walls. Bird indicated that giant cell was a commonly used term; coenocyte is however a more correcte term for a multinucleate structure formed without cell wall breakdown, but as some cell wall breakdown may also occur syncytium is a preferred descriptive term. In reply to a question, Bird said that the giant cells had outgrowths of the cell wall which are characteristic of transfer cells, but they did not always behave physiologically as typical transfer cells. The cell wall protrusions increase the surface area, allowing more rapid transfer of materials.

There were several references to observations indicating differences in giant cells formed in response to *M. hapla* and *M. incognita*. Mankau said that he had seen sieve plates on the surface of giant cells associated with *M. hapla*; Gommers referred to differences in the wall protrusions of giant cells in pea roots associated with *M. incognita*; Dalmasso reported that in *M. hapla* infections the giant cell nuclei tend to cluster rather than to be widespread and suggested that the host is important in relation to giant cell morphology.

Evans remarked that it was interesting that Bird found the peak metabolic activity of the giant cells or syncytia at the start of egg laying and asked if the nematode acquired all its food energy before laying eggs. Bird thought they did which led Evans to speculate that if the roots of trap crops are not removed the nematode may continue to lay eggs. Asked if the nematode can programme the host (giant) cells to produce its exact needs, Bird said that the systems seem very efficient and there is little waste matter eliminated during feeding. Triantaphyllou postulated that instead of wastes being eliminated they are accumulated within the body and hence the eggs have to be extruded as there is no space to retain them. Evans pointed to the behaviour of aphids feeding on phloem which void excess food materials via honey dew and suggested that the sub-crystalline layer in *Heterodera* may be a comparable device; Coomans pointed to the error of comparing *Meloidogyne* females with aphids which feed on substrates rich in carbohydrates and low in proteins.

10 LIFE CYCLE OF *MELOIDOGYNE* SPECIES AND FACTORS INFLUENCING THEIR DEVELOPMENT

G. de Guiran and M. Ritter

I.N.R.A., Station de Recherches sur les Nématodes, Antibes, France

Introduction

When Cornu (1879) described *Anguillula marioni* he gave some indications of the biological features of a root-knot nematode, but the first definitive description of the life-cycle of a species *Meloidogyne* must be attributed to Müller (1883) who also provided illustrations of the various larval and adult stages (Fig. 2). However, Müller was unaware of the work of Cornu and named his species *Heterodera radicicola*.

Other authors (Atkinson, 1899; Stone and Smith, 1898; Bessey, 1911; Nagakura, 1930; Goodey, 1932; Tyler, 1933a,b) further contributed to the knowledge of this important group of plant parasite nematodes. However, the value of their observations is limited somewhat as there is some uncertainty about the species they were working with, due to the confused taxonomical status of the genus until Chitwood's revision (1949).

The role of many *Meloidogyne* species as agricultural pests has stimulated, in the past two decades, much work on their biology. Consequently there is much information on the life-cycles, and on the environmental factors that affect it, of the most widespread and economically important species.

G. DE GUIRAN AND M. RITTER

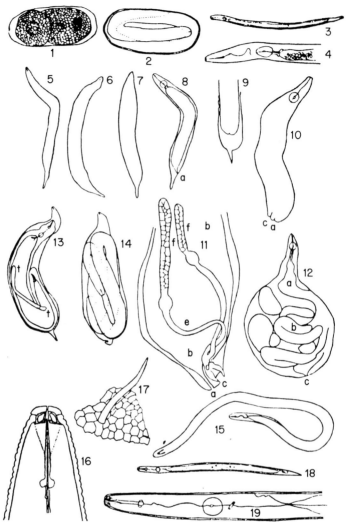

Fig. 1. Stages of the root-knot nematode. 1 and 2 eggs; 3 and 4, free-living larvae; 5-8, developing larvae; 9-13, developing female; 14 and 15, developing male; 16, head of male; 17, larva entering root; 18 and 19, larva of *H. schachtii* (cf. 3 and 4). (from Bessey, 1911).

General features of the life cycle

The life cycle of species belonging to the genus *Meloidogyne* can briefly be generalised as follows:

Eggs may be free in the soil or embedded in a gelatinous matrix which may still adhere to the root tissue of

the plant host. Second stage larvae hatch from the eggs
and invade new rootlets usually near the tip, just above
the root cap where there is intense meristematic activity.
They penetrate the cortex and establish themselves with
the anterior end in contact with the vascular cylinder
where, in susceptible hosts, they induce the formation
of giant cells upon which they feed; at this stage a
gall generally forms (as discussed by Bird in these
Proceedings). The nematode grows slightly in length and
much in width; at the same time the oesophageal glands
enlarge, the cells of the genital primordia divide and
six rectal glands, which secrete the ovisac, develop in
the posterior end of the female larval stages (Taylor
and Sasser, 1978).

During their further development the larvae gradually
assume a flask shape and undergo three further moults
(Taylor and Sasser, 1978). The last moult is a true metamorphosis for the male, which appears as a long filiform
nematode folded inside the cuticle of the fourth larval
stage; the adult female at first retains the same shape
of the last larval stage, but as it matures it enlarges
and becomes pyriform. After escaping from the last
larval exuviae the males copulate, in the amphimictic
species. The females secrete a gelatinous matrix into
which they extrude a large number of eggs, usually 500-
1,000 but sometimes more (Tyler, 1933a, recorded 2,882
eggs). Larvae hatching may migrate to another rootlet
or, when the egg masses are embedded in the root tissue,
stay in the same gall. After a few generations the gall
will contain a large number of females of various ages.
Bessey (1911) illustrated the various morphological
stages in the life cycle (Fig. 1).

Many species of *Meloidogyne* are parthenogenetic,
obligate or facultative, but this is not correlated with
presence or absence of males.

The number of generations per year varies according

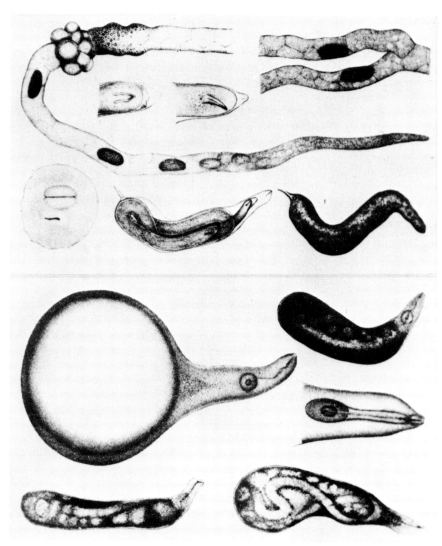

Fig. 2. Reduced reproduction of two of the four plates from Müller (1883) illustrating the life cycle of "*Heterodera radicicola*."

to species. Usually there are many, but in some species there is only one, e.g. *M. naasi* which attacks cereals in temperate climatic conditions. It is interesting to note

:hat with this species, which attacks only annual crops
in temperate conditions, the life cycle resembles a
:lassical *Heterodera* cycle and that of a tropical *Heter-*
ɔdera such as *H. oryzae* resembles a *Meloidogyne* life
:ycle, with short and continuous generations.

There is much information on the duration of life
cycles of various species of *Meloidogyne* under controlled
conditions.

Tarjan (1952) compared the life span of five *Meloid-*
ɔgyne species on the same host (*Antirrhinum majus* L.) at
21°C (Table I). Other authors have attempted to define
the optimal development conditions (i.e. the shortest
duration of the life cycle) for the most common species
by rearing them at several different temperatures.

TABLE I

Development of five *Meloidogyne* spp. on *Antirrhinum*
majus L. at 21°C (from Tarjan, 1952)

	Gelatinous matrix produced	Start of egg production	Host invasion by L2s
M. *javanica*	35[1]	39	63
M. *arenaria*	37	39	63
M. *hapla*	37	39	63
M. *incognita*	37	39	57
M. *incognita acrita*	37	37	59

[1]No. days after inoculation of plants

Tyler (1933b) carried out precise experimental investigations although the exact species she worked with
is not known. She infected each seedling with a single
larva obtained from a population of *Meloidogyne* originating from one female and reared for several generations
on the same tomato cultivar grown in nutrient solution.
She considered the threshold of development to be

about 9°C. The rate of development increased with temperature up to 28°C with a marked retardation above. For example, the minimum time required for the life cycle was 87 days at 16.5°C and 25 days at 27°C, development from gall formation to egg laying taking 79 and 15 days respectively. Development to egg laying was calculated to require 6,500-8,000 heat units (a heat unit being an hour-degree calculated from 10°C minimum) with the development of the different stages (as described by Bessey, 1911) according well with theoretical heat-unit parabola (Tyler, 1933b).

More recently Bird and Wallace (1965) investigated the influence of temperature on some aspects of the life cycle of *M. hapla* and *M. javanica*, the former being considered as more frequently occurring in cooler climates than the latter. Experiments at five temperatures in the range 15-35°C indicated that *M. hapla* had a thermal optimum of 25°C for hatching and 20°C for mobility in the soil, but these optima were respectively 30°C and 25°C for *M. javanica*. In other experiments it was shown that in both species, stages exposed to the soil environment have lower thermal optima (for hatching) than those occurring in root plant tissues (for growth).

Laughlin *et al.* (1969) demonstrated with *M. graminis* that high temperatures which allow the quickest growth of females also induce the highest proportion of males. Over 80% of adults were males at 27°C or 32°C, but males were absent at 16°C and 21°C.

The range of experiments by different authors indicate that for each particular situation, the duration of one generation of a particular *Meloidogyne* species depends on external factors which act differently on each stage. Temperature is an example, but many other factors can affect development, such as humidity, light or quality of the host in relation to its age and nutritional status.

Embryogenesis

Eggs are usually laid at the one-cell stage and a few
hours after deposition cell division starts. Tyler (1933b)
investigated embryonic development at temperatures varying between 16.5 and 31.5°C and showed that between
those limits development increases with temperature.
Hatching required 31 days at 16.5°C and 9 days at 29.5°C.
Godfrey and Oliveira (1932) obtained similar results
with another unidentified *Meloidogyne* species galling
cowpea.

Bird (1974) used cinephotomicrography to study the influence of temperature on the rate of hatching and on
the embryogenesis of *M. javanica*. He concluded that the
optimum is between 25 and 30°C. Embryogenesis was
slightly more rapid at 30°C (9-10 days), but more eggs
completed their development at 25°C (11-13 days). At
20°C development took longer, 16-48 days, and at 15°C,
46-48 days. The thermal optimum seems to be the same for
different stages of embryonic development.

Bird (1974) also investigated the influence of a
short period of exposure of eggs to high temperature.
When eggs were exposed for 10 min to 46°C, a sublethal
temperature, hatching was delayed by about 11 days but,
when the temperature stress occurred at an early stage
of egg development (two cells or gastrula), the effect
on embryogenesis itself was less pronounced.

Hatching and diapause

Hatching mechanism   After the first moult within the
eggs, the young L_2 obtains energy from its food reserves,
mostly lipids, within the intestine. Before emergence,
the empty space within the egg increases in volume and
the larva becomes more active, thrusting its stylet repeatedly at one end of the egg shell until a slit-like
aperture is produced. Bird (1968) studied ultrastructural modification of the egg shell and of the oesophageal

glands before and during the process of hatching. He
observed that as the activity of the larva continues
there is a corresponding decrease in the inner lipid
layer of the egg shell. The shell is thus reduced to the
outer, proteic layer, which is plastic and probably
elastic. The proteic granules present in the subventral
oesophageal glands and in their efferent ducts possibly
are responsible for the elimination of the lipid layer.
Thus hatching is believed to be initiated by an enzymatic
process, but nothing is known about the nature of the
enzymes or the mechanism of the process.

Hatching factors   Egg hatch in *Meloidogyne* spp., as with
other nematodes, could either be spontaneous, or controlled by an external stimulus (Rogers and Sommerville,
cited by Bird, 1968). According to the literature there
is almost total emergence from the egg masses in the (sub)
tropical "species" such as *M. arenaria, M. incognita* and
M. javanica under optimal temperature and humidity re-
gimes. Hatching would thus appear to be spontaneous.

Several workers have exposed the eggs of *Meloidogyne*
spp. to root exudates expecting stimulation of hatching
as observed with some *Heterodera* species. Viglierchio
and Lownsbery (1960) observed only a slight stimulation.
Other results are less convincing and sometimes conflict-
ing and Hamlen *et al.* (1973) concluded that a stimulating effect of root exudates is weak, if not absent, under
natural conditions. Nevertheless, all the experiments
could have been performed under conditions in which a
hatching factor was ineffective, or was inhibited. A
co-factor could be involved; Wallace (1966) observed a
partial inhibition of larval emergence in non-sterile
soils, and this inhibition was ineffective in the presence of a host plant. This observation has not been
confirmed by further experimentation but it seems likely
that soil microorganisms play a role in this as in many
other biological processes.

Quiescense   Hatching is suppressed when egg masses are
subjected to stress conditions such as high or low temperatures, low atmospheric humidity, high osmotic pressure,
or lack of oxygen.
     If the stress is limited in time and intensity, embry-
onic development continues and hatching resumes rapidly
when suitable conditions are restored. In this example
of quiescense it has frequently been shown that the
gelatinous matrix plays a protective role. When the
stress is intense, e.g. high osmotic pressure (Wallace,
1966) or high temperature (Bird, 1972, 1974), embryonic
development may be suspended, sometimes definitively or
apparently so.
     Hatching in the soil is governed by many biotic and
abiotic factors, sometimes interacting. For example,
when soil moisture is very low (at the plant wilting
point) hatching is temporarily suspended. This also
occurs when soil moisture is high (near saturation), but
in this case lack of oxygen is responsible. Wallace (1968)
has shown *in vitro* that if lack of oxygen is maintained,
hatching is delayed and decreased.
     The ability to endure climatic stresses usually
varies among species e.g. *M. hapla* withstands low temperatures better than *M. arenaria*, *M. incognita* and *M. javanica*, and is found more frequently in colder climates.
However, there are races within the other species which
have more or less adapted to the climatic stresses existing in geographic regions where they have been transported (Daulton and Nusbaum, 1961).
Diapause   Most experiments concerned with hatching generally focus on the number of larvae emerging and investigators fail to consider the status and fate of
unhatched eggs. A cause of non-hatching under optimal
conditions could be due to diapause affecting some or all
of the eggs laid. This phenomenon is known in *M. naasi*
and Watson and Lownsbery (1970) have shown that a

chilling period is necessary to stimulate hatching; this
explains why *M. naasi* has only one generation per year
(Gooris and D'Herde, 1972). In the group *M. arenaria*, *M.
incognita*, *M. javanica* there is almost total hatching
under optimal environmental conditions. In an experiment
involving egg masses, one of us (G. de G.) noted that in
M. incognita there was always a small proportion (about
10% but occasionally 60-80%) of eggs remaining alive
with their development blocked at an early stage.
Ishibashi (1969) considered that old or poorly nourished
females lay brown egg masses containing dormant eggs,
which are resistant to environment stresses and to nematicides, and which hatch under the stimulus of root
diffusates. Conversely, young or well nourished females
lay white egg masses with eggs susceptible to environmental stresses but hatching spontaneously.

Work being conducted in France (de Guiran, unpublished)
suggests that the phenomenon of diapause is complex. It
has been found that white egg masses are produced in dry
soil but brown ones in moisture-saturated soil. In both
cases ovisacs contained eggs in diapause, even in such
reputed good conditions as monoxenic cultures using
excised roots with only one female per gall. Observations
have also shown that diapause is not restricted to *M. incognita*, but is also present in *M. arenaria* and *M. javanica*. The proportion of eggs in diapause is not always
the same, but depends on the nematode-host plant interaction with an intraspecific variation at the nematode
level. The variation of the proportion of eggs in diapause among different egg masses of the same population
is transmitted from one generation to the other, irres-
pective of whether the female originates from an egg mass
with high or low proportion of eggs in diapause.

The percentage of eggs in diapause can be experimentally changed by physical stresses. In the ovisacs of
M. incognita kept in water-saturated soil, the

development of the eggs was suppressed in direct proportion to the exposure time. Chilling the eggs of a tropical population of this species blocked their embryonic
development. In both cases eggs remained alive at a non-
differentiated stage after favourable conditions had
been restored.

Whether artificially induced or spontaneous, diapause
is necessary in these species to enable them to survive
during the period between harvest and the beginning of
climatic stress (winter in temperate countries and dry
season in the tropics). If hatching occurred during
this period of favourable conditions it would result in
the exhaustion of the hatched larvae and consequently
of the population. Therefore, even if diapause is limited
to a small proportion of the eggs it is an important biological feature because of the high reproductive capacity
of the nematodes and especially if the breaking of diapause is governed by the presence of host plant which
increases the chances of multiplication of the emerged
larvae.

Diapause can persist for a long time - from weeks to
several months - but the ability of larvae to penetrate
host roots after different durations of the dormancy is
unknown.

Migration and penetration of host root

To complete its life cycle an emerged larva must reach
the root tip of its host plant, which it then penetrates.
This short exophytic phase is the only part of the complete life cycle in which the larva is not protected
either by the gelatinous matrix or by the root tissues.
Thus many studies have been undertaken with the aim of
finding a trapping process to eliminate the parasite
during its migratory phase. Many factors affect the survival during this free-living phase: soil water potential, size of the soil particles, oxygen and other gases

in the soil atmosphere, but mostly the presence of the
host plant. Prot (personal communication) observed a repulsive effect on *M. incognita* and *M. javanica* larvae by
salt concentration in the soil. The attraction of a larva
towards a root could thus result from its repulsion from
an area of high salt concentration towards the area of
the root where the concentration is lower because of the
uptake of mineral nutrients by the root. Janati (personal
communication) has demonstrated that the rate of penetration of invasive larvae is related to the host plant,
the soil temperature and the species of the root-knot
nematode (Table II).

TABLE II

Larval penetration of the roots of two tomato
cultivars (susceptible and resistant) at two temperatures

Mean percent penetration[1]

	M. incognita (Seville 36)	*M. arenaria* (Seville)[2]
Tomato cv. Marmande		
18–23°C	37	50
12°C	55	57
Tomato cv. Rossol		
18–23°C	30	42
12°C	50	52

[1]200 larvae per plant in a glass tube;
24 replications per treatment

[2]aggressive for cv. Rossol

Development in the root

The duration of the parasitic phase within the host
plant is also a specific feature of the root-knot nematode life cycle. It is affected by the genetic qualities
of the host (species and cultivar) and by environmental
conditions which act both on the nematode and on the
plant and which in total constitute the host-nematode

inter-relationship complex.

According to Oteifa (1951, 1952) the rate of reproduction of *M. incognita* on lima bean grown on nutrient
solutions increases with the potassium concentration of
the medium either with a light or a heavy initial infestation of the nematode. Marks and Sayre (1964) confirmed
these results using a cucumber cultivar, but found that
M. hapla and *M. javanica* were not influenced by potassium.

It is difficult to establish clearly the role of all
mineral elements as many of them interact with each
other. From the work of Bird (1960, 1970) and Davide and
Triantaphyllou (1967) it can be concluded that the rate
of growth of *M. javanica* mostly depends on the level of
initial infection: there is an acceleration of its
development in mineral-deficient plants at low level of
infection (40 larvae per plant) and a decrease at higher
nematode densities (400 and 4,000 larvae per plant).
These effects were more marked in tomato plants deficient
in nitrogen or potassium. Haque *et al.* (1975) obtained
different results with *M. incognita* infesting various
crops with different levels of inoculum and rates of
potassium.

Dropkin and Boone (1966) used a single *M. incognita*
larva on each excised root and showed that high levels
of potassium favour egg production, but Ishibashi *et al.*
(1963) obtained quite different results when the nematode
was reared on entire plants. We have done an experiment
in an attempt to clarify this point and to obtain
information on the subsequent diapause. Cultures were
reared on excised roots and only egg masses formed in
galls with 1-3 females per gall were considered. Egg
masses were collected weekly for 4 weeks so the last
ones were produced by old females. The results showed no
effect or potassium deficiency or age of the females on
the weekly rate or total egg production. The age of female
also had no effect on diapause in a complete medium but

an increase in diapause eggs was observed in the medium
without potassium (de Guiran and Villemain, unpublished).

Relation between host status and males occurrence

It is well known that the proportion of males in a
population, at least in parthenogenetic species, varies
according to the host plant and to the environmental
conditions. Sex determinism has been intensively investigated. Tyler (1933a) recorded that males are more
abundant under adverse conditions for development and
suggested that food supply was an important factor. This
opinion is supported by Linford (1941) who found a high
percentage of males in *Meloidogyne*-inoculated mature
leaves of cowpea (*Vigna sinensis* Endl.) where nutrition
does not allow females to develop.

Separating and incubating root tips infected with *M.
incognita*, Triantaphyllou (1960) found that intraspecific competition and damaged plants induce a stress
which convert many intrinsic female larvae into males
and that under these conditions male development is determined before the 15th day after penetration, i.e.
during the second larval stage. A similar mechanism
exists in *M. javanica*, but inversion takes place later
and may give rise to true intersexes (males with a vulva).
Fassuliotis (1970) found that *M. incognita* produced only
small galls on the resistant cucurbits *Cucurbita faci-
folia* Bouche and *Cucumis metuliferus* E. May and that
larval development was delayed with a stimulation toward
maleness. Webber and Fox (1971) studied the cycle of
three *Meloidogyne* spp. on *Cynodon* sp., *Phalaris arundinacea* L. and tomato and at two temperatures, 26° and
32°C., *M. incognita* produced 10-20 times more males in
Cynodon and 4-5 times more in *P. arundinacea* than in
tomato. At 32°C there were 9-14 times more males in tomatoes than at 26°C. Tomato is not a host for *M. graminis*
but the number of males produced was 3-4 greater than

in *Cynodon* sp. or *P. arundinacea* which are hosts.

By dissecting larvae out of the root after short or long periods of feeding, Trudgill (1972) has demonstrated the influence of nutrition on sex determination of *M. incognita*.

Results from these experiments indicate that many factors are involved in the development of males and as Bird (1959) has stated "the energy necessary for completion of the third and fourth moults must be obtained by the animal before the second moult because it is unable to feed on its host from the start of the second moult to the completion of the fourth moult".

Recently Bergé *et al.* (1974) showed that genetic characteristics of *Meloidogyne* spp. also play an important part in sex determination. In experiments using cucurbits, they noticed that the sex-ratio in *M. hapla* is really host dependent, but that there are several degrees of sex inversion according to the population. They considered that males having only one testis are inverted earlier than those having two and that maleness is correlated with the amount of lignin in the host plant. Orion and Minz (1971) had previously found an increase in the number of males associated with applications to the host plant of chlorofluorenol (Morphactine), a substance which promotes the development of ligneous tissues.

Bergé and Dalmasso (personal communication) observed in *M. hapla*-infected tomato that females are more numerous in the proximal part of the gall, males usually occur in the middle part, and that in the distal part larvae often fail to complete their development.

Male-promoting substances have still to be isolated. Probably several factors are involved in the mechanism, as with plant resistance; perhaps some of them are involved in both phenomena as maleness occurs more often in resistant hosts.

References

Atkinson, G.F. (1889). Nematode root-galls. A preliminary report on the life history and metamorphosis of a root-gall nematode, *Heterodera radicicola* (Greeff) Müll., and the injuries produced by it upon the roots of various plants. *Bull.Ala.agric.Exp.Sta* 9, 176-226.

Bergé, J.B., Dalmasso, A. and Ritter, M. (1974). Influence de la nature de l'hôte sur le développement et le déterminisme du sexe du nématode phytoparasite, *M. hapla. C.r.Acad.Agric* 60, 946-952.

Bessey, E.A. (1911). Root-knot and its control. *U.S. Dept Agric.Bur. Pl.Ind.Bull* 217, 1-88.

Bird, A.F. (1959). Development of the root-knot nematodes *Meloidogyne javanica* (Treub) and *Meloidogyne hapla* Chitwood in the tomato. *Nematologica* 4, 31-42.

Bird, A.F. (1960). The effect of some single element deficiencies on the growth of *Meloidogyne javanica. Nematologica.* 5, 78-85.

Bird, A.F. (1968). Changes associated with parasitism in nematodes. *J.Parasitol.* 54, 475-489.

Bird, A.F. (1970). The effect of nitrogen deficiency on the growth of *Meloidogyne javanica* at different population levels. *Nematologica.* 16, 13-21.

Bird, A.F. (1972). Influence of temperature on embryogenesis in *Meloidogyne javanica. J.Nematol.* 4, 206-213.

Bird, A.F. (1974). Suppression of embryogenesis and hatching of *Meloidogyne javanica* by thermal stress. *J.Nematol.* 6, 95-100.

Bird, A.F. and Wallace, H.R. (1965). The influence of temperature on *Meloidogyne hapla* and *M. javanica. Nematologica.* 11, 581-589.

Chitwood, B.G. (1949). Root-knot nematodes. *Proc.Helminth.Soc.Wash.* 16, 90-104.

Cornu, M. (1879). Etudes sur le *Phylloxera vastatrix.* Mém. présentés par divers savants à l'*Acad.des Sc. Paris.* 26, 163-175.

Daulton, R.A.C. and Nusbaum, C.J. (1961). The effect of soil temperature on the survival of the root-knot nematodes *Meloidogyne javanica* and *M. hapla. Nematologica,* 6, 280-294.

Davide, R.G. and Triantaphyllou, A.C. (1967). Influence of the environment on development and sex differentiation of root-knot nematodes. *Nematologica,* 13, 111-117.

Goodey, T. (1932). Some observations on the biology of root gall nematode, *Anguillulina radicicola* Greeff (1872). *J.Helminth.* 10, 33-44.

Dropkin, V.H. and Boone, W.R. (1966). Analysis of host parasite relationships of root-knot nematodes by single-larva inoculations of excised tomato roots. *Nematologica,* 12, 225-236.

Fassuliotis, G. (1970). Resistance of *Cucumis* spp. to the root-knot nematode *Meloidogyne incognita acrita. J.Nematologica.* 2, 174-178.

Godfrey, G.H. and Oliveira, J.M. (1932). The development of root-knot nematode in relation to root tissues of pineapple and cowpea. *Phytopathology,* 22, 325-348.

Gooris, J. and D'Herde, C.J. (1972). Mode d'hivernage de *Meloidogyne naasi* Franklin dans le sol et lutte par rotation culturale. *Rev. Agric.,* 4, 659-664.

Haque, Q.A., Khan, A.M. and Saxena, S.K. (1974). Studies on the effect of different levels of certain elements on the development of root-knot. *Indian J.Nematol.*, 4, 25-30.

Hamlen, R.A., Bloom, J.R. and Lukezic, E.L. (1973). Hatching of *Meloidogyne incognita* eggs in the neutral carbohydrate fraction of root exudates of gnotobiotically grown alfalfa. *J.Nematol.*, 5, 142-146.

Ishibashi, N. (1969). Studies on the propagation of the root-knot nematode *Meloidogyne incognita* (Kofoid and White) Chitwood, 1949. *Rev.Plant.Protec.Res.*, 2, 125-129.

Ishibashi, N., Kegasawa, K. and Kunii, Y. (1963). Egg spawning of root-knot *Meloidogyne incognita* var. *acrita* in physiological solution of sodium chloride. *Zool.Meg.Dobutsugaku.Zasshi*, 72, 170-176.

Laughlin, G.W., Williams, A.S. and Fox, J.A. (1969). The influence of temperature on development and sex differentiation of *Meloidogyne graminis*. *J.Nematol.*, 1, 212-215.

Linford, N.B. (1941). Parasitism of the root-knot nematode in leaves and stems. *Phytopathology*, 31, 634-648.

Marks, C.F. and Sayre, R.M. (1964). The effect of potassium on the rate of development of the root-knot nematodes *Meloidogyne incognita*, *M. javanica* and *M. hapla. Nematologica*, 10, 323-327.

Muller, C. (1883). Neue Helminthocecidien und derren Erreger. 5-50. Inaugural dissertation zur erlangung der philosphischen doctorwurde der philosophisher Facultat der Friedrich-Wilhems Univ. zu Berlin, Berlin.

Nagakura, K. (1930). Uber den Bau und die Lebensgeschichte der *Heterodera radicicola* (Greeff) Müller. *Jap.J.Zool.*, 3, 95-160.

Orion, D. and Minz, G. (1971). The influence of morphactin on the root-knot nematode *Meloidogyne javanica* and its galls. *Nematologica*, 17, 107-112.

Oteifa, B.A. (1951). Effects of potassium nutrition and amount of inoculum on rate of reproduction of *Meloidogyne incognita. Journ. Wash.Acad.Sci.*, 41, 393-395.

Oteifa, B.A. (1952). Potassium nutrition of the host in relation to infection by a root-knot nematode *Meloidogyne incognita. Proc. Helminthol.Soc.Wash.*, 19, 99-104.

Stone, G.E. and Smith, R.E. (1898). Nematode worms. *Bull.Hatch.exp. St.Massachussetts.Agric.*, 55, 1-67.

Tarjan, A.C. (1952). Comparative studies of some root-knot nematodes infecting the common snap dragon, *Antirrhinum majus* L. *Phytopathology*, 42, 641-644.

Taylor, A.L. and Sasser, J.B. (1978). Biology, identification and control of root-knot nematodes (*Meloidogyne* species). 111. Dept. Pl. Pathol. N.C. State Univ., Raleigh.

Triantaphyllou, A.C. (1960). Sex determination in *Meloidogyne incognita* Chitwood, 1949 and intersexuality in *M. javanica* (Treub, 1885) Chitwood, 1949. *Ann.Inst.Phytopath,Benaki,N.S.* 3, 12-31.

Trudgill, D.L. (1973). Influence of feeding duration on moulting and sex determination of *Meloidogyne incognita. Nematologica*, 18, 476-481.

Tyler, J. (1933a). Reproduction without males in aseptic root cultures of the root-knot nematode. *Hilgardia*, 7, 373-388.

Tyler, J. (1933b). Development of the root-knot nematode as affected by temperature. *Hilgardia*, 7, 389-415.

Viglierchio, D.R. and Lownsbery, B.F. (1960). The hatching response of *Meloidogyne* species to the emanations from the roots of germinating tomatoes. *Nematologica*, 5, 153-157.

Wallace, H.R. (1966). The influence of moisture stress on the development, hatch and survival of eggs of *Meloidogyne javanica*. *Nematologica*, 12, 57-69.

Wallace, H.R. (1968). The influence of aeration on survival and hatch of *Meloidogyne javanica*. *Nematologica*, 14, 223-230.

Watson, T.R. and Lownsbery, B.F. (1970). Factors influencing the hatching of *Meloidogyne naasi* and a comparison with *M. hapla*. *Phytopathology*, 60, 457-460.

Webber, A.J. Jr. and Fox, J.A. (1971). Parasitism of "Tif-green" Bermuda grass and reed canary grass by root-knot nematodes. *Virginia Acad.J.Sci*. 22, 87.

Discussion

Dr. Coolen questioned the use of the term diapause and suggested that in some circumstances the period of quiescence or inactivity would be better described as dormancy. This was particularly the case with *M. naasi* where larvae extracted from eggs can penetrate the host immediately. Ritter explained that diapause or dormancy occurred earlier in the species at the pre-larval stage. Evans suggested that the term diapause should be reserved for a definite pause in the nematode's development which is not simply broken by the restoration of suitable environmental conditions. Arrested development can be seen as a different physiological condition from that of diapause. It was pointed out, however, that there are many instances of egg masses from which some of the eggs apparently hatch without diapause. Dalmasso suggested there was probably a difference between amphimictic and parthenogenetic species and said that more detailed study of presence or absence of diapause in root-knot nematodes in general and in *M. naasi* in particular was needed before conclusions could be made satisfactorily.

The effect of temperature on the development of *M. hapla* was discussed with particular reference to its distribution in different climates. The species is

generally recognised as having a preference for a cooler
climate but it was suggested that where it developed in
warmer climates it had adapted to high temperatures and
could develop as well as and similarly to *M. javanica*.
Orion added that sometimes such populations of *M. hapla*
may infect poor hosts or hosts normally resistant to
M. javanica.

11 INFLUENCE OF MOVEMENT OF JUVENILES ON DETECTION OF FIELDS INFESTED WITH *MELOIDOGYNE*

J.C. Prot and C. Netscher

*Laboratoire de Nematologie, O.R.S.T.O.M.,
Dakar, Senegal*

Introduction

Numerous observations indicate that although there have been considerable improvements in nematode extraction from soil, discrepancies still exist between the number of juveniles of *Meloidogyne* extracted from field soil and the degree of infection observed on host plants grown in these fields. Thus while classical extraction methods frequently do not reveal the presence of *Meloidogyne*, susceptible crops grown in fields from which the soil samples have been taken are often heavily parasitized by the nematodes (de Guiran, 1966a).

In extraction methods currently used (Seinhorst, 1956, 1962; de Guiran, 1966a; Gooris and d'Herde, 1972; Demeure and Netscher, 1973) relatively small soil samples (100 or 250 cm^3) are analyzed. Sample size together with the uneven distribution of root-knot nematodes in the soil, related to the presence of egg-masses, could partly explain the discrepancies referred to. The only way to improve the situation is to increase the size of the samples or if this is not possible, to increase the number of samples.

Root-knot nematodes appear to be capable of moving over relatively large distances in short periods of time. Johnson and McKeen (1973) demonstrated that a population

of *M. incognita* localized at a depth of 120-125 cm induced galling of roots of tomatoes situated in the hori-
zon between 0 and 15 cm at the time of the first harvest.
Attacks of crops grown in seemingly uninfested land could
be explained by assuming that juveniles present in large
volumes of soil are capable of reaching and infecting
plants.

To test this assumption a series of experiments was
conducted to determine the degree of migration of 2nd-
stage juveniles of *Meloidogyne* and the relation to the
detection of these nematodes in the field.

Experimental results

Field experiments

Five experimental plots previously infested respectively with different field populations of: *M. javanica*;
a mixture of *M. javanica*, *M. incognita* and a form intermediate between *M. arenaria* and *M. incognita*; *M. incognita*
(two populations) were kept fallow during the dry season
(mid-August until November) to decrease nematode numbers.
In November, in each plot 20 soil samples (0-20 cm hori-
zon) were taken at 50 cm intervals in a transect and 250
cm^3 of soil from each site were processed to recover
Meloidogyne (Demeure and Netscher, 1973). At alternate
sampling sites in each plot a 4-week-old *Meloidogyne*-
susceptible tomato (*Lycopersicon esculentum* cv. Roma)
was planted. From each of the other 50 sampling sites,
two dm^3 of soil were taken and placed in plastic pots,
into each of which a 4-week-old Roma tomato plant was
transplanted, and the pots placed in the soil at each
planting site. Eight days after transplanting, all the
tomato plants were removed from the field, the root systems stained with cold cotton blue-lactophenol (de Guiran,
1966b) and the numbers of juveniles in the roots were
counted (Table 1). No differences were detected between
populations of *Meloidogyne*.

TABLE 1

Assessment of numbers of *Meloidogyne* juveniles and samples infested,
using different techniques (from Prot and Netscher, 1978).

Technique	Mean no. *Meloidogyne*	% samples with *Meloidogyne*
Soil extraction	19	47
Plants in pots	33	64
Plants *in situ*	109	96

If the soil samples had been analyzed by Demeure and
Netscher's method only, it would have been concluded that
the field was slightly infested with large areas free of
Meloidogyne (Table 1). Examination of the roots of the
plants grown in pots showed an increase in the efficiency
of detection. The two techniques are analogous since they
represent different methods of analyzing the number of
juveniles present in a given volume of soil.

Examination of roots of tomato plants transplanted
directly in the field revealed a much higher infection
and the field appeared to be almost completely infested,
only 4% of the sampling sites being free of *Meloidogyne*.
These results can be explained by assuming movement of
juveniles of *Meloidogyne* to the indicator plants followed
by penetration. The results could also be explained by
assuming that the indicator plant had induced eggs close
to the roots to hatch, but this hypothesis is difficult
to reconcile with the results when tomatoes were grown in
pots of soil.

Glasshouse experiments

Movement of juveniles of *M. javanica* in the presence of
absence of a susceptible tomato plant (cv. Roma) was studied using the experimental set up illustrated in Fig. 1.
A glass tube 25, 50, 75 or 200 cm long and 1.2 cm internal diameter was attached either at the side (Fig. 1A) or

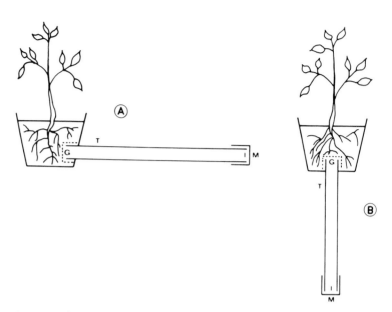

Fig. 1. Experimental set up for studying horizontal (A) and vertical (B) movement of juveniles of *Meloidogyne javanica*. G: stainless steel screen; I: inoculation site; M: polyethylene film; T: glass tube.

in the centre of the base (Fig. 1B) of 6 cm diameter pots with a capacity of 100 cm^3. Pots and tubes were filled with sterile sandy soil and 300 juveniles were placed at a vertical distance of 0, 25, 50, 75 and 100 cm from the roots. Pots were watered daily. Nine days later the roots were stained with cold cotton blue lactophenol (de Guiran, 1966b) and juveniles inside the roots counted. Soil samples from pots planted with tomatoes and from unplanted controls were extracted by elutriation and the number of juveniles counted. Fig. 2 shows the relation between distance of inoculation site and number of juveniles recovered in roots and pots.

An analogous experiment was made, placing 300 juveniles either horizontally or vertically at a distance of 25 and 50 cm from the roots. Counts were made at 1, 3, 5, 7, or 9 days after inoculation (Fig. 3).

The two experiments allow the following conclusions to

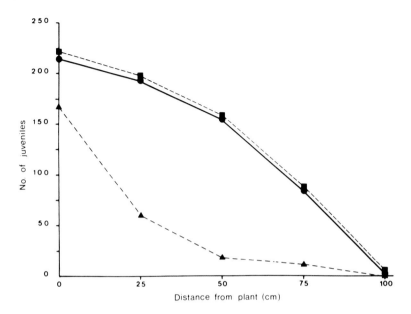

Fig. 2. Vertical movement of juveniles placed at different distances
from root systems of tomato plants.●——●: juveniles counted in roots;
■---■: total number of juveniles in roots and soil; ▲—·—▲ : Number of
juveniles recovered from soil of controls.

be made:

- Juveniles of *M. javanica* are capable of moving as
much as 75 cm and penetrating roots within a period of 9
days.

- Migratory movements are relatively fast as a considerable proportion of the juveniles were able to move vertically over a distance of 50 cm in three days.

- Movement over large distances is not exceptional; in
9 days 60% of the juveniles moved 25 cm, 50% 50 cm and
25% 75 cm.

With the same technique, vertical migration of four
field populations was studied: one population of *M. javanica*, one of *M. incognita*, one with a perineal pattern
intermediate between *M. arenaria* and *M. incognita* and one
population containing a mixture of these species.

Migration was studied as a function of distance by

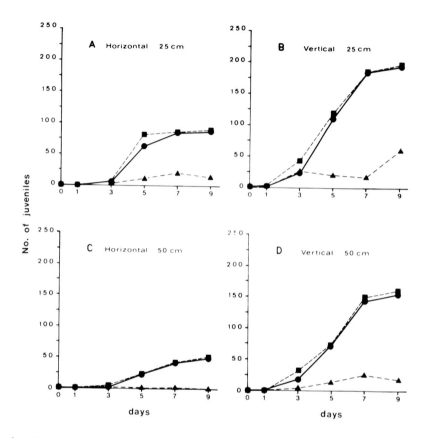

Fig. 3. Vertical and horizontal movement of *M. javanica* as a function of time. A: horizontal movement, inoculation site 25 cm from roots; B: vertical movement, inoculation site 25 cm from roots; C: horizontal movement, inoculation site 50 cm from roots; D: vertical movement, inoculation site 50 cm from roots; ●——●: juveniles counted in roots; ■--■: juveniles counted in roots and soil; ▲--▲: juveniles counted in soil of controls.

introducing the juveniles at 0, 5, 10, 25 and 50 cm from the root systems and leaving the plant in position for nine days. Migration was also studied as a function of time; juveniles were placed 25 cm from the roots and the experiment terminated 1, 3, 5, 7 and 9 days after introduction. Each treatment was replicated five times.

At the end of an experiment the tomato plant was removed from the pot and the root system stained with cold

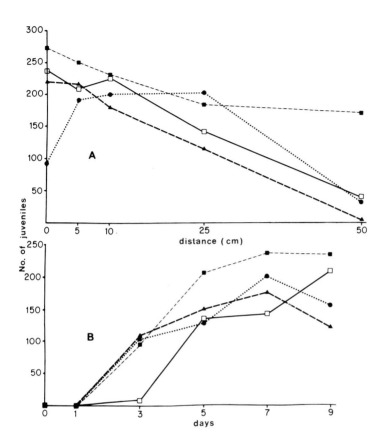

Fig. 4. Vertical migration of juveniles of four natural populations of *Meloidogyne*. ■--■ *M. incognita*;□—□ *M. javanica*;▲-▲population intermediate between *M. arenaria* and *M. incognita*;●·····●population including the three previous species. A: vertical migration as a function of the distance between root systems and inoculum (abscissa); ordinate, number of juveniles in the roots after 9 days. B: vertical migration of juveniles placed 25 cm from root systems as a function of time; abscissa: time of experiment in days; ordinate: number of juveniles found in the roots, (from Prot, 1977, *Rev.Nematol*.**1**, 109-111; with permission).

cotton blue lactophenol (de Guiran, 1966b). Only juveniles found in the roots were counted. Nematodes that entered the pots but did not infect were ignored.

The mean numbers of juveniles in the roots 9 days after inoculation and placed 0, 5, 10, 25 and 50 cm from

the root systems are shown in Fig. 4A. Fig. 4B represents
the mean penetration of juveniles when placed 25 cm from
the roots 1, 3, 5, 7 and 9 days before examination.

In the experiments in which distance between the point
of introduction of nematodes and roots was varied, the
results were approximately the same for the four populations at distances less than 25 cm. The sole exception
occurred with the heterogeneous population at a distance
of 0 cm (in direct contact with the roots). In this case
in three of the five replicates many of the roots died
resulting in an unrealistically low penetration figure.
When nematodes were introduced 5 or 10 cm from the roots,
penetration was similar to that obtained when the same
populations were introduced at a distance of 0 cm.

At a distance of 25 cm, mean percentages of the four
populations varied between 50 and 66%. At a distance of
50 cm, approximately 50% of the *M. incognita* population
penetrated the tomato roots within 9 days (Fig. 4A).
This figure is similar to that obtained with a clone of
M. javanica (Prot, 1977b). Less than 15% penetration was
observed at this distance with the other three populations.

In the experiment in which the distance was constant
(25 cm) the rapid rate of movement is evident, i.e. in
three of four populations approximately 33% of the juveniles reached and penetrated the roots within three
days. Within seven or nine days a penetration of more
than 50% was achieved with each population.

On the basis of these results it is concluded that
the capability to migrate over relatively large distances
is common within populations from West Africa. It seems
logical to assume that the same ability exists within
Meloidogyne populations from other geographical areas as
well. It should be noted that populations do vary in this
character, e.g. higher penetration figures were obtained
in both experiments with the *M. incognita* population than

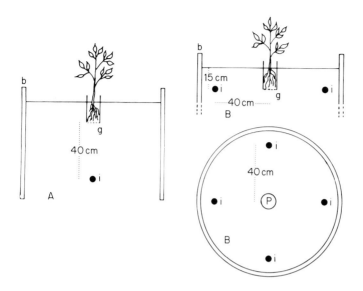

Fig. 5. Arrangement of plants and inoculation sites in microplots.
A = inoculation 40 cm below plant; B = inoculation 40 cm from plant
horizontally, b = wall of micro plot, g = stainless steel screen,
p = 4 week-old Roma tomato in polyvinyl chloride cylinder; i =
point of inoculation, (from Prot and Netscher, 1978, with permission).

with the others tested.

Experiments in microplots

Vertical and horizontal movement of juveniles of *M.
javanica* was studied in cylindrical micro plots of 1m
diameter. Before beginning the experiment, absence of
Meloidogyne was verified by growing a susceptible tomato
plant in the centre of each plot for 10 days, after which
the plants were uprooted and examined for the presence of
Meloidogyne. A 4-week-old tomato seedling (cv. Roma) was
placed in sterile soil contained in a tube 6 cm in diameter and 20 cm tall, closed at the bottom by a stainless
steel screen with 35 μm mesh. One tube was placed in the
centre of each of 21 micro plots. Two thousand juveniles
of *Meloidogyne* were inoculated 40 cm under the screen
(Fig. 5A) or four aliquots of 500 juveniles were

TABLE 2

Infection of roots of tomato cv. Roma 10 days after inoculation with
2000 juvenile *M. javanica* 40 cm from roots
(from Prot and Netscher, 1978)

Root apices		No. juveniles	
Infected	Non-infected	In apices	other root parts
A. 40 cm horizontal[1]			
49	275	194	15
B. 40 cm vertical[2]			
126	202	624	58

[1]10 replicates   [2]11 replicates

inoculated at a depth of 15 cm and 40 cm from the tubes
(Fig. 5B). Tubes were removed 10 days after inoculation
and the tomato roots were stained and the number of juveniles that had penetrated were counted (Table 2). The
results confirm those obtained in glasshouse studies and
demonstrate that under conditions approximating to those
occurring in the field, juveniles of *M. javanica* are
capable of moving vertically or horizontally and subsequently infesting tomato plants.

References

Demeure, Y. and Netscher, C. (1973). Méthode d'estimation des populations de *Meloidogyne* dans le sol. *Cah.ORSTOM, sér.Biol.* **21**,
85-90.
Gooris, J. and d'Herde, C.J. (1972). A method for the quantitative
extraction of eggs and second stage juveniles of *Meloidogyne*
spp. from soil. State Nematology and Entomology Research Station,
Ghent.
Guiran, G. de (1966a). Infestation actuelle et infestation poten-
tielle du sol par les nématodes phytoparasites du genre *Meloidogyne*. *C.r.Acad.Sci.Paris*, **262**, 1754-1756.
Guiran, G. de (1966b). Coloration des nématodes dans les tissus
vegetaux par le bleu cotton à froid. *Nematologica* **12**, 646
Johnson, P.W. and McKeen, C.D. (1973). Vertical movement and dis-
tribution of *Meloidogyne incognita* (Nematodea) under tomato in
a sandy loam greenhouse soil. *Canad.J.Pl.Sci.* **53**, 837-841.
Prot. J.C. (1977a). Amplitude et cinetique des migrations de nématode
Meloidogyne javanica sous l'influence d'un plant de tomate. *Cah.
ORSTOM.ser.Biol.* (1976) **11**, 157-166.

Prot, J.C. and Netscher, C. (1978). Improved detection of low population densities of *Meloidogyne*. *Nematologica* **24**, 129-132.
Seinhorst, J.W. (1956). The quantitative extraction of nematodes from
soil. *Nematologica* **1**, 249-267.
Seinhorst, J.W. (1962). Modifications of the elutriation method for
extracting nematodes from soil. *Nematologica* **8**, 117-128.
Rode, H. (1962). Untersuchungen über das Wandervermögen von Larven
des Kartoffelnematoden (*Heterodera rostochiensis*); Modelversuchen
mit verschiedenen Bodenarten. *Nematologica* **7**, 74-82.
Viglierchio, D.R. (1961). Attraction of parasitic nematodes by plant
root emanations. *Phytopathology* **51**, 136-142.

Discussion

Viglierchio (1961) reported that the horizontal movement of *Meloidogyne hapla* and *M. incognita acrita* was only
a few centimetres but our experiments show that juveniles
of *M. javanica* are capable of moving over distances up to
75 cm in 9 days. These migrations over relatively long
distances are not exceptional as 50% of the juveniles
moved 50 cm vertically in 9 days; Rode (1962) observed
movement of *Globodera rostochiensis* juveniles over distances of 45 cm occasionally.

The great capability by juveniles of *Meloidogyne* may
explain partially why it is so difficult to determine the
level of infestation of slightly infested soils. If the
juveniles can migrate 50 cm in 10 days and infect roots,
the potential inoculum represents all juveniles located
in a hemisphere of soil with a radius of 50 cm or 261.8
dm^3 of soil. According to Poisson's law, the probability
of finding juveniles in 1 dm^3 of soil at population densities of 1 juvenile per 1 or 2 dm^3 of soil, assuming random distribution of the nematodes, is 0.37 to 0.61 respectively; yet at these densities there may be 131 or 262
juveniles in the hemisphere capable of reaching and infecting a susceptible plant. Egg masses will give a more
contagious distribution, thus increasing the frequency
of negative samples. It is therefore not surprising that
Meloidogyne infections occur in fields thought from routine sampling to be free of these nematodes.

H. Ferris and S.D. Van Gundy

*Department of Nematology, University of
California, Riverside, USA*

Introduction

In considering the ecology and host interrelationships
of *Meloidogyne*, a systems analysis approach to the agro-
ecosystem is useful to develop a holistic overview of
the interacting components. "System" denotes a series
of interacting components with a defineable boundary,
for example, an agricultural field. Each component has
its own dynamic attributes which can be examined at
various levels of interest and maintains behavioural con-
sistency with its biotic and abiotic environment (Caswell
et al., 1972; Patten, 1971). Systems analysis involves
investigation and characterization of the component composition and behaviour of the system relative to its environment. Holistic consideration of the system is
achieved by integrating knowledge gained through reductionist approaches to biological research within and
across levels of organization, to describe behaviour
of the system as a whole (Clawson, 1969; Zadoks, 1971).
Understanding of the system may be improved and areas
where knowledge is weak will be delineated (Ferris,
1976a).

Generally, ecological studies on *Meloidogyne* spp.
have involved examination of effects of extrinsic or
intrinisc factors on individual components of the system.

Such reductionist experiments can be interpreted without
ambiguity, although possible interactions among the factors may be overlooked or not investigated because of
interpretation and experimentation difficulties. When
discussing his two-variable experiments on nematode and
plant subsystems and their interaction, Wallace (1969)
underscored the complexity of the system and problems
of interpretation. Extrinsic factors are influences
from outside the system which are not reciprocally influenced by the system, e.g. temperature, Intrinsic
factors arise within the system and are influenced by
the system, e.g. hatching factors, CO_2 levels.

For this discussion, we choose to limit the system
under consideration to a plant and the interacting *Meloidogyne* population. These are actually two interacting
subsystems of a much larger system - the field as a whole,
or ultimately, the universe (Rosenbluth and Weiner,
1945). Other subsystems within this system include all
members of the soil flora and fauna interacting with
each other and with the plant, at varying levels of intensity and directness, and the above-ground organisms.
Such a system may be too complex in initial studies for
for holistic understanding, and boundary restrictions must
be imposed due to lack of information on many components.
Ultimately, consideration of all components and subsystems is necessary, even if only to ensure that they have
negligible effect on movement of the system as a whole,
or to determine whether they warrant consideration in
agricultural management of the system. From an agricultural standpoint, crop productivity is the major output
of the system and is a reflection of the integration of
the interaction of all subsystems and components with
each other and with their environments.

In the systems approach, components are identified
at the level of interest and their interactions dia-
grammed as a conceptual model. Such models provide a

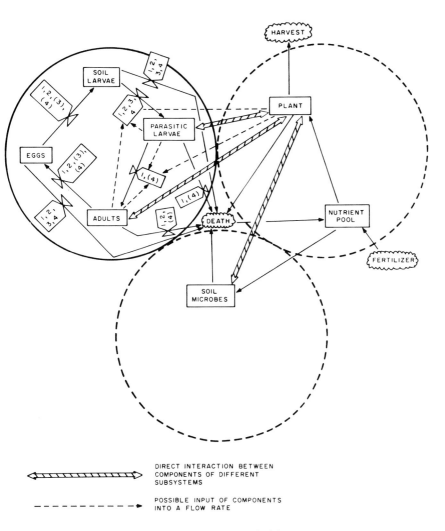

Fig. 1. Schematic representation of a *Meloidogyne*-plant system, with emphasis on details of the nematode subsystem. Extrinsic effects on rates, shown in the "valves", are: 1) soil temperature, 2) soil moisture, 3) soil oxygen, and 4) nematicides. A number in parentheses indicates uncertainty of the effect.

framework for analysis of the system (Tummala *et al.*, 1975) and review of the available literature. The *Meloidogyne*/plant system modelled by Ferris (1976a) will be discussed in detail and will serve as a specific example

of the systems analysis approach. In Fig. 1, solid
arrows indicate the flow of individuals among components
by development or death, and broken arrows denote direct
interaction between the two subsystems. Under this framework, extrinsic and intrinsic factors influencing movement along the arrows can be identified by literature
research and biological intuition, and ecological relationships can be considered.

Nematode ecologists, and ecologists generally, use
the term "optimum conditions" to denote the range of
conditions under which a process functions at its maximum rate, usually in single variable experiements. These
optima, however, may not be favourable for longevity and
survival of the population or species as a whole; they
merely specify the set of conditions under which the
life cycle proceeds most rapidly. Conditions under which
development and reproductive rate of the nematode proceed
at a maximum may result in decreased longevity and
vigour of the nematode or plant host. Definition of the
most favourable conditions from a population standpoint,
involving integration of the interacting subsystems and
the effects of environmental conditions upon them, may
only be possible through a simulation approach.

Heat, an extrinsic factor, is the primary driving
force for the system by its determination of metabolic
rates. All other extrinsic and intrinsic variables (soil
moisture, aeration, pH, soil texture, root exudates, host
variety, food availability, light intensity, etc.) have
a regulating effect on the basic response of the system
to temperature level. The magnitude of the effects of
these secondary factors can be quantified on a 0-1 scale
relative to their level at any point in time. Since most
of the factors are dynamic, the magnitude of their
effects is constantly changing and contributing to the
dynamic interactions of the system. Quantifying these
effects on a 0-1 scale allows their incorporation as

multipliers in systems model equations (Ferris, 1976a).
If a factor is in the optimum range for a process, it
will have a value 1 in the equation describing the rate
of that process and allow the process to proceed at the
maximum rate possible under prevailing conditions of all
other factors. Conversely, if one of the secondary factors is so far out of the optimal range as to be completely limiting to the process, it will have the value
0, and the process described by the equation will stop.
All gradations, between 0 and 1, of the effects of each
factor on the process are possible. This somewhat sim-
plistic initial approach assumes multiplicative effects
of the factors. Systems simulation and further experimentation will elucidate interactions among the effects.

Conceptually, then, the flow down an arrow depicted
in Fig. 1, for example a portion of the nematode life
cycle, can be represented as a function of temperature:
$\lambda_i = \Phi t_i$. This flow at time i will be modified and regulated by various factors, each having an effect in a
0-1 range:

$$\lambda = (\Phi t_i)(m_i p_i h_i \text{ etc.})$$

where m = soil moisture, p = particle size, h = host
variety, etc. - all at the time point i. In quantifying
the ecological system it is necessary, then, to determine
the relationship between development and temperature,
and the shape of the effect curve of the regulating factors. Note in Fig. 1 that as extrinsic and intrinsic
factors become sub-optimal, they may not only slow or
stop processes in the nematode subsystem, but also increase flow rate out of the main cycle and toward death.
Many of the data necessary for formulating such a model
for *Meloidogyne* are present in the literature (Ferris,
1976a) although not always in appropriate form. Areas of
weakness become apparent in the process and some will
be outlined in this presentation. It is not the objective of this paper to review all literature on *Meloidogyne*

ecology, most of which is readily available. Rather, we
attempt to present a philosophical approach for the study
and understanding of *Meloidogyne* in a complex and dynamic
ecosystem. Further, we illustrate the process and under-
score the dynamic attributes of the system, using published data, within the conceptual framework of a model.

The development-temperature relationship

In the nematode subsystem, some intrinsic and extrinsic factors may not affect all stages of the life cycle,
or the effects may be differential. Hence, it is useful
to examine the temperature relationship for each component
stage. A convenient measure of the temperature effect is
the heat unit approach, degree-hours or degree-days above
a developmental threshold (Ferris, 1976b; Milne and Du-
Plessis, 1964; Starr and Mai, 1976; Tummala *et al.*, 1975).
Above a minimum threshold, *Meloidogyne* development approaches linearity with temperature (Ferris *et al.*, 1978;
Milne and DuPlessis, 1964; Tyler, 1933) and can be expressed in physiological time (heat units) rather than chronological time. Tyler (1933) found that 7000-8500 heat units
(degree-hours) were required from larval penetration of
tomato roots to first egg production. A further 5000 heat
units were required for development and hatch of eggs.
Studies with *M. arenaria* showed a requirement of 4200 heat
units for egg development (Ferris *et al.*, 1978). Tyler
(1933) calculated a requirement of 12 heat units per egg
for egg production and 250-500 units for larval penetration. Thus, in an experimental situation with all other
intrinsic and extrinsic factors held in the optimum range,
the life cycle of the nematode was completed in 13000 heat
units. This would be the shortest possible development,
the first observation life cycle completion under those
conditions. In fact, all larvae do not penetrate roots at
the same rate, due to genetic and spatial variances; similarly there are differences in larval development and

female egg production and egg development (Bird, 1972;
Ferris *et al.*, 1978). Tyler (1933) presented data on larval penetration of roots with time which can be manipula-
ted by liberal interpretation into a penetration frequency
distribution relative to physiological time (Fig. 2).
Similar curves can be derived for the probability of penetration of *M. hapla* into alfalfa roots (Griffin and Elgin,
1977). Another distribution can be determined from Tyler's
(1933) data on larval development in the root. These probability distribution patterns are important in the epidemiological analysis of nematode population dynamics and in
simulation efforts. Each probability curve is specific for
a host-parasite relationship, in this case under ideal
conditions of other factors, and reflects parasite genotype variability and phenotype history, root tip distribution, and relative suitability of infection sites. These
kinds of data from ecological studies are necessary for
generation of realistic quantitative population models.

Heat units required for completion of the life cycle
may increase as other environmental conditions move out
of the optimum range. It is useful to consider the concept

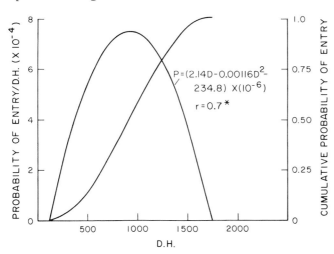

$$P = (2.14D - 0.00116D^2 - 234.8) \times (10^{-6})$$

$$r = 0.7^*$$

Fig. 2. Probability of entry of a tomato root per degree-hour above
10°C for second-stage larvae of a *Meloidogyne* sp. Data from Tyler
(1933).

of "effective heat units". If any factor has an effect on
the development less than 1 on the scale previously discussed, effective heat unit accumulation will be reduced
accordingly (Jones, 1975; Ferris, 1965b). Host status of
the plant is another factor involved in development time.
Although Pyrowolakis (1977) found little difference in
generation time of *M. javanica* on hosts from six families,
Milne and DuPlessis (1964) measured a 9000 heat unit requirement from penetration to first egg production on tobacco compared with 14,000 units on alfalfa, and differences were detected among grape varieties (Hackney and
Ferris, 1975). Host effects on this type strongly influence population dynamics of the nematode. Other reflections of host status include larval penetration patterns
and female egg production relative to effective heat
units. Measures of these parameters in cultivars of a
plant species would indicate levels and mode of expression
of horizontal resistance as defined by Vanderplank (1963)
and would be useful in simulation studies for pest management decisions (Ferris, 1976a).

An area of *Meloidogyne* ecology where fundamental data
are lacking is the temperature requirements for egg production. Tyler's (1933) calculations involved individual
infections and ideal conditions. Host interactions are
involved here, including carrying capacity per unit of
root and competitive effects of multiple infections.
These interactions may also be involved in larval penetration, length of development and production of males and
will be discussed later. Other data, e.g. Wallace (1969)
are often ambiguous as they may reflect both nematode
development in the root and egg production.

Estimates of optimum temperature for phases of the
life cycle of *M. javanica* are: embryogenesis - 25-30°C
(Bird, 1972); hatching - 30°C; mobility - 25°C; growth -
25-30°C (Bird and Wallace, 1965); and egg production -
32-34°C (Thomason and Lear, 1961). The egg production

measurements were made on a California population and the
other measurements on Australian populations. It is reasonable to assume that populations become acclimated and adapt
to local conditions and there is a danger of generalizing
about environmental responses with populations of different geographic origin. Daulton and Nusbaum (1961) showed
that a population of *M. javanica* had temperature survival
capabilities consistent with their geographic origin. A
Netherlands population of *M. incognita* had a temperature
threshold for infection and reproduction of 5°C lower than
a Venezuelan population (Dao, 1970). It is likely that
nematodes, esepcially those with several generations a
year, are similar to other organisms (Tummala *et al.*,
1975) in having similar temperature response optima and
thresholds in different developmental stages.

There are some consistent differences in species response to temperature. In general, optima for *M. hapla* are
some 5°C lower than those for *M. javanica*, reflective of
its existence and survival in temperate regions and high
altitudes. Isolated populations of *M. hapla*, however, do
occur and survive in warm regions, for example Palo Verde
Valley of Southern California (Radewald *et al.*, 1969).
Temperature responses of this desert population have not
been studied. Generally in California, *M. hapla* is most
common in the cool coastal areas. *M. naasi* and *M. thamesi*
are found along the coast and in the cool northern areas,
primarily restricted to barley and potatoes, respectively.
M. incognita and *M. javanica* are more commonly found in
the hot interior valleys (Siddiqui *et al.*, 1973). In East
Africa, *M. hapla* occurred in several areas but flourished
only above 1800 m altitude. *M. incognita* and *M. javanica*
were restricted to areas below 2000 m (Whitehead, 1969).
This pattern is typical of other observations on the
world-wide distribution of *Meloidogyne* spp.

As well as being determinant in the nematode
development rate, heat energy has the capacity to shunt

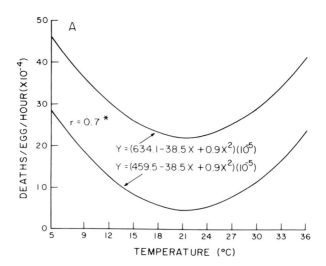

Fig. 3a-b. Effect of temperature and time on the death rate (egg deaths/egg/hr) of eggs of *Meloidogyne arenaria*. a) Death rate during first week of exposure; upper curve based on actual deaths measured, lower curve based on assumption that death rate for first week at 21°C was the same as for subsequent weeks.

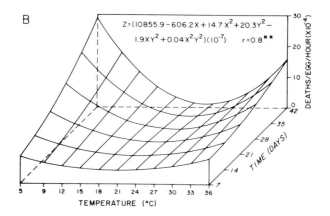

b) Death rate of eggs after first week of exposure. (From Ferris *et al.*, 1978)

various components out of the main path of the life cycle
and toward death or non-viability. The death rates generated usually increase as the temperature and other environmental factors get further from their optimum ranges.
There is a temperature-dependent death rate of eggs, larvae and adults. Egg death rates for *M. arenaria* were higher during the first week than during subsequent weeks of
exposure to temperatures outside the optimum range (Fig.
3). This differential was due to greater sensitivity of
younger eggs to the stress, either genetically based or
due to lack of acclimation by previous exposure (Ferris
et al., 1978). Data on temperature sensitivity of second-
stage larvae (Van Gundy *et al*., 1967) can be converted to
death rates (Fig. 4); however, there is little information
on temperature death rates of adults. Davide and Triantaphyllou (1967) provided insight into parasitic larval
death and development of males relative to temperature
extremes (Fig. 5). Development of a male is equivalent to

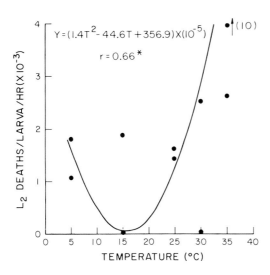

$$Y = (1.4T^2 - 44.6T + 356.9) \times (10^{-5})$$

$$r = 0.66^*$$

Fig. 4. Death rate relative to temperature of second-stage *Meloidogyne javanica* (larvae/larva/hr). Data from Van Gundy *et al*., (1967).

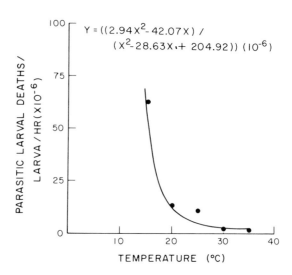

Fig. 5. Death rate relative to temperature of parasitic *Meloidogyne incognita* (larvae/larva/hr). Larvae developing into males are considered "dead". Data from Davide and Triantaphyllou (1967).

death in terms of the quantitative direction of the nematode subsystem as there is no further productive input, particularly in parthenogenetic species.

Temperature response and survival of *Meloidogyne* eggs and larvae is also a function of their physiological age. Larvae of *M. javanica* hatch from the egg with 30% of their body wieght in lipids as a stored energy reserve. The use of this reserve depends upon the rate of metabolism which is in turn influenced by temperature, moisture, aeration, etc. (Van Gundy *et al.*, 1967; Wallace, 1966a). The more food reserves the nematode has when the host root enters its environment, the greater are its chances of getting to the infection site and entering the root. Extended survival and infectivity in soil is related to conservation of food reserves and physiological condition of aged larvae rather than to chronological age. The susceptibilit of larvae to temperatures between 0 and 10°C is related to temperature-induced phase transitions in lipid position in the biological membranes of the nematode. *M. javanica*

exhibits a phase transition at 10°C while *M. hapla* does
not, possibly accounting for, or reflecting, differences
in geographical distribution (Lyons *et al.*, 1975). *M. hapla*
(Sayre, 1964) and *M. naasi* (Franklin *et al.*, 1971) exhibit
cryophilic tendencies and can survive subzero soil temperatures. In fact, egg hatching in *M. naasi* is stimulated
by chilling. Storage of infested soil at 0°C or 10°C for
1 week prior to incubation at 20°C resulted in three times
as many emerging larvae as unchilled soil (Franklin *et al.*,
1971). On the other hand, *M. incognita*, *M. javanica*, *M.
arenaria* (Thomason and Lear, 1961), and possibly *M. exigua*,
are normally subtropical or tropical region species
(thermophilic) incapable of extended survival in soils
below 10°C. Since nematodes are continuously moved about
by man, distribution ranges of *Meloidogyne* spp. overlap
resulting in the occurrence of temperate species in many
subtropical and tropical areas (Siddiqui *et al.*, 1973;
Whitehead, 1969).

Effects of other extrinsic factors

Soil is a three-phase system: solid, liquid, and gas.
The solids are organic and inorganic and include sand,
silt and clay particles; the liquid contains a variety of
ions and gases; the gas phase is inversely proportional
to the liquid phase. Nematodes are dependent upon water
for most of their activities. In moist soils (40-60% of
field capacity) nematodes are active and move through
the soil in water films. In dry soils they become inactive
and may die through desiccation. In saturated soils nematodes are unable to move because of water film thickness
and may show reduced hatch and activity rates from lack
of oxygen (Wallace, 1966b; Van Gundy and Stolzy, 1961).
It is difficult to separate the interacting effects of
soil texture, moisture and aeration on nematode movement
and survival (Stolzy and Van Gundy, 1968). In terms of
the conceptual *Meloidogyne*/plant system used as a

framework for this discussion (Fig. 1), soil moisture
would have a 0-1 effect on the accumulation of effective
heat units for various flows between components. It might
also cause quantitative limitations on larvae penetrating
roots, or be restrictive in egg hatch. Effects on parasitic stages of the life cycle are indirect, through its
effect on host physiology.

Direct effects of moisture can be expressed in terms
of suction potential, a measure of the accessibility of
water to the nematode. There is an optimum range which
Wallace (1963, 1966a) generalizes as being in the region
of field capacity. In this range, the moisture effect
would be 1; as suction increases, points are reached at
which larval movement, egg hatch, or egg development become increasingly inhibited. The effect on these processes then drops below unity until the processes are
completely limited when the effect of the moisture factor
is numerically zero. The shape and origin of the effect
curve probably differs for each process.

At low suction potential, soil pores are filled and
oxygen deficiency may result. Exposure of embryonated
eggs to anaerobic conditions for one week is lethal due
to lactic acid accumulation in the cells. At low oxygen
levels, the metabolism and movement of larvae, and
their infectivity, are reduced. Continued low oxygen
levels around larvae and females within plant roots
cause correspondingly reduced growth and reproduction
(Van Gundy and Stolzy, 1961; Van Gundy *et al.*, 1967).
The effect of O_2 on invasion, movement and hatch of *M.*
hapla in organic soil appears to be modified by the
level of CO_2 present. Wong and Mai (1973) found that
activity was not adversely affected at 10% CO_2 but was
at 30% CO_2. High suction potentials directly, or indirectly because of increased aeration, induce high metabolic activity in *M. javanica* larvae causing rapid body-
reserve depletion and reducing survival time (Van Gundy

et al., 1967). Maximum hatch of *M. javanica* eggs occurs
when the soil pores are drained to field capacity (Wallace
(Wallace, 1966b, 1968b). The relationship between rate of
hatch of *M. javanica* eggs and oxygen concentration is
linear over a range of 5% to 20% O_2. No hatch occurs in
the absence of O_2 (Wallace, 1968a).

There are compounding problems in consideration of
moisture availability. Simultaneous consideration of the
effects of soil solution is necessary. Dissolved salts,
hydrogen ion concentration, root exudates, fertilizers,
organic compounds, and various pesticides all vary in
concentration with soil moisture level. Differences in
osmotic potential inside and outside the nematode body
occur with the changes in concentration of dissolved
salts in the soil water. Hatch of *M. javanica* eggs is
unaffected below osmotic potentials of 2.5 atmospheres
and decreases to zero at 12.5 atmospheres (Wallace,
1966b), although osmotic potentials in agricultural soils
seldom exceed two atmospheres. However, the actual stress
on the nematode involves some combination of the effects
of suction and osmotic potential. Wallace (1966b) uses
"soil moisture stress" to describe the total potential
of the soil water energy, with main components suction
and osmotic potential. Suction and osmotic potential
may be interchangeable in terms of quantitative availability of water to eggs, but differ in effects on
larvae which are physically limited by thin water films
at high suction. In quantifying the model, either the
rate of larval entry into the roots and/or the number
entering would be regulated by the soil moisture stress.

Embryos and first-stage larvae are more resistant to
water loss than unhatched second-stage larvae because
of changes in the egg membrane after the first moult in
the egg. The egg sac appears to give more protection to
water loss from the unhatched second-stage larvae than
the egg shell. In drying environments the gelatinous

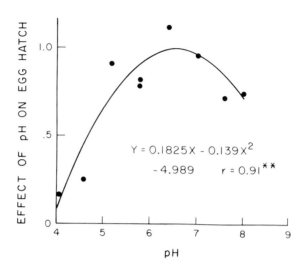

Fig. 6. Effect of pH on hatch of *Meloidogyne javanica* eggs. (After Wallace, 1966a).

matrix of the egg sac maintains a high moisture level and provides a barrier to water loss from the eggs (Wallace, 1966b). Since the nematode cuticle allows passage of water rapidly back and forth, solutions of high osmotic potential cause second-stage larvae to shrink in size and sometimes become inactive or quiescent. *Meloidogyne* larvae are able to osmoregulate and become active when returned to favourable osmotic environments. Little is known about osmoregulatory mechanisms but studies on *Aphelenchus avenae* indicate that the average residence time of a water molecule in a nematode is approximately 0.9 sec; thus water movement in and out of nematodes is a very dynamic process (Marks *et al.*, 1968).

Meloidogyne spp. survive, hatch and reproduce over a wide pH range, 4.0-8.0. Hatch of *M. javanica* eggs was maximum between pH 6.4 and 7.0 and inhibited below pH 5.2 (Wallace, 1966a). As with other soil solution factors, the pH fluctuates, increasing and decreasing with changes in soil moisture level. The effect of pH on egg

hatch can be transformed to a 0-1 scale (Fig. 6). Further
analysis may be necessary to determine whether the effect
is on length of egg development or on numbers hatching.
Organic chemicals in the soil solution may be toxic to
nematodes; decomposition products of rye residues immo-
bilized *M. incognita* larvae in three hours and killed
them after 24 hours (Patrick *et al.*, 1965).

Many reports associate differences in distribution
and disease severity of *Meloidogyne* with soil type and
texture. *M. incognita* and *M. hapla* occur in greater infestations in sandy loam soils than in heavy clay soils
in Maryland (Sasser, 1954). In Arizona, heaviest infestations of *M. incognita* occur on the coarse textured
soils (O'Bannon and Reynolds, 1961). On the other hand,
Whitehead (1969) found the relative frequencies of *M.
javanica* and *M. incognita* in East Africa were not correlated with soil texture. Generally, *Meloidogyne* spp.
occur on a wide range of soil types but crop damage is
often accentuated on sandy soils or sandy patches within
a field. In these cases, the plant damage is often a
reflection of a series of stresses, including low fert-
ility, poor moisture holding capacity and poor management,
which are compounded by the presence of the nematode
(Ferris and McKenry, 1977). Soil particle size affected
hatch rate of larvae, with highest numbers at 150-250 μm.
Reproduction rate was highest in the coarse sand while
root penetration was greatest in fine sand (Wallace,
1969). Moisture distribution in pore spaces is a confounding factor in studies of this type. The effects of
these physical constants of the soil environement can
be transformed to a scale reflecting their influence on
various stages of the nematode life cycle (Fig. 7).

Host-parasite interrelations

Host plant roots and their emanations are major
factors affecting the distribution of plant-parasitic

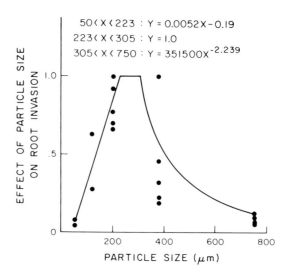

Fig. 7. Effect of particle size on root invasion by second-stage larvae of *Meloidogyne javanica*. Data from Wallace (1966a).

nematodes in the soil (Jones, 1960). The spatial distribution of *Meloidogyne* spp. is closely associated with root distribution in grape vineyards (Ferris and McKenry, 1974). Grape roots and galls have been found 510 cm below the surface (Raski and Lear, 1962). Although this is largely a response to food source, there are other factors that influence nematode behaviour. Viglierchio and Lownsbery (1960) have shown a significant increase in hatch of *M. hapla*, *M. incognita acrita* and *M. javanica* due to root emanations. Wallace (1966a) found that hatching of *M. javanica* was inhibited in nonsterile soil. Inhibition disappeared when the soil was either sterilized or planted to a tomato seedling. It was further observed that larvae stored in a porous cup embedded in the soil near a tomato seedling used their body reserves much faster than those not associated with a seedling (Van Gundy *et al.*, 1967). The relationship between numbers of *Meloidogyne* in the soil, rate of reproduction and growth of host plant is a complex ecological

situation. Environmental factors may influence both plant
and nematode. They may influence the nematode in the soil,
soil water, and in the plant. Thus, the nutrient level
supplied to the plant influenced the development and egg
production of parasitic stages (Wallace, 1969).

Weakness in our knowledge of the system are revealed
by attempts to quantify the host-parasite interrelations
from a holistic standpoint (Fig. 1). Major points of
direct interaction of the nematode and plant subsystems
involve the parasitic larval stages (3rd and 4th) and
adults. Interactions also occur with the second-stage
larvae through the infection process, and through any
root exudate effects on egg hatch and larval attraction.

In the infection process, the host-parasite interaction effect on the number of larvae able to enter and
become established at each infection site is important.
There is some evidence that root tip entry is not random
(Wallace, 1973) but presumption of a carrying capacity
related to infection site size and host physiological
status is conceptually logical. The epidemiological concepts of Vanderplank (1963) and Seinhorst (1965) are useful in quantifying the ecological aspects of the host/
parasite interaction. Thus, if d represents the proportion
of a root system of unit size damaged by one female nematode, and V is the current volume of infection sites, then

$$X = (1-d/V)^{(A + \beta P)}$$

where X is the proportion of the root infected, $(1-d/V)$
represents the proportion of the root system not damaged
by one nematode, and a parasitic larva (P) has an effect
only β as great as an adult (A). Using this relationship,
$(1 - X)$ represents the proportion of the root system
still available for infection; further, it is a measure
of the proportion of the root system available for active
plant and nematode growth, and hence of plant vigour
(Ferris, 1976a). The value $(1 - X)$ is a dynamic,

constantly changing proportion, its magnitude being influenced by the rate of plant and nematode growth, and
by the rate of infection; reciprocally, it affects each
of these rates.

Changes over a period of time in the parasitic larval
component of the nematode subsystem (Fig. 1) are represented by flows into minus flows out of the compartment.
This change can be represented by:

$$\Delta P = S(1 - X)R_1 - P(1 - X)R_2 - PR_3$$

Here S represents the number of infective second-stage
larvae in the soil, $(1 - X)$ measures available infection
sites, and R_1 is the rate of penetration. However, Wallace
(1969) found no difference in larvae penetrating relative
to inoculum density so that availability of sites may
not be a factor. The rate of penetration is a function
of the penetration probability heat unit relationship
(Fig. 2), with numbers penetrating being influenced by
effects of soil physical factors on larval movement and
activity, and by any effect of host resistance expressed
through penetration. For flow out of the compartment to
adulthood, P is the number of parasitic larvae, $(1 - X)$
a measure of food availability, host vigour and intraspecific competition, and R_2 a function of the develop-
ment-heat unit relationship and effects of host resistance to development. Wallace's (1969) data show a slow-
ing effect on development under crowded conditions. R_3
represents death rate of parasitic larvae due to temperature, competition or other conceivable stress. Larval
death rate includes development to males and can be
related to temperature (Fig. 5) and to conditions of
crowding when up to 50% of larvae may develop to males
(Davide and Triantaphyllou, 1967). There is little data
available to define crowded conditions; Ferris (1976a)
arbitrarily used a value of X = 0.9, i.e. $(1 - X) = 0.1$,
in a simulation model.

A similar equation can be analyzed for change in the

number of adult females, with flow in from the parasitic
larval compartment and flow out represented by a natural
death rate and a death rate due to competitive effects
and starvation. Few data are available on either of these
death rates. The effects of plant vigour on egg output of
females is unknown, and there are indications of host
resistance factors being expressed through egg production
rates (Hackney and Ferris, 1975). These parameters of
interaction have important bearing on the population
dynamics of *Meloidogyne* spp.

Interspecific competition in Meloidogyne

We have presented a somewhat superficial discussion of
two subsystems of the complex agro-ecosystem. There are
many other subsystems, both in and out of the soil, which
are interacting directly with each other, or indirectly
through the plant - the subsystem of primary interest
from an agricultural standpoint.

Some of the interacting subsystems contribute to the
buffering effects of species diversity (Nusbaum & Ferris,
1973) and some might be considered by pest managers or
system manipulators as falling into the sphere of biological control.

Plant-parasitic nematodes occur in polyspecific communities in agricultural soils (Oostenbrink, 1966). In
the conceptualized system, competitive interactions in
terms of food availability are expressed through (1 - X),
the measure of host vigour and physiological activity.
As niches of competitive species become closer, in terms
of infection and feeding sites, other aspects of the
competition including larval agressiveness and vigour
in infecting and colonizing available sites become important. Such aggression would depend on the adaptation
of the nematode species to the environment and on plant
host status relative to its effect on penetration, development and egg production of that species.

There has been limited work on intrageneric competition
of *Meloidogyne*, primarily due to problems of species
identification in any quantitative study. At least four
species of *Meloidogyne* were reported from a vineyard
(Hackney, 1974; 1977). Two species were more common than
the others, they occurred together on more than 50% of
the vines samples and showed no differences in distribution throughout the root zone (Ferris, 1976c).

Conclusions

We have attempted to emphasize the dynamic nature of
Meloidogyne ecology, not only in terms of response of the
nematode to various levels of the environmental factors,
but also with regard to the confounding complexity of
the dynamic, interacting nature of the environmental
factors themselves. An important consideration is the
lack of homogeneity of the soil environment, both with
regard to textural and structural anomalies and in terms
of those environmental factors which vary in a vertical
continuum. Superimposed on the complexity of the soil
physical environment is the spatial distribution of the
interacting biological subsystems - in this case a plant
root system and a *Meloidogyne* population. At each vertical
level in the soil the interacting subsystems are subjected to different magnitudes of the extrinsic physical
factors. Further, the dynamics of fluctuation in the
magnitude of these physical factors varies with soil
depth. Holistic comprehension of the system would involve
consideration of this environmental variability. Essentially this could be achieved by considering the soil as
a multi-layered medium and determining the effects of
the environmental conditions at each depth of components
of the biological subsystems at those depths. Thus the
dynamics of the subsystem will vary with relative
location in the soil. The real benefit of the approach
at this level of complexity is probably through the

regimentation of model construction in its promotion of understanding of the system. Computer simulations, using the derived models, on a multi-layered soil system are possible, but the expense in terms of machine time at the capability level of existing computer hardware would make their use impractical in the pest management decision process. A compromise can be achieved by simulating the system at the depth of maximum feeder root production of the plant, that is the root requiring protection for promotion of high yields.

References

Bird, A.F. (1972). Influence of temperature on embryogenesis in *Meloidogyne javanica*. *J.Nematol.* **4**, 206-212.

Bird, A.F. and Wallace, H.R. (1965). The influence of temperature on *Meloidogyne hapla* and *M. javanica*. *Nematologica* **11**, 581-589.

Caswell, H., Koeing, H.E., Resh, J.A. and Ross, Q.E. (1972). *In* "Systems Analysis and Simulation in Ecology" (Ed. B.C. Patten), II, 3-78, Academic Press, New York.

Clawson, M. (1969). *In* "Physiological aspects of crop yield" (Eds. J.D. Eastin, F.A. Haskins, C.Y. Sullivan and C.H.M. Van Bavel) 1-14, Am.Soc.Agron., Madison.

Dao, D.F. (1970). Climatic influence on the distribution pattern of plant parasitic and soil inhabiting nematodes. *Meded.Landbouwhogeschool Wageningen* **70**, 181.

Daulton, R.A.C. and Nusbaum, C.J. (1961). The effect of soil temperature on the survival of the root-knot nematodes *Meloidogyne javanica* and *M. hapla*. *Nematologica* **6**, 280-294.

Davide, R.G. and Triantaphyllou, A.C. (1967). Influence of environment on development and sex differentiation of root-knot nematodes. *Nematologica* **13**, 102-110.

Ferris, H. (1976a). Development of a computer simulation model for a plant-nematode system. *J.Nematol.* **8**, 255-263.

Ferris, H. (1976b). A generalized nematode simulator based on heat unit summation. *J.Nematol.* **8**, 284.

Ferris, H. (1976c). Observations on the distribution of *Meloidogyne* spp. in a vineyard soil. *J.Nematol.* **8**, 184-185.

Ferris, H., DuVerney, H.S. and Small, R.F. (1978). Development of a soil-temperature data base on *Meloidogyne arenaria* for a simulation model. *J.Nematol.* **10**, 39-42.

Ferris, H. and McKerny, M.V. (1974). Seasonal fluctuations in the spatial distribution of nematode populations in a California vineyard. *J.Nematol.* **6**, 203-210.

Ferris, H. and McKerny, M.V. (1977). Vineyard management and nematode populations. *Calif. Agric.* **31**, 6-7.

Franklin, M.T., Clark, S.A. and Course, J.A. (1971). Population changes and development of *Meloidogyne naasi* in the field.

Nematologica 17, 575-590.

Griffin, G.D. and Elgin, J.H., Jr. (1977). Penetration and development of *Meloidogyne hapla* in resistant and susceptible alfalfa under differing temperatures. *J.Nematol.* 9, 51-56.

Hackney, R.W. (1974). The use of chromosome number to identify *Meloidogyne* spp. on grapes. *J.Nematol.* 6, 141-142.

Hackney, R.W. (1977). Identification of field populations of *Meloidogyne* spp. by chromosome number. *J.Nematol.* 9, 248-249.

Hackney, R.W. and Ferris, H. (1975). Infection, development, and reproduction of *Meloidogyne incognita* in eight grapevine cultivars. *J.Nematol.* 7, 323.

Jones, F.G.W. (1960). Some observations and reflections on host finding by plant nematodes. *Deel* 25, 1009-1024.

Jones, F.G.W. (1975). Accumulated temperature and rainfall as measures of nematode development and activity. *Nematologica* 21, 62-70.

Lyons, J.M., Keith, A.D. and Thomason, I.J. (1975). Temperature-induced phase transitions in nematode lipids and their influence on respiration. *J.Nematol.* 7, 98-104.

Marks, C.F., Thomason, I.J. and Castro, C.E. (1968). Dynamics of the permeation of nematodes by water, nematicides and other substances. *Exp.Parasitol.* 22, 321-337.

Milne, D.L. and DuPlessis, D.P. (1964). Development of *Meloidogyne javanica* (Treub.) Chit., on tobacco under fluctuating soil temperatures. *S.Afr.J.Agric.Sci.*, 7, 673-680.

Nusbaum, C.J. and Ferris, H. (1973). The role of cropping systems in nematode population management. *Annu.Rev.Phytopathol.*, 11, 423-440.

O'Bannon, J.H. and Reynold, H.W. (1961). Root-knot nematode damage and cotton yields in relation to certain soil properties. *Soil Sci.* 92, 384-386.

Oostenbrink, J. (1966). Major characteristics of the relation between nematodes and plants. *Meded.Landbhogesch.Wageningen.* 66, 46.

Patrick, Z.A., Sayre, R.M. and Thorpe, H.J. (1965). Nematicidal substances selective for plant-parasitic nematodes in extracts of decomposing rye. *Phytopathology* 55, 702-704.

Patten, B.C. (1971). *In* "Systems Analysis and Simulation in Ecology." (Ed. B.C. Patten) I, 3-121, Academic Press, New York.

Pyrowalakis, E. (1977). Einfluss von Bodentemperaturen und Wirtzpflanzen auf de Generationsdauer von *Meloidogyne javanica*. *Nematologica* 23, 47-50.

Radewald, J.D., Mowbray, P.G., Paulus, A.O., Shibuya, F. and Rible, J.M. (1969). Preplant soil fumigation for California head lettuce. *Plant Dis. Rep.*, 53, 385-389.

Raski, D.J. and Lear, B. (1962). Influence of rotation and fumigation on root-knot nematode populations on grape replants. *Nematologica* 8, 143-151.

Rosenbleuth, A. and Wiener, N. (1945). The role of models in Science. *Philos. Science* 12, 316-321.

Sasser, J.N. (1954). Identification and host-parasite relationships of certain root-knot nematodes (*Meloidogyne* sp.). *Univ.Md.Agric. Exp.St.Bull.* A-77, 30.

Sayre, R.M. (1964). Cold-hardiness of nematodes. *Nematologica* 10, 168-179.
Seinhorst, J.W. (1965). The relation between nematode density and damage to plants. *Nematologica* 11, 137-154.
Siddiqui, I.A., Sher, S.A. and French, A.M. (1973). Distribution of plant parasitic nematodes in California. Dept. Food and Agric. Sacramento, California. 324.
Starr, J.L. and Mai, W.F. (1976). Predicting on-set of egg production by *Meloidogyne hapla* on lettuce from field soil temperatures. *J.Nematol.* 8, 87-88.
Stolzy, L.H. and Van Gundy, S.D. (1968). The soil as an environment for microflora and microfauna. *Phytopathology* 58, 889-899.
Thomason, I.J. and Lear, B. (1961). Rate of reproduction of *Meloidogyne* spp. as influenced by soil temperature. *Phytopathology* 51, 520-524.
Tummala, R.L., Ruesink, W.A. and Haynes, D.L. (1975). A discrete component approach to the management of the cereal leaf beetle ecosystem. *Environ.Entomol.* 4, 175-186.
Tyler, J. (1933). Development of the root-knot nematode as affected by temperature. *Hilgardia* 7, 391-415.
Vanderplant, J.E. (1963). Plant Diseases: Epidemics and Control. 349, Academic Press, New York.
Van Gundy, S.D., Bird, A.F. and Wallace, H.R. (1967). Ageing and starvation in larvae of *Meloidogyne javanica* and *Tylenchulus semipenetrans*. *Phytopathology* 57, 559-571.
Van Gundy, S.D. and Stolzy, L.H. (1961). Influence of soil oxygen concentrations on the development of *Meloidogyne javanica*. *Science* 134, 665-666.
Viglierchio, D.R. and Lownsbery, B.F. (1960). The hatching response of *Meloidogyne* species to the emanations from the roots of germinating tomatoes. *Nematologica* 5, 153-157.
Wallace, H.R. (1963). The Biology of Plant Parasitic Nematodes. 280. Edward Arnold, London.
Wallace, H.R. (1966a). Factors influencing the infectivity of plant parasitic nematodes. *Proc.Royal Soc.B.*, 164, 592-614.
Wallace, H.R. (1966b). The influence of moisture stress on the development, hatch and survival of eggs of *Meloidogyne javanica*. *Nematologica* 12, 57-69.
Wallace, H.R. (1968a). The influence of aeration on survival and hatch of *Meloidogyne javanica*. *Nematologica* 14, 223-230.
Wallace, H.R. (1968b). The influence of soil moisture on survival and hatch of *Meloidogyne javanica*. *Nematologica* 14, 231-242.
Wallace, H.R. (1969). The influence of nematode numbers and of soil particle size, nutrients and temperature on the reproduction of *Meloidogyne javanica*. *Nematologica* 15, 55-64.
Wallace, H.R. (1973). Nematode Ecology and Plant Disease. 228 Edward Arnold, London.
Whitehead, A.G. (1969). The distribution of root-knot nematodes (*Meloidogyne* spp.) in tropical Africa. *Nematologica* 15, 315-333.
Wong, T.K. and Mai, W.F. (1973). *Meloidogyne hapla* in organic soil: Effects of environment on hatch, movement, and root invasion. *J.Nematol.* 5, 130-138.
Zadoks, J.C. (1971). Systems analysis and the dynamics of epidemics.

Phytopathology **61**, 600-610.

Discussion

Asked about the practical application of the model
Ferris said that it had not yet been used as a basis for
a guide to farmers but it had been used extensively in
suggesting and planning glasshouse and laboratory experiments and for testing possible treatments to be used in
field experiments. Replying to a question about limitations on the use of factors in the model, he explained
that there are factors which completely inhibit processes
in the model, even though other factors are within a
suitable range. For example, soil moisture level can
have the effect of preventing egg hatch or movement of
larvae either directly or through interaction with factors
such as aeration. This can occur while temperature conditions are quite suitable for egg hatch. Asked whether
the presence of other microorganisms would interfere
with the model, Ferris replied that conceptually these
would be handled as models of separate sub-systems within
the total system. Interactions between these sub-systems
would have to be quantified just as interactions between
the nematode and plant sub-system are quantified. So
far these other sub-systems have not yet been considered.

13 NEMATODES AND GROWTH OF PLANTS: FORMALIZATION OF THE NEMATODE-PLANT SYSTEM

J.W. Seinhorst

Instituut voor Plantenziektenkundig Onderzoek, Wageningen, The Netherlands

Introduction

The ultimate aim of almost all research on plant parasitic nematodes is to find ways of preventing damage and yield reductions caused by these nematodes that give maximum returns within the narrowing limits set by the necessary protection of man and environment. This means that in relation to the economics of crop protection nematological investigations must provide the basis for two predictions to be made in time to plan control measures of the proper kind and extent:
1) the yield reduction expected without control measures,
2) the effect on this yield reduction of the various possible protective measures (dosages of nematicides, number of years without host plant) (Seinhorst, 1973b). The first prediction requires the measurement of a variable, on which to base the prediction, that is related to yield sufficiently closely at the time of measurement. The relation between this variable and yield, the values of the parameters to be applied in this relationship and their variability must be supplied by nematological investigations. In a great majority of the cases of nematodes attacking crops, the nematodes were present in the soil before sowing or planting. Then the only variable that has to be considered as a basis for the prediction

of crop losses is nematode population density in the soil
at the time of sowing or planting. It differs mainly from
attacks by nematodes already present in the plant material
in that in this case the total initial nematode population
is much smaller but concentrated in or on the first roots.
Therefore infested plant material is particularly harmful
if the nematode species involved multiplies relatively
fast (e.g. *Meloidogyne*) and in perennial plants.

Nematodes as damaging organisms and the nature of their effects on plants

   In general the amount of damage caused by a noxious
organism to a plant depends on the density of the population of that organism on the plant. This density depends
on the multiplication of the noxious organism on a food
source, which may be the ultimately damaged plant or another source in a previous period. Most organisms that
plant pathologists and entomologists deal with are capable
of rapid increase in numbers or total volume if external
conditions allow. Very often an increase from hardly de-
tectable to devasting densities takes place in a matter
of days. In comparison with such rates of increase the
rate of growth of the plant becomes irrelevant; there
merely is destruction of the plant, or parts of it, at
one or more moments. However, if the rate of increase of
the population of the noxious organism is inherently
small (of the same order as, or smaller than, the rate
of growth of the plant) as it is in almost all plant
parasitic nematodes (bud and leaf nematodes and stem nematodes are the only ones that can "overrun" a crop in one
season) the nematode density on the plant differs little
from that at the time of sowing or planting, especially
during the first months of the growth of the plant. The
effect of the nematode attack on the ultimate size of the
plant is largely determined during this period. Therefore,
a close, though not necessarily simple and unique,

relation exists between nematode density in the soil at
the time of sowing or planting and the size of the plant
at the end of the growing season. Reduction of plant size
does not result from repeated destruction by the nematodes
but from a reduction of the growth rate during a certain
period. Varying environmental factors seem to be of little
importance in the plant/nematode relationship. Moreover,
here the soil acts as an effective buffer.

Whereas the study of the dynamics of the system of
association between most parasites and plants concentrates
on the ecology of the parasite, that of the nematode-plant
system must concentrate on the growth of the plant. The
results of ecological studies have not been particularly
helpful in understanding nematode problems in crops and
nematode control. Also, the inadequate understanding of
the dynamics of the nematode-plant system on the one hand
made interpretation and practical application of the ecological findings practically impossible, whereas on the
other it has led to blaming puzzling results of experiments on the effect of environmental factors, without
production of the relevant evidence, and to entangling
ecology and system dynamics in a most confusing manner
(e.g. in Barker and Olthof, 1976). Seinhorst (1965, 1972)
has shown that in many cases the relation between relative
plant weight y (plant weight at nematode density P divided
by that in the absence of nematodes) at the end of an
experiment and nematode density P at the time of sowing
or planting is well described by the arbitrary equation
$y = m+(1-m)z^{P-T}$ for $P \geqslant T$ and $y = 1$ for $P \leqslant T$ (1) in
which m may vary from 0 to 0.9 (mostly 0 to 0.5) and z^{-T}
is conveniently chosen to be 1.05 ($z^T = 0.95$). T is the
so-called tolerance limit (Seinhorst, 1965), the nematode
density below which no or practically no yield reduction
occurs. It is also a measure of the tolerance of a plant
to a particular nematode species. So is z but, as its
value depends on the density scale, it is more convenient

depends on the density scale, it is more convenient to
choose as a parameter a nematode density at which z^P has
a convened value. The parameter m is a measure of the
proportion of the root activity that escapes from the
effect of nematode attack even at very high nematode
densities. Therefore, it has a higher value the more
irregular the distribution of the nematodes in the soil
(Seinhorst, 1973a, Seinhorst and Kozlowska, 1977). If eq.
(1) always applied and, moreover, parameter values were
constant during the development of the plant there would
not be much reason, besides curiosity, to investigate
the relation between nematode density and plant weight
further. All that would be needed would be estimates of
the values of the parameters T and m for different combinations of plant and nematode species and perhaps
under different conditions. However, eq. (1) does not
always apply nor are T and m constant in a single experiment (Seinhorst and den Ouden, 1971, Seinhorst and
Kozlowska, 1977). Moreover, estimating T and m in field
experiments is difficult. The required range of nematode
densities is difficult to achieve without creating differences in growth conditions resulting from the manipu-
lations applied to obtain particular nematode densities
(differences in previous crops, chemical control, e.g.
see Hijink, 1969) or to natural differences in population density (patchy occurrence of nematodes). If the
field situation could be simulated by using inoculated
soil this would greatly simplify the determination of
the relevant system parameters. Such a simulation could
be made with all characteristics that could affect the
nematode-plant-system parameters equal to those in the
field, or with changed characteristics (as in a pot
experiment) and translation of the observations into the
parameters operative under field conditions. In both
cases the factors that affect and may change the relation
between nematode density at sowing or planting and plant

weight during the development of the plant must be known.

General description of the nematode-plant system

If plants are sown or planted in soil that is infested with nematodes which can attack the plants the nematodes will pass from a passive existence in the soil to an active one when they come in contact with a root and will try to enter it and/or feed on it. The subsystem nematodes-in-soil is discussed by Ferris in these Proceedings. It differs in one important way from the subsystem nematodes-in/on-plants: all changes in nematode numbers in the sub-system nematodes-in-soil are reductions, the rates of which are independent of nematode density, whereas changes in nematode numbers in or on plants are density dependent at the higher nematode population densities.

The possible developments of root, nematode attack of them and the associated growth of above ground parts are illustrated in Fig. 1. It is assumed that the nematodes are distributed throughout the whole soil mass and that they do not move towards roots over great distances. Therefore, roots are always exposed to the same nematode density where they penetrate in the soil. Fig. 1A represents the growth of a plant in the absence of nematodes; Fig. 1B that in a large quantity of soil with nematodes, as in the field; Fig. 1C that in pots with nematodes; Fig. 1D that when the nematodes produce a second generation which is much more numerous than the first; and Fig. 1E that of a resistant plant attacked by nematodes. It is assumed that nematode attack reduces the activity of the root system and consequently the rate of growth of the plant, and further, that the nematodes increase in number continuously but slowly in Figs 1A to C. In Fig. 1D the new generation is considered to appear in much greater numbers than the parent generation at some time after sowing or planting, first at the base of the plant and then proceeding to deeper layers. The nematodes in

resistant plants (Fig. 1E) are presumed to die some time
after penetration.

Figs 1A and B illustrate the usual exponential growth
initially which decreases later. However, plants grow slower in B than in A. Moreover, in the final stages shown in
Fig. 1B root density near the plant continues to increase

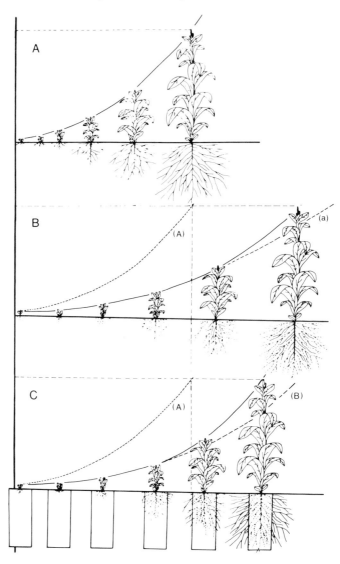

Fig. 1A-C. (Legend opposite)

although all nematodes have already entered roots that
developed earlier; this leads to an increase in the rate
of growth. It follows that the greater the initial reduction in growth the later this stage will be reached.
Fig. 1C illustrates a similar phenomenon. Initial root
growth is as in an unlimited soil volume. However, when
the roots reach the pot wall they are forced to grow on
in the limited volume of soil in the pot already occupied

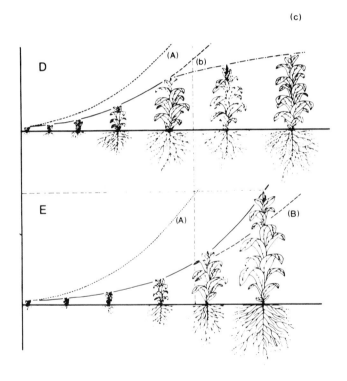

Fig. 1D-E Growth of plants in the absence (A) and presence of nematodes (B to E). B: in a large volume of soil, C: in small pots
(roots shown outside the pot in fact grow inside after initial root
growth has reached the pot wall), D: with a second generation of
the nematode, and E: when plants are resistant to nematode but sensitive to damage caused by penetration. Roots: —————— , without
nematodes; ——————— , with nematodes. (A), (B): growth curves
according to A and B, (a) in B: growth curve if nematode number per
g root had been constant, (b) in D: growth curve if there had been
only one generation.

by other roots. By that time the latter roots have at-
tracted all or almost all nematodes in the pot. Therefore,
roots developing after that (shown here outside, but in
fact growing inside the pot) grow on free of and uninhib-
ited by nematode attack, if the nematodes become immobile
after attaching themselves to or having penetrated into
roots (*Meloidogyne, Heteroderidae*). Migratory nematodes
might redistribute themselves resulting in a decrease
in the number of nematodes per unit weight of roots and
therefore in an increase of the rate of growth of the
root system. In both cases the rate of growth of the
plant will increase compared to that in an unlimited
volume of soil with the same nematode density (line (B)
in Fig. 1C) when a certain root density in the pot is
exceeded. In *Heteroderidae* and *Meloidogyne* only second
stage juveniles penetrate into roots; it takes them
three to six weeks to become adults and produce eggs. If
the eggs hatch soon after completing their development
a flush of new second stage juveniles, much larger in
number than the initial population, starts to attack
the roots from three to six weeks after sowing or planting. This will result in an additional reduction of the
rate of growth of the plant (Fig. 1D).

Nematodes penetrate into the roots of resistant plants
and damage the tissue, resulting in growth reduction as
in susceptible plants (good hosts). However, the roots
fail to provide adequate food for the nematodes, which
therefore die sooner or later. Although the damage
caused by the penetration and the resulting growth reduction may be severe, it is temporary. It decreases
relatively with expansion of the root system and growth
approaches more and more to that of plants without nematodes (Fig. 1E). All that remains is a delay in the development of the plants.

Fig. 1 demonstrates that the size of a plant at any
moment is determined by its size at a certain time

$t = 0$ and the subsequent rates of growth of the plant.
Nematode attack reduces these growth rates, but not by a
constant proportion. It changes with the nematode density
associated with the root system. This density may be
affected in two ways. Root systems expand and then become denser at some distance from their perimeter, and
nematode numbers or sizes of individual nematodes increase
on good hosts. Increase in number may be slower or faster
than increase in root mass per unit volume of soil. Figs
1B and C illustrate the result of a slower, Fig. 1D
that of a suddenly faster increase of nematode numbers
than of root mass. Further we may distinguish here
between an increase in size and perhaps in numbers without redistribution, so that new roots remain nematode
free, and increase in numbers with redistribution. There
are no indications whether the difference is of practical
importance.

First steps in the formalization of the description of the nematode-plant system

Nematode attack may do more to the plant than change
its rate of growth and, therefore, its ultimate size.
Certainly, heavily attacked plants are not only relatively small but may also look unhealthy. However, less
severely attacked plants cannot be distinguished from
plants of the same size without nematodes, grown under
the same conditions, nor does the literature provide
any indications that there are considerable differences
in chemical composition. Therefore, in a nematode-plant
system the only relevant characters of the plant are
size and growth rate, and of the nematode the reduction
of this growth rate per individual, and how the separate
reductions must be summed.

To construct an operative dynamic model of the nematode-plant system we need:
1) the size of the plant at time $t = 0$ (the moment

nematode attack begins),
2) the growth rates of the plant at times $t = 0$ to x,
3) the general relation between nematode density and
reduction of the growth rate.
4) the values of the parameters in this relation at
times $t = 0$ to x, and
5) nematode densities on or in the roots at times $t = 0$
to x.

Fig. 1 suggests that 4 and 5 may be different in different parts of the root system. Theoretically nematode
densities in or on different parts of the root system can
be derived from initial nematode density in the soil if
the size of the root system at different times, the relation between this size and the proportion of the soil
from which nematodes have started feeding in or on the
roots, the mortality rate of nematodes in the soil in the
absence of food, the multiplication or increase in size
and, in the former case, degree of redistribution are
known. However, the rate of growth of the root system is
affected by nematode density, but the nematode density
roots meet with is also affected by root growth if the
mortality rate of the nematodes in the soil is high (the
sooner the roots reach the nematodes the more are still
alive) or their multiplication rate on the roots is high
compared to the rate of increase in size of the root system (many nematodes are added to those already present in
the region of recent expansion of the root system). Root
growth and increase of nematode numbers are mutually dependent in perennial plants. Such feed-back systems require a step by step calculation of the progress of growth
and calculations and (computer) simulation require formal-
ization of the general description given above. Unlike
that for nematode attacks on perennial plants, a simple
approach may be satisfactory for most nematode attacks on
annual plants if supported by sufficient experimental evidence (Seinhorst & den Ouden, 1971, Seinhorst & Kozlowska,

(1977). The latter is due to the overwhelming effect of the growth reduction in the first growth phase, during the expansion of the root system, on the final plant weight. The growing period can conveniently be divided into two periods. In the first the root system expands without increasing much in density, and nematode density on or in the roots decreasing noticeably. In the second period the root system rapidly increases in density and, therefore, nematode density per unit weight of root decreases at about the same rate, if the nematodes multiply relatively slowly. Growth during a good part of the first phase can be approximated by an exponential relation between plant size and time. If Y is the size of the plant at time t then $dY/dt = rY$ (2) in which the parameter r is the so-called growth constant (the increase of Y per unit time and per unit weight of Y). Then $Y = Ce^{rt}$ (3) in which C is the integration constant, here the weight of the plant at time $t = 0$. In actual plant growth the rate of growth per unit weight of plant soon starts to decrease as a decreasing proportion of the plant tissue is productive and an increasing proportion serves to support and protect it. In time the productive part of the plant does not increase further. Therefore, a still better approximation for the first period of growth might be the logistic curve. Then

$$dY/dt = rY(Y_{max} - Y)Y_{max} \qquad (4)$$

and

$$Y_t = e^{rt}C\ Y_{max}\ /\{(e^{rt} - 1)C + Y_{max}\} \qquad (5)$$

Contrary to the use of these equations in population dynamics (e.g. Seinhorst, 1967), Y_{max} is not a true maximum, but only an auxiliary constant to create a growth rate that decreases with increase of Y and the equation should be used only for values of Y not exceeding about $0.5Y_{max}$. The only parameter in eqs (2) to (5) that could be affected by nematode attack is the growth rate

r. Now if the growth rate r_0 of a plant in the absence
of nematodes is reduced by a nematode population of density P to $r_p = pr_0$ this indicates that a proportion $(1-p)$
of the productivity of the plant necessary for a growth
rate r_0 is put out of action by the nematodes. If we now
assume that the effects of a nematode on root tissue are
local and nematodes attack the parts of the roots they
feed on (whole roots, root tips, the portion of roots just
behind the tip) in a random fashion, then increased density of the nematode population will result in more overlapping of the area affected by single nematodes. The area
or volume *not* affected by a nematode population of density
P then is conveniently and sufficiently accurately described by Nicholson's competition model as a proportion
$y = z^P$ ($z < 1$) (6) of the total area or volume which could
be exploited by the nematodes. Strictly, this area or volume depends on the shape of the area or volume affected
by a single nematode, but differences that might arise
lie well within the limits of variation even in very
accurate experiments. Observations in experiments have
shown that very often it seems to be difficult or even
impossible to reduce growth to below a certain level.
This may have to do, at least partly, with irregular distribution of the nematodes after inoculation, as suggested
by Seinhorst and Kozlowska (1977). Also the phenomenon is
less pronounced with nematodes moving about very actively,
e.g. *Heterodera* larvae, than with more sluggish species,
e.g. *Pratylenchus penetrans* and *Rotylenchus uniformis*.

If a proportion k ($0 < k < 1$) of the part of the root
system that is suited to attack by a particular nematode
species is out of reach of these nematodes eq. (6) becomes $y = k + (1 - k)z^P$ (7). The growth rate of plants
attacked by a nematode density P leaving a proportion
y of the production capacity of the root system intact
then becomes $r_0 = r_0 k + (1 - k)z^P$ (8) (Seinhorst and den
Ouden, 1971, Seinhorst and Kozlowska, 1977). Apparently

it is not important in what way the growth reduction is
brought about by the nematode. What counts is how the
effects of individual nematodes add up to a total effect.
If the same nematode causes growth reduction in different
ways (e.g. Heteroderidae by penetration into root tips
and by giant cell formation, *Meloidogyne* in these two
ways and also by causing a storage of material in unpro-
ductive gall tissue) these different growth reductions
must be evaluated separately (Seinhorst and den Ouden,
1971). If a nematode reduces the rate of growth of a
plant first in one way and then in another (e.g. by
penetrating into a root and then causing the development
of a giant cell) the rate of growth of a plant is reduced
according to the equation:

$$r_P/r_o = i_1 i_2 = \{k_1 + (1 - k_1)z_1^P\}\{k_2 + (1 - k_2)z_2^P\} \quad (9)$$

Values of r_P/r_o, according to eq. (9) for different values of P and z_1, z_2, k_1 and k_2 independent of P, only
deviate materially from those according to

$$r_P/r_o = k_3 + (1 - k_3)z_3^P \quad \sqrt{(k_3} = k_1 k_2, \; z_3 = z_1 z_2)$$

if k_2 is considerably larger than k_1 and at the same
time $z_2 < z_1$ (or vice versa). No indications were found
in experiments that this ever was the case as long as
the different causes of growth reduction operated simultaneously. For example, *Heterodera* larvae first cause
penetration damage (Seinhorst and den Ouden, 1971) and
then cause the development of a giant cell, but meanwhile
new larvae have penetrated into the root tips. There will
be a shift in the ratio of the types of growth reduction
and eventually changes of k_3 and z_3 depending on P (de-
synchronizing effect of nematode attack) resulting in
certain deviations of the relation between nematode density and growth rate from eq. (8). Seinhorst and den Ouden

(1971) derived the effect of giant cells on the growth of potato from the final weights of the plants at a range of nematode densities. These weights suggest that a second cause of growth reduction operated after a first had ceased to do so. Of this second cause only the final result was considered and assumed to be according to eq. (1) with a high value of m.

Substituting eq. (8) in eqs (3) and (5) results in

$$Y_P = C \, e^{r_0 t \{k+(1-k)z^P\}}$$

and

$$Y_P = e^{r_0 t \{k+(1-k)z^P\}} / \{(e^{r_0 t \{k+(1-k)z^P\}} -1)C + Y_{max} \tag{11}$$

respectively. Y_P / Y_0 then becomes

$$y = e^{(r_p - r_0)t} = e^{r_0 t \{k+(1-k)z^P -1\}} = e^{r_0 t (1-k)(z^P -1)} \tag{12}$$

and

$$y = e^{r_0 t (1-k)(z^P -1)} \times (e^{r_0 t} + x) / (e^{r_0 t \{k+(1-k)z^P\}} + x) \tag{13}$$

for

$$Y_{max} = (x-1)C$$

respectively.

The relation between y and P according to eqs (9) and (10) is practically identical to $y = m + (1 - m)z^P$ (14) with z and m decreasing with increase of t (Figs 2, 3). An even more important character of eqs (3), (5), (12) and (13) is that they become invariant for P if t is substituted by $t' = tr_0 / r_P$. This is the same as the statement that sizes (weights) of plants are the same at times $t' = tr_0 / r_P$. (15). The constancy of the ratio r_0 / r_P can be established by checking that of t'/t, provided growth conditions before time t do not differ from those between times t and t'.

According to the models discussed above and any other model satisfying eq. (15) the effect of differences in nematode density can be described as a desynchronization of the growth of the plants exposed to these different nematode densities. It leads to a considerable change of the values of z and m in eq. (13) if curves according to this equation are fitted to the results of eq. (11) for different values of e^{rt} and k. However, the shape of

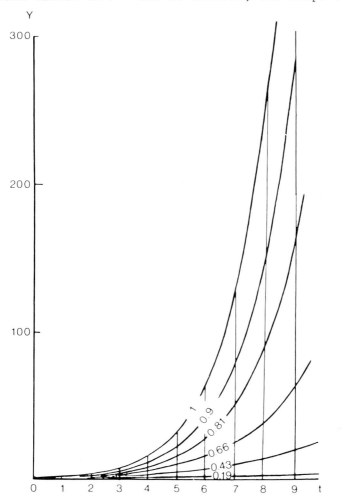

Fig. 2. Exponential growth of plants with and without nematodes. Effect of nematodes according to $r_p \; r_0 z^{t^p}$ (eg. (8)) with $k=0$ and $e^r=2$. Numbers in lines: z^P. $Y=1$ for $t=0$.

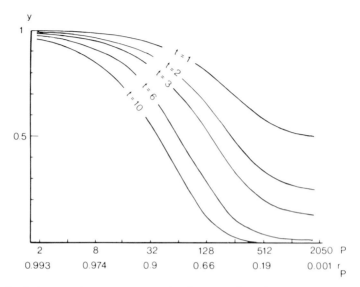

Fig. 3. Relation between nematode density P and relative plant weight y at different times t according to Fig. 2 (= according to eq. (12)). The lines also represent the relations between y and P for the combinations of $e^{r_o t_k}$ and k listed in the following table ($e^{r_o} = 2$ in all cases).

t	$e^{r_o t_k}, k$	$e^{r_o t_k}, k$	$e^{r_o t_k}, k$	$e^{r_o t_k}, k$
1	2,0	4,0.5	8,0.67	
2	4,0	16,0.5	64,0.65	128,0.72
3	8,0	64,0.5	128,0.57	1024,0.7
6	64,0	128,0.14	1024,0.4	4100,0.5
10	1024,0			

In general: $(k-1)t_k=t$, in which t_k is the time at which the graph for a given t and $k = 0$ applies for a given value of k.

curves according to eq. (11) if y is plotted against log P is the same for the scale of $(1-m)$ (Fig. 3). As a result exponential growth with constant r_P suggests a decrease of apparent tolerance of the plant at successive measuring dates. This decrease is smaller when k is larger. It is also smaller according to eq. (13) than to eq. (12). Growth rates are assumed to decrease continuously in Fig. 4 but times t at which plants reach a certain size at different nematode densities are in accordance with eqs (8) and (15). This probably is the most realistic of all

growth models discussed here. Contrary to the relation
between nematode density and plant weight at different
times according to eqs (12) and (13) the apparent tolerance of the plants decreases little at first and increases afterwards (here from t = 8 on, according to Fig. 5).
The relation between nematode density and plant size is

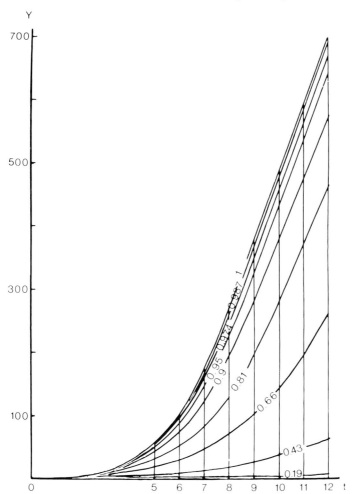

Fig. 4. Weights of plants Y at different times t if growth rate decreases with increase of t but eq. (15) applies (t' = tr_p/r_o; t' and
t times at which plants reach a certain weight at nematode density
P and in the ansence of nematodes), and $r_p = r_o z^P$ (eq. (8) with
k = 0). Numbers in lines: z^P. For equivalent relative nematode densities see Figs 3 and 5. Y = 1 for t = 0.

again very close to that according to eq. (14) at high
nematode densities with older plants. The deviations are
too small to be demonstrated experimentally. In Figs 4
and 5 k is assumed to be 0. For $k > 0$, m decreases with
increase of the age of the plants according to Figs 2 and
3, but increases slightly with increase of the age of the
plants according to Figs 4 and 5 (e.g. from 0.26 to 0.38
between $t = 5$ and $t = 12$ for $k = 0.66$).

Figs 2 to 5 indicate that considerable reductions in
the weight of the plant can be expected even if the nematodes reduce the growth rate by a small proportion, but
continuously from the beginning of growth. To provide
some orientation on growth rates it may be noted that
tripling the weight per week is a normal growth rate as
for example with cereals grown at optimum temperature and
that theoretical initial plant weight of oats was calculated to be about 15 mg dry matter per plant (weight of
seed with hulls 33 mg). Therefore plant weight 100 times
the initial weight at sowing time, assuming exponential
growth, is still a small plant. Growth curves as in Fig.
4 cannot be distinguished from those according to

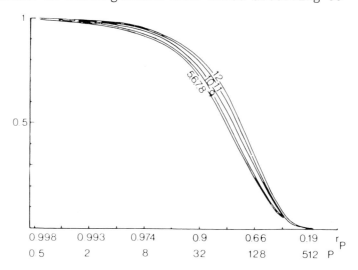

Fig. 5. Relation between nematode density P and relative plant weight
y at different times t (numbers in lines) according to Fig. 4.

exponential growth on the basis of observed plant weights
up to about 100 times the initial weight.

Comparison of theory and experiment

The small amount of information on weights of plants
at different initial nematode densities in the soil during
the first weeks after sowing is reasonably in accordance
with eq. (1) with z^T increasing during the experiment from
1 (eq. (14)) to 1.05, and to eq. (15) for the higher densities (determination of t'/t for the lower densities is
too inaccurate to be of much use). Therefore, we may assume that eq. (8) gives a fair representation of the relation between nematode density and growth reduction. Only
the increase of T with increase of t requires explanation. It is not due to the decrease of the growth rate of
the plants with increase of age, as demonstrated by Figs
4 and 5. According to Seinhort and Kozlowska (1977) it
results from a decrease of nematode density caused by an
increase of root density in the soil which starts earlier
at low than at high nematode densities (Figs 6 and 7). It
is much more difficult, but fortunately also unimportant
for our purposes, to know whether eq. (12), eq. (13) or
even another relation for which $t' = tr_0/r_P$ applies.

Approximate extension of constructive specification of the system for later stages of growth of plants

Decrease of growth rate with increase of age of plants
does not much affect the relation between nematode density and relative plant weight as long as eq. (15) applies (Figs 4 and 5). Therefore, changes in this relation
as reported by Seinhorst and den Ouden (1971) and Seinhorst and Kozlowska (1977) may indicate a change of the
t'/t ratio. Increase of this ratio at a given initial
nematode density may indicate an increase of population
density per unit productive root tissue, decrease a decrease of this density, e.g. because of a strong increase

of root density in the soil or of death of attacking nematodes on a (resistant) plant. Both increase and decrease
will be greater at low than at high nematode densities,
increase because nematodes multiply faster and decrease
because roots grow faster at low than at high nematode
densities. Seinhorst and Kozlowska (1977) explain how,
with nematodes that multiply slowly, the latter leads to
a relation between y and P more similar to eq. (1) than
to eq. (14) in certain cases (Figs 6 and 7) and different
from both in others. Increase of root density also occurs
in the field, but it generally starts later and is less
than in pots as a much larger volume of soil is available
per plant. Final yields at different nematode densities in
field experiments with nematodes that multiply relatively
slowly may therefore be expected to be according to eq. (1)

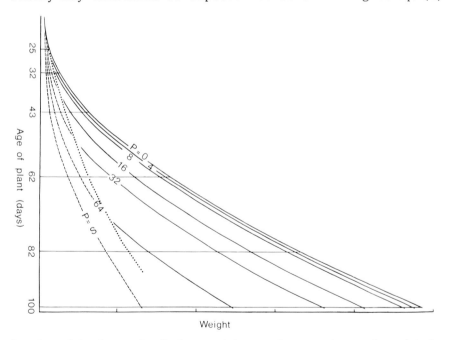

Weight

Fig. 6. Model of growth of plants with growth rates decreasing with increase of age of the plants and growth reduction according to eq. 8 wit
k = 0.53 and eq. 15 up to a certain size of the plants, after which the
is no growth reduction by nematodes. (---) growth rate reduced by nemat
attack; (——) growth after cessation of reduction by nematode attack.
P = 1,2,4 - nematode densities (From Seinhorst and Kozlowska, 1977).

Those in pot experiments may be when $m > 0.25$, but then T
increases with the age of the plants (Fig. 7). When
$m < 0.1$ the range of P between $y = 0.95$ and $y = 0.15$ will
decrease with increase of the duration of the experiment
(Seinhorst and Kozlowska, 1977). The same is to be expected in the field with a range of nematode densities
attacking resistant plants. Here the death of the nematodes after contact with the plant causes the necessary
decrease of nematode density on the root system.

No data are available on the effects of increasing
nematode numbers per unit volume or weight of roots on
the growth of plants. It is expected that this will lead
to an extension of the range of densities between those
at which $y = 0.95$ and $y = 1.1\ m$, as when a second generation, which is considerably more numerous than the first.
starts to appear some time after sowing or planting.

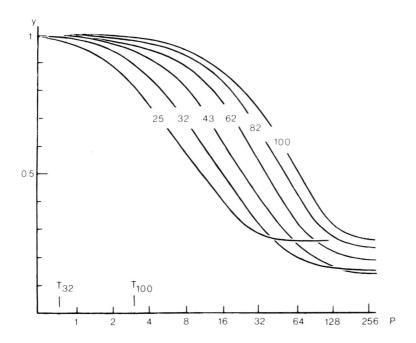

Fig. 7. Relation between nematode density P and relative plant
weight Y of plants of different ages according to Fig. 6. Numbers in
lines: ages of plants in days (After Seinhorst and Kozlowska, 1977).

Indirect effects of nematode attack on plant weight

Not all developments of plants (haulm formation in cereals, flowering, ripening) are closely correlated with the size (weight) of the plant. The assumption by Sein-horst and Kozlowska (1977) that carrot weight and leaf weight in carrots were correlated in the same way in plants with and without nematodes (though not linearly) was consistent with observed final plant weights. However, *Heterodera avenae* delayed flowering of oats but less than the time at which a certain plant weight was reached (Seinhorst, unpublished). As a result the initial deceler-ation of growth resulted in lower seed yields than with plants without nematodes. In an experiment by Seinhorst and den Ouden (1971) *Globodera rostochiensis* attack shif-ted the development of potato plants towards a period with longer days, resulting in a larger maximum haulm weight and, possibly, a greater ultimate productivity of the plant.

Technical problems in investigations on the relation between nematode density and plant growth

The models discussed so far are more or less arbitrary relationships derived from more definite ones applying to early stages of growth. Without further detailed informa-tion they still allow a great variety of (theoretical) realizations. In the absence of such information the models proposed by Seinhorst and den Ouden (1971) and Seinhorst and Kozlowska (1977) lean heavily on hypotheses necessary to construct the required growth curves. Much of the speculation can be eliminated by obtaining experi-mental data from measurements at short time intervals of plant sizes (weight) and with plants such as potato and carrot of storage organs. However, here the experimenter meets with some difficulty. A typical experiment on the relation between nematode density and plant growth may contain a range of nematode densities $a \times 2^n$ with $n=0$ to

14. If there are five replications per density, this means 70 pots per treatment. If the experiment were to last 10 weeks and plant weight to be determined every week after taking the plants from the pots and perhaps drying them, the experiment would require 700 pots at the start. Comparison of treatments would then be virtually impossible. Therefore, it would be of great help if non-destructive methods to estimate plant weight could be applied. Seinhorst and den Ouden (1971) convey an impression of the growth of potato plants in their experiments in graphs showing haulm lengths at different dates but they did not determine a relation to weight. Seinhorst and Kozlowska (1977) assumed that leaf weight of carrots was proportional to the third power of the length of the longest leaf of a plant. They derived this relation from measurements of plants of the same age but of very different lengths.

Seinhorst (unpublished) found that the weight at the end of the growing season of second year shoots of apple seedlings (a single shoot left to grow per plant) attacked by *Pratylenchus penetrans* was proportional to the square of their length and that this was most probably so during the whole growing season. The weight of *Tagetes erecta* plants attacked by *P. penetrans* was approximately proportional to their length. Water consumption during four days to one week was approximately proportional to plant weight in oats attacked by *Heterodera avenae*, white clover with *H. trifolii* and *Lolium perenne* with *Tylenchorhynchus dubius*. However, in oats with *T. dubius*, and tobacco inoculated with *P. penetrans* some time after sowing, water consumption per g plant per day was less the greater the nematode density. In all cases nematode attack did not affect the shoot weight to root weight ratio. However, *Longidorus elongatus* reduced root weight without change of shoot weight in *Lolium perenne*, a most unusual phenomenon; water consumption was proportional to root weight.

These examples indicate that often plant weight can be
derived from other measurable properties of a plant provided that it is possible to check the relationship between
these properties and plant weight in the course of the
experiment. Where nematode attack only decelerates the
growth of the plant without nematodes, plants of different
age can be used to establish the relation at the different
measurement dates between plant weight and the property to
be measured.

Another problem still to be solved is the uniform distribution of fairly sluggish nematodes (*Pratylenchus*,
Rotylenchus) in the soil. Injection into the soil, as
described by Seinhorst and Kozlowska (1977) is quite
satisfactory for eggs of Heteroderidae (low minimum
yields) but not for *Pratylenchus* and *Rotylenchus*. It is
impossible to inoculate soil with *Longidorus* species in
this way.

Final considerations

A simple growth model for plants associated with an
equally simple relation between nematode density and
growth rate leads to a model of the growth of plants in
the presence of nematode populations at different densities which is reasonably in accordance with growth
observed in experiments during the first weeks after sowing. For later periods of plant growth less well specified
considerations on changes of root density and nematode
density on the roots must be added. So far these changes
have not been studied and, therefore, are purely hypothetical. From a practical point of view this is not too
great a shortcoming for annual crops attacked by nematodes producing only one generation per season or multi-
plying slowly. Most probably the relation between nematode density and crop yield in the field can, in these
cases, be described by a simple equation with two parameters and it is expected that the relation between the

values of these parameters and those applying during the
first stage of growth of the plants does not vary much
from case to case.

However, we are still unable to construct specifica-
tions for the final stages of a model for plant growth
in the presence of a nematode that multiplies quickly or
produces more than one generation active in the same season. This is even more so for a model for perennial
plants attacked by nematodes. Consequently our knowledge
of the extent of nematode damage and our judgment of the
importance of different nematode species on perennial
plants is vague and the economic efficiency of control
uncertain.

References

Barker, K.R. and Olthof, T.H.A. (1976). Relationship between nematode population densities and crop responses. *Ann.Rev.Phytopathol.*
  14. 327-353.
Hijink, M.J. (1969). Groeivermindering van finjspar veroorzaakt
  door *Rotylenchus robustus: Meded.Rijksfaculteit Landb.wetensch.*
  Gent. **34**, 539-549.
Seinhorst, J.W. (1965). The relation between nematode density and
  damage to plants. *Nematologica* **11**, 137-154.
Seinhorst, J.W. (1967). The relationships between population
  increase and population density in plant parasitic nematodes.
  Nematologica **12**, 157-169.
Seinhorst, J.W. (1972). The relationship between yield and square
  root of nematode density. *Nematologica* **18**, 585-590.
Seinhorst, J.W. (1973). The relation between nematode distribution
  in a field and loss in yield at different average nematode
  densities. *Nematologica* **19**, 421-427.
Seinhorst, J.W. (1973). Principles and possibilities of determining
  degrees of nematode control leading to maximum returns. *Nematol.*
  medit. **1**, 93-105.
Seinhorst, J.W. and Kozlowska, J. (1977). Damage to carrots by
  Rotylenchus uniformis, with a discussion on the cause of increase of tolerance during the development of the plant.
  Nematologica **23**, 1-23.
Seinhorst, J.W. and Ouden, H., den (1971). The relation between
  density of *Heterodera rostochiensis* and growth and yield of two
  potato varieties. *Nematologica* **17**, 347-369.

Discussion

Seinhorst explained that so far he had used only

annual crops to develop his mathematical model in order
to start with a relatively simple situation in which the
growth of whole plants can be assumed to relate to nematode infestations. Yields of fruit are derived as a
final product from plant growth through a series of
growth parameters and these could be fitted to the model
but not as simply as with whole plants. It was also possible to develop the model to account for attack by more
than one nematode species provided the action of the different species was independent of each other. Following
some discussion about initial and sustained attacks,
Seinhorst remarked that nematodes enter roots as they
become available. The success of the second generation of
Meloidogyne and *Heterodera* relates to the availability of
new roots, whereas *Pratylenchus* and some other species
enter almost any type of root; thus there was need for
observations for each species.

Asked about his definition of the "tolerance limit",
Seinhorst said it is better not to use this term without
reference to the regression equation $y = m + (1 - m)z^{P-T}$
and explained that T is just a symbol introduced to make
the formula fit better to the data obtained. Ferris asked
if it were possible to predict maximum yields and if an
accurate measurement of T were necessary: Seinhorst replied that predictions are difficult to make if many factors are involved.

J.N. Sasser

*Dept. of Plant Pathology, North Carolina State
University, Raleigh, North Carolina, U.S.A*

Introduction

Root-knot nematodes, *Meloidogyne* species, constitute a
major group of plant-pathogenic nematodes affecting crop
production. Their worldwide distribution, extensive host
ranges and involvement with fungi and bacteria in disease
complexes rank them among the top five major plant pathogens affecting the world's food supply. Collectively,
the various species of root-knot nematodes attack nearly
every crop that is grown. Not only are yields greatly
affected, but quality is also reduced especially for root
crops such as potato, yams and peanuts. Much of the extensive research completed to date on these pests has
been summarized in a recent book (Taylor and Sasser, 1978)
as an aid to co-operators participating in the "Inter-·
national *Meloidogyne* Project". Although we have known
about these nematodes for more than a century, much remains to be learned before effective and economical control measures can be implemented.

In considering the present status and progress that
has been made in recent years in understanding the root-
knot nematodes, it is helpful to visualize the situation
as it existed in the late 1940's to early 1950's. A few
excellent monographs had been published such as those of
Neal (1889), Atkinson (1889), Stone and Smith (1898) and

Bessey (1911). These studies were exhaustive and remark-
ably accurate for the time and in many ways laid the
groundwork which later led to progress in our understanding of the taxonomy and behaviour of root-knot nematodes.

Pathogenicity

The first evidence that root-knot nematodes differed
in their host preference was probably that presented by
Sherbakoff (1939). He reported considerable root-knot
injury to cotton, *Gossypium hirsutum* L., grown on land
previously planted with tomatoes, *Lycopersicon esculentum* Mill., even though the tomatoes had not been severely
injured by root-knot nematode. At that time, all root-
knot nematodes were taxonomically classified as *Heterodera marioni* (Cornu) T. Goodey. Christie and Albin
(1944) and Christie (1946) established experimentally
that populations of *H. marioni* differed in their host-
parasite relationships. Among 14 populations studied,
these investigators demonstrated that host specialization
may be shown by:
(i) distinct host preference, or
(ii) the character of root galling on susceptible hosts.
Chitwood (1949), after a morphological study, removed
the root-knot nematodes from the genus *Heterodera* and
re-established the genus *Meloidogyne* (Goeldi, 1887).
Although emphasis was on morphology, Chitwood made
important observations with reference to host differences. For example, he observed that *M. hapla* Chitwood
was the only root-knot nematode known to reproduce on
strawberry and that *M. incognita* (Kofoid and White)
Chitwood failed to infect peanut but reproduced readily
on cotton.

Although a number of publications prior to Chitwood's
revision of the genus (Bessey, 1911; Marcinowski, 1919;
Buhrer *et al.*, 1933; Buhrer, 1938; Tyler, 1941) listed
host plants found to be susceptible, resistant, tolerant

and immune, it became apparent that these studies had
limited value as it could not be determined which species
were involved in the various reports. Similarly, much
of the research on the effects of temperature, moisture,
soil type and other environmental factors on survival,
development, and pathogenicity, was of little value in
developing control practices for the newly described
species. The need, therefore, for new host range studies
using single species populations became obvious.

Host ranges

Sasser (1954), using single species cultures identified according to Chitwood's morphological characters,
studied the suitability of 50 plant species and cultivars
as hosts for the known root-knot nematode species at
that time, except for *M. exigua* Goeldi. These investigations, among other things showed that some species
were more host specific than others and that for each
species there were certain crops which were non-hosts.
These differences gave encouragement to the possibility
of identifying species on the basis of host response of
certain plant species when grown in soil infested with
a single root-knot nematode species (Sasser, 1952). This
method, which employed the use of differential hosts,
proved useful in identifying the common species within
a given geographical region but was found to be less
reliable when populations of the various species were
obtained from widely separated geographical regions
(Sasser, 1966, 1972).

Host range studies conducted in more recent years
(Gaskin and Crittenden, 1956; Winstead and Sasser, 1956;
Thomason and McKinney, 1959; Winstead and Riggs, 1959;
McGlohon *et al.*, 1961) using known species have added
much to our knowledge concerning susceptible and resistant crops. Additional studies are needed, however, to
extend our knowledge of the host ranges of the

individual species. Furthermore, information pertaining
to sources of resistance to individual species for use
in the development of resistant cultivars, is greatly
needed.

Pathogenic variation

Some of the early evidence of pathogenic variation
within a species involved differences in parasitism on
cotton among isolates of *M. incognita*, ranging from non-
parasitism to severe parasitism (Martin, 1954). Other
investigators (Allen, 1952; Sasser and Nusbaum, 1955;
Linde, 1956; Powell, 1957; Goplen *et al.*, 1959) found
that within a species, certain populations react differently on a given host. These have been referred to as
races, biotypes or pathotypes. More recently, Netscher
(1978) has given a comprehensive review of published
morphological and physiological variability among species
of *Meloidogyne*. To date, within-species variation has
been reported for *M. incognita*, *M. javanica*, *M. hapla*,
M. arenaria and *M. naasi*.

Because of this pathogenic variation within and
among species, it became evident that progress in understanding this genus would be more rapid if populations
could be studied from widely separated geographical
regions. Thus, a worldwide collection of *Meloidogyne*
species and populations was begun at North Carolina
State University about 15 years ago and at present we
have more than 500 populations representing about half
of the known species. A determination of the extent of
pathogenic variation within and among species of *M. incognita*, *M. javanica*, *M. hapla*, *M. arenaria*, and, to a
limited extent, *M. exigua*, has been a major objective.

To study such variation it was necessary to continue
the use of "differential hosts" for testing against the
various species and populations within a species. When
involving a single-species population, the differential

TABLE I

Usual response of plant species to
attack by the more common *Meloidogyne* spp.

Meloidogyne species	Differential Hosts[1]					
	Tobacco	Cotton	Pepper	Water melon	Peanut	Tomato
. incognita	⊟ [2] (+)[3]	⊞ (-)	+	+	-	+
. javanica	+	-	⊟	+	-	+
. hapla	+	-	+	⊟	⊞	+
. arenaria	+	-	+	+	⊞ (-)	+

Plant cultivars include: Tobacco cv. N.C. 95; Cotton cv. Deltapine
6; Pepper cv. California Wonder; Watermelon cv. Charleston Grey;
eanut cv. Florrunner; Tomato cv. Rutgers. See Taylor and Sasser (1978)
or testing procedure.

Indicates key differentials for that species.

Some populations attack the differential host whereas others do not.

hosts have been used somewhat successfully in distinguishing between species and for detecting pathogenic variation
among populations of the same species. This method is
based on the assumption that species and populations
within a species will always react on given hosts according to previous tests. When populations react differently
on the differentials this constitutes evidence of pathogen variation within the species being tested or that
the population is a mixture of two or more species
(Table I).

After a study of a large number of populations of
Meloidogyne from various regions of the world, it became
evident that the most widespread and most important agronomic species were those originally described by Chitwood
(1949), namely, *M. incognita, M. javanica, M. hapla* and
M. arenaria, in that order. *M. exigua* occurs primarily
in Central and South America where it causes extensive
damage on coffee. The frequency of occurrence of these
species in 180 populations studied at the time by Sasser

and Triantaphyllou (1977) was as follows: *M. incognita*, 56%; *M. javanica*, 24%; *M. hapla*, 12%; *M. arenaria*, 5%; and others, 3%. In the 3% are such species as *M. exigua*, *M. naasi*, *M. graminis* and *M. microtyla*. It should be pointed out that a large number of the populations studied were collected from tropical or subtropical regions. Additional collections from the cooler regions of the world will undoubtedly contain a higher proportion of the species *M. hapla* and *M. naasi*, which occur in cooler climates or at high altitudes in the tropics.

As of January 1, 1977, 37 species of *Meloidogyne* had been described (Esser, *et al.*, 1976). Observations indicate that with the exception of *M. incognita*, *M. javanica*, *M. hapla*, *M. arenaria*, *M. exigua*, and *M. naasi*, the remainder of the species are highly specialized parasites of certain plants and are not widespread, and probably are of little economic importance except for the one crop they attack in a restricted area (Sasser, 1977). Further studies are needed, however, to clarify the real significance of these uncommon species in agriculture. In the meantime, those species which are widespread and economically important should have high priority with respect to research directed toward control since these probably account for 95% or more of the losses caused by *Meloidogyne* species. A study of the less common species is essential, however, to clarify phylogenetic relationships, to determine biological phenomena leading to host specializations, and to determine other complexities of the genus.

In attempts to detect pathogenic variation among and between species of *Meloidogyne* collected from widely separated geographical regions of the world through use of the standard differential host tests used at North Carolina State University, it has become apparent that the *M. incognita* complex is made up of at least 4 host races; *M. arenaria*, 2 host races; and *M. javanica* and

TABLE II

Differential host test identification

Meloidogyne species and race	Differential host plant cultivars[1]					
	Tobacco	Cotton	Pepper	Water Melon	Peanut	Tomato
M. incognita						
Race 1	−	−	+	+	−	+
Race 2	+	−	+	+	−	+
Race 3	−	+	+	+	−	+
Race 4	+	+	+	+	−	+
M. arenaria						
Race 1	+	−	+	+	+	+
Race 2	+	−	−	+	−	+
M. javanica	+	−	−	+	−	+
M. hapla	+	−	+	−	+	+

[1]Tobacco (*Nicotiana tabacum*) cv.NC 95; cotton (*Gossypium hirsutum*) cv.Deltapine 16; pepper (*Capsicum frutescens*) cv.California Wonder; watermelon (*Citrullus vulgaris*) cv.Charleston Gray; peanut (*Arachis hypogaea*) cv.Florrunner; tomato (*Lycopersicon esculentum*) cv.Rutgers. See Taylor and Sasser (1978) for testing procedure.

M. hapla, 1 race each (Table II). The race designations for *M. incognita* are based on parasitism of cotton, cv. Deltapine 16, and resistant tobacco, cv. N.C.95. A total of 222 populations of *M. incognita* have been studied. From this number, 63% (139 populations) were Race 1; 22% (49 populations) were Race 2; 10% (22 populations) were Race 3, and 5% (12 populations) were Race 4. For *M. arenaria*, the key differential is peanut cv. Florrunner. Generally, but not always, those populations which fail to reproduce on peanut also fail to reproduce on pepper cv. California Wonder. The host races of *M. incognita* and *M. arenaria* are distributed throughout the world and are morphologically indistinguishable.

Discussion and conclusions

A study of over 400 populations of *Meloidogyne* species
from the United States and 6 major geographical regions
of the world outside the United States indicate that root-
knot nematodes are widespread throughout the world. Although approximately 37 species have been described, 6
probably cause most of the damage to economic food crops;
these include *M. incognita, M. javanica, M. hapla, M. arenaria* and *M. naasi*. Evidence thus far indicates that the
other species are highly specialized parasites of only a
few and sometimes only one host and known distribution at
this time for many of these species is that indicated in
the original reports. Frequency of occurrence of the
various species indicate that the *M. incognita* group,
comprised of at least four races, is the most widespread.
M. javanica, often found coexisting with *M. incognita*, is
second, followed by *M. hapla* and *M. arenaria*. Differential
host tests have provided the best evidence of "host races"
within some species. They have also revealed that there
is considerable uniformity in host response among species
collected from widely separated geographical regions. For
example, *M. incognita* populations do not reproduce on
peanut, *M. hapla* populations do not reproduce on watermelon, and *M. arenaria, M. javanica* and *M. hapla* populations do not attack cotton. Furthermore, there is considerable evidence that a large number of "resistance-
breaking races" do not occur in cultivated fields. It
implies that any crop cultivar resistant to one or more
of the four races of *M. incognita*, two races of *M. arenaria*, one race of *M. javanica* or one race of *M. hapla*
may be useful in all parts of the world. However, the
possibility of resistance-breaking races has been demonstrated in some cases (Graham, 1968; Nishizawa, 1974)
and therefore resistant cultivars should not be planted
continuously (year after year) in the same field but

instead should be included as part of a well designed crop rotation system. The need for more extensive tests with the various species and races to determine their host ranges is urgently needed if rotation of nonhost or resistant crops with susceptible ones is to be an effective part of the integrated control programme. More emphasis should also be given to looking for new sources of resistant germ-plasm in wild species, plant introductions and breeding lines, and incorporation of such resistance into local cultivars. This approach is especially important since there are so few plants with natural resistance to use for reducing the population density, and because of the limitations being imposed on growers in the use of nema-ticides.

Acknowledgement

This is paper number 5609 of the Journal Series of the North Carolina Agricultural Experiment Station, Raleigh, North Carolina. The "International *Meloidogyne* Project," involving approximately 70 cooperators from developing countries is funded by the United States Agency for International Development through a contract with North Carolina State University.

References

Allen, M.W. (1952). Observations on the genus *Meloidogyne* Goeldi 1887. *Proc.Helminthol.Soc.Wash.* **19**, 44-51.
Atkinson, G.F. (1889). A preliminary report upon the life history and metamorphosis of a root-gall nematode, *Heterodera radicicola* (Greeff) Mull., and the injuries caused by it upon the roots of various plants. *Ala.Agric.Exp.Stn.Bull.* **1**, 54.
Bessey, E.A. (1911). Root-knot and its control. *U.S. Dept. of Agric. Bull.* **217**, 89.
Buhrer, E.M. (1938). Additions to the list of plants attacked by the root-knot nematode *Heterodera marioni*. *Plant Dis.Rep.* **22**, 216-234.
Buhrer, E.M. Cooper, C. and Steiner, G. (1933). A list of plants attacked by the root-knot nematode *Heterodera marioni*. *Plant Dis. Rep.* **17**, 64-69.

Chitwood, B.G. (1949). Root-knot nematodes. *Proc.Helminthol.Soc.Wash.* **16**, 90-104.

Christie, J.R. (1946). Host-parasite relationships of the root-knot nematode, *Heterodera marioni*. *Phytopathology* **36**, 340-352.

Christie, J.R. and Albin, F.E. (1944). Host-parasite relationships of the root-knot nematode, *Heterodera marioni*. *Proc.Helminthol. Soc.Wash.* **11**, 31-37.

Esser, R.P., Perry, V.G. and Taylor, A.L. (1976). A diagnostic compendium of the genus *Meloidogyne* (Nematoda:Heteroderidae). *Proc. Helminthol.Soc.Wash.* **43**, 138-150.

Gaskin, T.A. and Crittenden, H.W. (1956). Studies on the host range of *Meloidogyne hapla*. *Plant Dis.Rep.* **40**, 265-270.

Goeldi, E.A. (1887). Relatoria sobre a molestia do cafeeiro na provincia do Rio de Janeiro. Apparently an advance separate of: *Arch. Mus.Nac. Rio de Janeiro* **8**, 7-121.

Goplen, B.P., Standord, E.H. and Allen, M.W. (1959). Demonstration of physiological races within three root-knot nematode species attacking alfalfa. *Phytopathology* **49**, 653-656.

Graham, T.W. (1968). A new pathogenic race of *Meloidogyne incognita* on flue-cured tobacco. *Tobacco* **168**, 42-44.

Linde, W.J. (1956). The *Meloidogyne* problem in South Africa. *Nematologica* **1**, 177-183.

McGlohon, N.E., Sasser, J.N. and Sherwood, R.T. (1961). Investigations of plant-parasitic nematodes associated with forage crops in North Carolina. *N.C.Agric.Exp.Stn.Tech.Bull.* **148**.

Marcinowski, K. (1919). Parasitisch und semiparasitisch an pflanzen lebende nematoden. *Arbeiten aus der Kaiserlichen Biologischen Anstalt fur Land und Forstwirtschaft, Berlin.* **7**, 1-192.

Martin, W.J. (1954). Parasitic races of *Meloidogyne incognita* and *Meloidogyne incognita acrita*. *Plant Dis.Rep.Suppl.* **227**, 86-88.

Neal, J.D. (1889). The root-knot disease of the peach, orange and other plants in Florida, due to the work of *Anguillula*. *U.S. Dept Agric.Div.Entomol., Bull.* **20**, 31.

Netscher, C. (1978). Morphological and physiological variability of species of *Meloidogyne* in West Africa and implications for their control. *Meded.Landbouwhogeschool Wageningen* **78-3**, 1-46.

Nishizawa, T. (1974). A new pathotype of *Meloidogyne incognita* breaking resistance of sweet potato and some trials to differentiate pathotypes. *Japanese J.Nematol.* **4**, 37-42.

Powell, W.M. (1957). Variation in host-parasite relationships of the root-knot nematode *Meloidogyne hapla* (Chitwood). *Unpublished Master's Thesis, North Carolina State University*.

Sasser, J.N. (1952). Identification of root-knot nematodes (*Meloidogyne* spp.) by host reaction. *Plant Dis.Rep.* **36**, 84-86.

Sasser, J.N. (1954). Identification and host-parasite relationships of certain root-knot nematodes (*Meloidogyne* spp.). *Univ.of Md. Agric.Exp.Stn.Tech.Bull.* **A-77**, 30.

Sasser, J.N. (1966). Behaviour of *Meloidogyne* spp. from various geographical locations on ten host differentials. *Nematologica* **12**, 98-99.

Sasser, J.N. (1972). Physiological variation in the genus *Meloidogyne* as determined by differential hosts. *OEPP/EPPO Bull.* **6**, 41-48.

Sasser, J.N. (1977). The distribution and ecology of the root-knot

nematodes in the potato-production areas of the world. *Organi-
zation of Tropical American Nematologists (OTAN), Annual Meeting
Lima, Peru, March 1977*. 16.
Sasser, J.N. and Nusbaum, C.J. (1955). Seasonal fluctuations and
host specificity of root-knot nematode populations in two-year
tobacco rotation plots. *Phytopathology* 45, 540-545.
Sasser, J.N. and Triantaphyllou, A.C. (1977). Identification of
Meloidogyne species and races. *J.Nematol.* 9, 283.
Sherbakoff, C.D. (1939). Root-knot nematodes on cotton and tomatoes
in Tennessee. *Phytopathology* 29, 751-752.
Stone, G.E. and Smith, R.E. (1898). Nematode worms. *Mass.Agric.Exp.
Stn.Bull.* 55, 67.
Taylor, A.L. and Sasser, J.N. (1978). Biology, identification and
control of root-knot nematodes (*Meloidogyne* species). *Department
of Plant Pathology, N.C. State University, Raleigh.* 111.
Thomason, I.J. and McKinney, H.E. (1959). Reaction of some cucur-
bitaceae to root-knot nematodes (*Meloidogyne* spp.). *Plant Dis.
Rep.* 43, 448-450.
Tyler, J. (1941). Plants reported resistant or tolerant to root-
knot nematode infestation. *U.S. Dept. Agric. Misc. Pub.* 406,
91.
Winstead, N.N. and Riggs, R.D. (1959). Reaction of watermelon
varieties to root-knot nematodes. *Plant Dis.Rep.* 43, 909-912.
Winstead, N.N. and Sasser, J.N. (1956). Reaction of cucumber
varieties to five root-knot nematodes (*Meloidogyne* spp.). *Plant
Dis.Rep.* 40, 272-275.

Discussion

Coomans asked whether the two chromosomal forms in
M. incognita, as recognised by Triantaphyllou, could be
correlated with any of the races detected on the basis
of the host differentials. Sasser replied that there
were no real correlations with such forms and also that
the various races could not be correlated with the sub-
species in *incognita* and *acrita*. Moreover, the races have
no special geographical distribution and are found as
such in all parts of the world. It was pointed out that
Golden (*in* Plant Parasitic Nematodes, Vol 1, eds. B.M.
Zuckerman, W.F. Mai and R.A. Rohde, (1971)) recognises
several sub-species and races within *M. incognita* but
Sasser said that he felt that this division would not
serve any useful purpose and said there was a need to
study more populations rather than a few populations with
peculiar characteristics. Coomans observed that a taxo-

nomist can describe races as sub-species and designate
them with trinomials; other workers would not have to
use these trinomials but could stick to the binomial
nomenclature for the species. However Sasser said that
if you cannot look at a specimen and decide what it is
and what it will do then it should not be separated as
a sub-species or a different form. Coomans expressed the
opinion that the description of a sub-species is a matter
of personal judgment but he basically agreed that naming
should have some biological significance. Asked whether
the host differentials were used primarily for race
identification, since the species had already been identified, Sasser said that at Raleigh both the perineal
pattern and host differentials were used for species
identification and in the case of *M. incognita* the host
range also identifies the race.

C. Netscher and D.P. Taylor

*Laboratoire de Nématologie, O.R.S.T.O.M.,
B.P. 1386, Dakar, Sénégal*

Introduction

Dependence of an obligate parasite upon its host is axio-
matic. Root-knot nematodes (*Meloidogyne* spp.) as plant
parasites are no exception and the energy for juvenile development, maturation and reproduction, can only be obtained by feeding on suitable host plants. As root-knot nematodes have no resistant stage, absence of hosts causes a
decrease in populations which in the extreme results in complete eradication.

Present research emphasis is being placed on reduced
dependence on chemicals for the control of nematodes, as
with other plant pests. This is because of today's high
cost of nematicides, usually petroleum derivatives or by-
products, and the emphasis on non-pollution of the environment by agricultural chemicals. Thus, to achieve nematode
control, approaches other than those employing chemicals
are now receiving considerable attention. To achieve this
end, it is logical to utilize non-hosts or resistant
cultivars of susceptible crops in the development of a
control programme designed to reduce root-knot nematode
populations and minimize crop damage.

Crop rotations to control root-knot nematodes have been
used since the early 1900's. Orton (1903) recommended that
oats and maize be used in a "starvation rotation" to

control *Meloidogyne*, and Bessey and Byars (1911) utilized groundnut for the same purpose. Nevertheless, in the search for an immediate or interim solution to a particula root-knot nematode problem, rotation with non-hosts is a technique often neglected.

Although these early recommendations offered consider- able promise, subsequent research has demonstrated that root-knot nematodes are extremely polyphagous. Thus, in 1953 the files of the United States Department of Agricul- ture listed 1865 plant species that were recorded as hosts of the root-knot nematodes (Sasser, 1954). So many crops were known to be hosts that it became difficult to find a sufficient number of non-hosts to use in a practical rotation. Further, it soon became evident that specific rotation recommendations could not be used universally because the same plant species often reacted differently to root-knot nematode populations from different areas. It was suggested that this variability of plant reaction to root-knot nematodes might be due to the existence of different physiologic races of these nematodes. This hypothesis was proven by Christie and Albin (1944), who demonstrated striking differences in susceptibility of cotton, groundnut and alfalfa to root-knot nematode pop- ulations collected from different regions of the United States. After a detailed morphological study of several root-knot nematode populations, Chitwood (1949) was able to distinguish certain of these populations and split the root-knot nematodes, that had been considered a single species, *Heterodera marioni*, into five species and one variety (subspecies). These he classified in the genus *Meloidogyne* Goeldi, 1887, which he re-established at that time. Since then many species of *Meloidogyne* have been described and in 1976, a total of 35 species were listed (Esser *et al.*, 1976; see also Mary Franklin in these Proceedings).

In principle, subdivision of root-knot nematodes into

species was a step forward in the development of crop rotations intended to control them. In theory, if different
plant species were inoculated with monospecific populations
of *Meloidogyne*, it would be possible to develop lists of
hosts and non-hosts of the various species of *Meloidogyne*.
Thus, if a field population of *Meloidogyne* could be
correctly identified to species, crop rotations could be
developed by utilizing non-hosts of that species as contained in such lists. However, compilation of such lists
has been difficult to achieve because even within morphological species, individual populations vary greatly in
host specificity as will be discussed in detail later.

Most populations of root-knot nematodes consist of one
or more of the species originally described by Chitwood
(1949). These species are all highly polyphagous and have
a wide geographical distribution with the exception of
the type species, *M. exigua*, which apparently has a more
restricted host range and distribution. This becomes
obvious when the world literature on *Meloidogyne* is con-
sulted. Nearly 93% of the references dealing with *Meloidogyne* in *Helminthological Abstracts* between 1949 and 1976
refer to Chitwood's species. Accepting the synonymization
of *M. incognita* and *M. incognita acrita* proposed by Triantaphyllou and Sasser (1960) one concludes that at present the most widely distributed and polyphagous species
are: *M. incognita*, *M. javanica*, *M. arenaria* and *M. hapla*.
The first three species are widely distributed in warm
regions whereas *M. hapla* occurs primarily in cooler climates. Most of the other species are more specialized and
generally confined to restricted geographical areas. *M.
brevicauda* well illustrates this as it is known to be a
parasite only of tea and is restricted to a few estates
in Sri Lanka and one in South India (Sivapalan, 1972).
Certain species, however, such as *M. naasi* which is found
in various regions of Europe and the United States, are
intermediate between very polyphagous and highly

specialized species.

Although *M. incognita*, *M. javanica*, *M. arenaria* and *M.
hapla* have overlapping host ranges which include a large
number of crops, several rotations have been developed
for their control. Thus Gramineae such as the Ermelo strai
of *Eragrostis curvula* Nees., certain cultivars of *Chloris
gayana* Kunth, and *Panicum maximum* Jacq. have been effectiv
in Rhodesia in reducing populations of *M. javanica*, in
rotations with tobacco on which the nematode is an important pathogen (Daulton, 1963). Experiments in North Carolina demonstrated that several grasses, including Bermuda
grass (*Cynodon dactylon* (L.) Pers.) and *Eragrostis curvula* were resistant to *M. incognita* (McGlohon *et al.*,
1961). Grasses and grain crops were non-hosts of *M. hapla*
in the Noord Oost Polder (the Netherlands) and no damage
by this nematode was observed on susceptible crops when
grains were incorporated in crop rotations (Kuiper, 1977).

  Groundnut is very resistant to all populations of root-
knot nematodes (consisting of *M. incognita*, *M. javanica*
M. arenaria or mixtures of these species) in West Africa
(Netscher, 1975) and Madagascar (de Guiran, 1970). Sasser
(1954) observed that *Crotalaria mucronata* Desv. and *C.
spectabilis* Roth. were resistant to all species of *Meloidogyne* tested. Similarly, de Guiran (1970) found that *Crotalaria fulva* Roxb. and *C. grahamiana* Wight & Azm. were
resistant to root-knot nematode populations in Madagascar.
Resistant varieties of alfalfa, beans, cowpeas, cotton,
pepper, tomato and tobacco have been developed that are
resistant to one or more species of *Meloidogyne* (Kehr,
1966).

  The few samples cited here demonstrate clearly that
crop rotations can be used to decrease root-knot nematode
populations. However, caution should be exercised in
applying the results obtained in one location to another,
as different populations of the same species of *Meloidogyne* may react differently on the same plant species.

This physiologic variability and its implications on control are discussed here.

Host-parasite relations

Sasser (1954) was the first to report the susceptibility or resistance of different plant species to species of *Meloidogyne*. He tested 50 plants against one population each of *M. incognita, M. incognita acrita, M. javanica, M. arenaria* and *M. hapla*. His data are summarized in Table I. This Table is grossly modified by the synonymization of *M. incognita* and *incognita acrita* and by the omission of 10 hosts which were not tested against all five species of root-knot nematodes. Moreover, only those plant species that might serve as differential hosts are listed in Table I. For those plants reacting in the same way to the different species of *Meloidogyne*, only the number of plant species is listed. Examination of Table I shows that the majority of plants tested have the same reaction towards *M. incognita, M. javanica* and *M. arenaria* (36 out of 40). In this group of plants, 15 show a different reaction to *M. hapla*, suggesting that this species differs considerably from the other three.

When different populations of the same species of root-knot nematodes are tested against the same host plants, differences sometimes become apparent in host reaction. Thus, van der Linde (1956) was one of the first to demonstrate such variability between populations of *M. incognita acrita* from South Africa. It soon became evident that when many populations of *Meloidogyne* were tested, differences in host reaction from those reported by Sasser (1954) were found. Without attempting to present an exhaustive list of these differences, a number of these exceptions involving Sasser's differential hosts will be cited.

Sasser (1954) rated groundnut (*Arachis hypogaea* L.) as a host for *M. arenaria*, the so-called "peanut root-knot

TABLE I

Reaction of 40 different plant species to *Meloidogyne incognita*,
M. javanica M. *arenaria* and M. *hapla*
(Data adapted from those of Sasser, 1954)

Number of plant species showing same reaction	M. *incognita*	M. *javanica*	M. *arenaria*	M. *hapla*
14	+[1]	+	+	+
7	−	−	−	−
14	+	+	+	−
1 (Strawberry)	−	−	−	+
1 (Cotton)	+	−	−	−
1 (Groundnut)	−	−	+	+
1 (Sweet pepper)	+	−	+	+
1 (Sweet potato)	+	−	−	+

[1] + = reproduction of *Meloidogyne* sp
  − = no reproduction of *Meloidogyne* sp

nematode" and *M. hapla* (see Table I). However, Minton
(1963) reported a population of *M. arenaria* that did not
reproduce on groundnut. Later work of Sasser (1966) demonstrated that when several populations of *M. arenaria*
were tested on groundnut cv. Florunner some populations
reproduced, whereas others did not. This same variation
was obtained when six populations of *M. arenaria* from
Florida were tested on groundnut (Kirby *et al.*, 1975).
Sasser (1954) considered groundnut as a non-host of *M.
javanica* and *M. incognita*. However, these observations
have not always been confirmed by other workers. For example, groundnut has been infected by *M. javanica* in
Rhodesia (Martin, 1956), Egypt (Ibrahim and El Saedy,
1976) and Georgia (Minton *et al.*, 1969). Oteifa *et al.*
(1970) and Taha and Youssif (1976) reported groundnut as
a host for *M. incognita* in Egypt.

Sasser (1966) found that cotton (*Gossypium hirsutum*
L.) was severely attacked by eight of 18 populations of
M. incognita; four populations did not attack cotton and

the remaining six populations produced an intermediate reaction. Kirby *et al.* (1975) reported that only four, out of 14 populations of *M. incognita* tested, attacked cotton.

Strawberry (*Fragaria ananassa* Duch.) was originally reported as a non-host for *M. incognita*, *M. javanica* and *M. arenaria* (Sasser, 1954); this was supported by later work of Sasser (1966) and Kirby *et al.* (1975). Netscher: (1970) tested the reactions to strawberry of 25 populations consisting of either *M. javanica*, *M. incognita* *M. arenaria*, or a mixture of these species, from Sénégal: none of the populations was able to parasitize this crop. Nevertheless, Martin (1962), Minz (1958), and Taylor and Netscher (1975) reported strawberry as a host for *M. javanica* in Rhodesia, Israel and Sénégal respectively, while Perry and Zeikus (1972) found a population of *M. incognita* that reproduced on strawberry.

Sweet pepper (*Capsicum frutescens* L.) is another differential host frequently utilized to distinguish between species of *Meloidogyne*. In the majority of cases *M. incognita* parasitizes pepper, whereas *M. javanica* does not (Sasser, 1966; Kirby *et al.*, 1975). Southards and Priest (1973) reported that one among 17 populations of *M. incognita* tested did not reproduce on pepper, and neither did two populations from Florida (Perry and Zeikus, 1972). On the other hand, Colbran (1958) in Australia reported parasitism of pepper by *M. javanica*; however, he did not specify to which group of *C. frutescens* the host belonged. A hot pepper plant (*C. annuum* L.) found in the field in Sénégal was parasitized by both *M. incognita* and *M. javanica* (Netscher, 1970).

Sweet potato (*Ipomoea batatas* (L.) Lam.) the last of the differential hosts mentioned in Table I, does not have a uniform reaction to different populations of the same species of *Meloidogyne*. Thus, contrary to the results of Sasser (1954), Kirby *et al.* (1975) observed that the:

cultivar Allgold was parasitized by some populations of
M. arenaria and by both populations of *M. javanica* tested.
Summarizing the data presented, one may state that although the differential hosts listed in Table I often may
serve to separate *M. incognita*, *M. javanica* and *M. arenaria*, enough populations exist within these species that
react differently to question the feasibility of using
differentials for the identification of *Meloidogyne*.

Variation in pathogenicity to hosts among populations
of the same root-knot nematode species may be very common.

TABLE II

The reaction of 17 isolates of *M. incognita* from Tennessee on six
hosts (+ = infection and reproduction; - = no infection and
reproduction). (Southards and Priest, 1973)

Isolate	Tobacco	Cotton	Cowpea	Water Melon	Pepper	Tomato
1	-	-	-	+	+	+
2	-	-	-	+	+	+
3	-	-	-	+	+	+
5	-	-	-	+	+	+
6	-	-	-	+	+	+
7	-	-	-	+	+	+
13	-	-	-	+	+	+
17	-	-	-	+	+	+
4	-	-	+	+	+	+
14	-	-	+	+	+	+
15	-	-	+	-	+	+
9	-	+	-	+	-	+
12	-	+	-	+	+	+
8	-	+	+	+	+	+
10	-	+	+	+	+	+
11	-	+	+	+	+	+
16	-	+	+	+	+	+

Southards and Priest (1973), studying 17 populations of
M. incognita from Tennessee, found striking differences
in the reactions of cotton, pepper, cowpea and watermelon
(Table II). They reported one isolate capable of parasitizing watermelon and another not attacking pepper. Most
isolates did not attack cotton and more than half of the
populations did not reproduce on cowpea, a plant which
was used for years at the O.R.S.T.O.M. Laboratoire de
Nématologie at Abidjan, Ivory Coast, to rear *Meloidogyne*
populations from West Africa, including many *M. incognita*
isolates (Table II). Physiologic differences between populations of *M. naasi* were observed by Michell *et al.*
(1973) who tested five different isolates of this nematode to *Agrostis palustris* Hucls., *Sorghum bicolor* (Jav.),
Stellaria media (L.) Vill. and *Rumex crispus* L. (Table III).

Differences in host-parasite relationships are not
necessarily due only to the variability of different populations of the species of *Meloidogyne*. They may also depend on differences between cultivars of the same host.
Thus, different degrees of resistance to a given *Meloidogyne* population may be observed when several cultivars
of a normally susceptible crop are tested. Sweet potato
is a good example of a plant species with extremely variable reaction to root-knot nematodes. This is illustrated
by the data in Table IV where the reaction of 4343 sweet
potato lines to one population of *M. incognita* are summarized (Struble *et al.*, 1966). These data suggest that resistance to *Meloidogyne* is determined by several genes.
In other cases resistance of otherwise susceptible plants
is determined by one or a few major genes, although the
situation may be more complicated than suggested here.

When different populations of a species of *Meloidogyne*
are tested against several cultivars of the same host,
complexity of host-parasite relationships are potentially
the greatest, as variability of both host and parasite are
are involved. This situation is illustrated in Table V

TABLE III

Differentiation of races of *Meloidogyne naasi* by reaction on
four plant species (Michell *et al.*, 1973)

Host status of key plant species[1]

Race No.	Location	Curly Dock	Sorghum "RS 610"	Creeping Bentgrass "Tomato C 15"	Common Chickweed[2]
1	Berkshire, England	−	−	−	+
2	Siskiyou Co, Cal.	+	−	−	+
3	Du Page Co, Ill.	+	−	+	+
4	Kenton Co, Ken	−	−	+	−
5	Leavenworth Co, Kan	+	+	+	−

[1] + = host and − = non host, based on presence or absence of egg-masses, 50 days after inoculation.
[2] Supplemental host key species only; not reliable at low nematode densities.

which summarises the reaction of five isolates of *M. hapla*, all collected in central or southern France, to different varieties of cucumber and melon (Bergé *et al.*, 1974). None of the varieties has the same reaction to all of the five *Meloidogyne* isolates and all isolates differ from each other regarding their reactions to the same set of culti- vars. The same phenomenon was observed when tomatoes pos- sessing different genes for resistance were tested against different populations of *Meloidogyne*. Resistance to *M. in- cognita* was incorporated into tomato (*Lycopersicon escu- lentum* (L.) Mill.) by developing a hybrid between tomato and the resistant *Lycopersicon peruvianum* (L.) Mill. (Smith, 1944). Commercial cultivars possessing resistance derived from this hybrid were subsequently developed (see Fassuli- otis in these Proceedings). Resistance was determined by the gene Mi (Gilbert and McGuire, 1956) that not only pre- vented parasitism of *M. incognita*, but also of *M. arenaria* and *M. javanica* (Barham and Sasser, 1956. Resistance break- ing biotypes of *M. incognita*, *M. arenaria* and *M. javanica* have been reported since that time by several authors (Riggs and Winstead, 1959; Triantaphyllou and Sasser,

TABLE IV

Distribution of 4343 sweet potato lines according to their reaction to *Meloidogyne incognita* in relation to the reaction of their parents (Struble *et al.*, 1966)

Parental combination[1]	Lines in each reaction class[1]					
	R		I		S	
	No.	%	No.	%	No.	%
R x R	322	53	227	38	57	9
R x I	284	55	175	34	58	11
R x S	564	40	537	38	301	22
I x I	45	30	54	37	48	33
I x S	219	29	287	38	250	33
S x S	132	14	209	23	573	63

[1]R = resistant; I = intermediate or tolerant; S = susceptible

1960; Sauer and Giles, 1959; Netscher, 1970, (1977). Recently Sidhu and Webster (1974, 1975) showed that resistance to $M.$ $incognita$ was determined by two dominant genes
$LMiR_1$ and $LMiR_2$ which were closely linked, and that $LMiR_1$
and Mi are either allelic or identical. During a study of
populations of $M.$ $incognita$ and $M.$ $javanica$ from Sénégal in
which these populations were inoculated on tomato cultivars possessing $LMiR_1$, $LMiR_2$ or both genes, evidence was
obtained that different types of resistance breaking races
of $M.$ $incognita$ and $M.$ $javanica$ existed, and it was suggeste
that a gene for gene relationship existed between races of
$Meloidogyne$ and resistant tomato cultivars (Netscher, 1978
Table VI).

Other observations on populations of $M.$ $javanica$ and
$M.$ $incognita$ from West Africa have demonstrated that
certain populations are able to develop resistance

TABLE V

Reaction of five populations of $Meloidogyne$ $hapla$ to different
cultivars of cucumber and melon (after Bergé et $al.$, 1974)

	Origin of $M.$ $hapla$ populations				
	Saulcy	St. Emilion	Juan les Pins	Lyon	Beaucaire
Cucumber:					
Anglais	+	++	+	+	+
Maraîcher	+	++	+	+	+
Genereux	+++	+	++	+	0
Melon:					
Charentais ordinaire	0	+	0	0	0
Charentais 3967	+	+	0	0	+
Ordinabel	+	+	++	+	++
Doublon Fl	+	+	0	0	0

+++ numerous egg-masses recovered from plants inoculated with 500
    juveniles
++ 10-20 egg-masses recovered from plants inoculated with 500
    juveniles
+   1-10 egg-masses recovered from plants inoculated with 500
    juveniles
0   no multiplication

TABLE VI

Reaction of populations of *Meloidogyne* to resistant
tomato varieties (Netscher, 1978)

Species	Population	Type	Variety and resistance genes			
			Roma (universally susceptible) $LMir_1$ $LMir_2$	Nematex $LMiR_1$	Small Fry $LMiR_2$	Rossol[1]
. javanica	11310	1.2	+	+	+	+
. incognita	11304	1.2				
. incognita	11575	2	+	−	+	−
o be found	−	1	+	+	−	−
. javanica	Majority of					
. incognita	populations	0	+	−	−	−

[1] As it is not known whether Mi and $LMiR_1$ are identical or allelic, and since the reactions observed make it not possible to deduce whether Rossol possesses $LMiR_2$, the following gene combinations are proposed for this variety: $LMiR_1LMiR_2$, $LMiR_1LMiR_2$, $MiLMiR_2$, and $MiLMir_2$

breaking races, whereas other do not (Netscher, 1977; Table VI). Indications were obtained that the process of adaptation to resistant cultivars was gradual. Further, it was observed that certain naturally occurring populations of *M. incognita* and *M. javanica* were capable of severely attacking resistant tomato cultivars.

As yet, no data are available to explain the causes of the physiological variability of species of *Meloidogyne*. However, without this knowledge, it is impossible to answer such crucial questions as: What determines the degree of polyphagy of a species? i.e. why are certain species very polyphagous and others more specialized in their parasitism; and what are the genetic mechanisms controlling variability within species of *Meloidogyne*? Present answers can only be highly speculative and almost certainly any hypothesis put forward now may subsequently prove to be wrong. Nevertheless, an attempt is made to develop a hypothetical model of the host-parasite

TABLE VII

Number of egg-masses recovered from tomato cv. Ronita inoculated with juveniles from 18 populations and one sub-culture of *Meloidogyne* from Sénégal. Data were collected after 1,2 and 6 generations (Netscher, 1977)

Sample Number	Location	Species[1]	Generation I J[2]	Generation I E.m.[3]	Generation II J	Generation II E.m.	VI J	VI E.m.	Number of egg-masses expressed as percentage of juveniles inoculated — Generation I	Generation II	VI
3 239	K. Mamadou N. déné	*incognita*	1 850	3	1 500	0			0.16	0	0
3 244	Kaolack	*javanica*	11 000	0					0	0	0
3 249	Bambey	*javanica*	9 500	0					0	4.29	8.64
3 257	Dakar	*javanica*	2 800	11	8 150	350	3 300	285	0.39	0	0
3 262	Koumoune	*javanica*	19 000	0					0	0	0
3 264	Koumoune	*incognita*	1 800	10	4 400	0			0.56	0	0
3 265	Koumoune	*incognita*	3 300	0					0	0	0
3 272	Kaniak	*incognita*	5 000	1	103	1	4 600	147	0.02	0.97	3.20
3 277	Tampe	*javanica*	2 000	1	790	0			0.05	0	0
3 280	Tilène	*javanica*	7 000	4	2 000	35	4 400	310	0.06	1.75	7.05
3 407	M'Boro	*inc + java*	4 800	0					0	0	0
3 418	M'Boro	*javanica*	8 000	0					0	0	0
3 414	M'Boro	*javanica*	4 500	0					0	0	0
3 424	Lampoul	*javanica*	4 700	0					0	0	0
3 433	K. Koura	*arenaria*	3 700	3	81	3	6 200	37	0.08	3.70	0.60
3 438	St Louis	*incognita*	7 000	0					0	0	0
3 448	Dak. Bango	*incognita*	1 800	0					0	0	0
3 488	Bambey	*incognita*	3 700	0					0	0	0
3 280C	Laboratory	*javanica*	7 000	7	12 700	>40	2 000	192	0.24	n.c.[4]	9.6

[1] Identified on the basis of perineal patterns. [2] J = Number of juveniles inoculated. [3] E.m. = Number of egg-masses recovered. [4] n.c. = Percentage not calculated.

relationship of *Meloidogyne* using available knowledge in an attempt to provide better comprehension of a complex situation.

The proposed model is based upon the following statements of Bird (1974):

1) "It seems that if *Meloidogyne* species are to reproduce normally, they must be able to initiate and maintain these syncytia"; and

2) "During syncytial formation the nematode induces specialized cells to be formed from unspecialized cells. These unspecialized cells would normally form specialized cells of different types with specific functions in

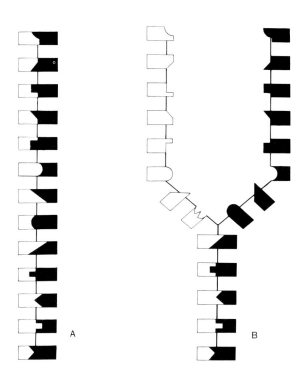

Fig. 1. Diagrammatic representation of host-parasite relationships. Each pair of interlocking teeth represent an action of the nematode and the corresponding response of the host, the white teeth representing the nematode physiological "input" and the black teeth the host response (see text).

response to various repressors and inducers operating
within the plant. The nematode also represses part of the
cell's genetic coding and activates other parts so that a
special type of cell is induced". In addition it should
be assumed that root-knot nematodes induce the same processes in the different host plants which they parasitize
and that they interfere in a group of functions that are
common to the plants.

In the model the host-parasite relationship is depicted
as a zipper, each pair of interlocking teeth of the zipper
representing an action of the nematode and the corresponding response of the host (Fig. 1). The left part of the
zipper composed of white teeth represents the total physiological "input" of the nematode, the right part characterized by black teeth the sum of host response ultimately leading to the formation of the syncytium (Fig.
1A). Thus a pair of teeth may represent an enzyme injected
by the nematode and a corresponding substrate produced
by the host. It also may represent a substance released
by the nematode and a gene of the host cell blocked by
it. Host plants of a polyphagous nematode will thus be
characterized by identical teeth in the zipper and conversely, relations between different populations of *Meloidogyne* will be characterized by the number of "teeth"
they have in common. Thus *M. incognita*, *M. javanica* and
M. arenaria may have almost all "teeth" in common except
for a few that are required to interlock with "teeth" of
the differential hosts mentioned in Table I.

It seems logical to assume that in such complex relationships as exist between *Meloidogyne* and a host plant,
relatively minor genetic differences in the host may up-
set the delicate balance between stimulus of the nematode
and response of the host resulting in malfunction or
repression of syncytia and hence resistance of the host.
Similar minor differences in the genotype of the nematode
may account for the existence of populations capable of

parasitizing resistant plants. This situation may be represented by a zipper with one or a few "teeth" that do not
fit the corresponding "teeth" of the host (Fig. 1B). In
cases of very specific host-parasite relationships, e.g.
M. ovalis and *Acer* spp., the host may have many "teeth"
of a particular nature and therefore *M. ovalis*, equipped
to match the special constitution of its host, will not
be able to attack plants commonly attacked by other root-
knot nematodes. This actually has been reported by Riffle
and Kuntz (1967) who observed that *M. ovalis* was not capable of parasitizing tomato, carrot and *Pelargonium* sp.

    Although the model may help to visualize the intricate
host-parasite relationships existing between *Meloidogyne*
and various plants, it does not explain the extreme variability encountered in certain species of the genus.
Actually most species discussed in this article have a
parthenogenetic mode of reproduction. *M. hapla* is characterized by meiotic parthenogenesis but as amphimixis is
possible (Triantaphyllou, 1966) genetic variation might
occur by meiosis and fertilization during amphimictic reproduction and segregation through the process of "selfing" during parthenogenesis; the same might occur in
M. naasi. However, *M. javanica, M. incognita* and *M. arenaria* are all characterized by a mitotic parthenogenetic
mode of reproduction and the situation is completely different. The mitotic nature of the reproduction of these
three species is difficult to reconcile with the diversity
observed in these organisms. However, in certain cases
mitotic reproduction may be more favourable for the establishment of certain specialized populations. As Toxopeus (1956) pointed out, races adapted to resistant cultivars whose creation is based upon a mutation from a
dominant to a recessive gene are difficult to realize in
amphimictic species as it involves the mating of females
and males both possessing the rare recessive gene. Mutation in parthenogenetic species generally results in the

creation of a new clone characterized by the mutant gene, if the mutant is viable and competitive. If a mutation in a parthenogenetic population provides the mutated individual with the potential capacity to parasitize a resistant cultivar, a resistance breaking race will automatically be selected when it parasitizes a resistant plant.

The fact that adaptation to resistant cultivars of tomatoes might be a gradual rather than an abrupt change (Netscher, 1977) is not in agreement with the hypothesis that the creation of a resistant breaking race of *Meloidogyne* was caused by a single mutational step. It rather seems that a selection among increasingly aggressive individuals was exerted by the resistant cultivar. The seemingly polygenic inheritance of resistance of sweet potato to *M. incognita* (Struble *et al.*, 1966) also indicates that relations between root-knot nematodes and host plants are extremely complex and that the model, proposed here, may be oversimplified. Nevertheless, it is hoped that the concepts expressed may help the comprehension of the wide variation of host-parasite relations in the genus *Meloidogyne*.

Practical implications

The considerable physiologic variability of the most commonly encountered species of *Meloidogyne*, previously discussed, seriously obstructs the application of crop rotations utilizing non-host plants and resistant cultivars designed to reduce populations of these parasites and minimize crop damage. It has been shown that it is impossible to predict with certainty whether or not a population consisting of a known species of *Meloidogyne* will parasitize a given host. Expressed more simply, if it is known that a particular field is infested with *M. incognita* only, it is impossible to know whether cotton would be attacked if it were planted in that field. Therefore, from a practical viewpoint, the value of making

species indentifications is questionable. It is suggested
that the only satisfactory approach is experimentally to
determine the reaction of particular plants to individual
field populations.

An additional complication is that many populations of
root-knot nematodes are composed of more than one species.
For example, 25% of the populations examined at the ORSTOM
Laboratoire de Nématologie in Dakar, Sénégal, consisted
of a mixture of species (Netscher, 1978). It is assumed
that the physiologic variation of mixed populations is
even greater than that of monospecific populations. The
following example may illustrate the difficulties that
are encountered in field experiments involving mixed populations of *Meloidogyne*.

In 1974, a rotation experiment was made at the Centre
de Développement d'Horticulture at Camberene, Sénégal.
The nematode population consisted of 57% *M. incognita*,
14% *M. javanica*, 9% *M. arenaria* and 20% of forms intermediate between these species (determinations based on
examination of 105 perineal patterns). On the basis of
these identifications, the following predictions might
have been made: strawberry would not be parasitized by
this population because of the absence of *M. hapla*, whereas groundnut could be parasitized because of the presence
of *M. arenaria*. Three months after planting, strawberry
and groundnut were examined and no evidence of galling
was found on either crop. In those plots containing the
crops, soil populations of *Meloidogyne* had decreased to
very low levels whereas in control plots in which tomato
had been grown populations were high. However, in five
of the 15 replications, strawberry was allowed to grow
for six months, after which time a population of *M. javanica* had developed on this host (Taylor and Netscher,
1975). *Meloidogyne* resistant cultivars of tomato growing
in the same field were initially resistant to the root-
knot nematode population present (Taylor, 1975). However,

from a few egg masses that developed on a resistant tomato plant, it was possible to establish resistance breaking races (Netscher, 1977).

For practical reasons identification of field populations should not be encouraged, especially in developing countries where such identification requires the expertise of a highly specialized nematologist and is very time consuming. Under these conditions it seems much more practical to have field populations of the nematodes tested by agronomists, extension specialists, etc., who are perfectly able to uproot plants and determine if roots are galled or not. It would be better if such tests were made in cooperation with a nematologist, but it is not absolutely essential. A wide range of plants, including cash, cover and food crops of interest to the grower or to the region under study, should be tested for susceptibility or resistance to the particular *Meloidogyne* population. Resistant cultivars of normally susceptible crops, if available, should also be included in such tests.

Even if little or no physiologic variation exists within *Meloidogyne* populations that are present in a given area, it is possible that routine testing will not detect a mixed population in which one component is rare. An example of such a situation is found in the work of Sauer and Giles (1959) who grew groundnut in a field thought to be infested only by *M. javanica* and inadvertently increased a population of *M. hapla* that had escaped detection.

If the concept is accepted that crop rotation recommendations should be based on plant reactions to particular *Meloidogyne* populations the question arises as to how work should be organized in a given region. Ideally, populations from each infested field should be tested, but such an approach is obviously not realistic. The best solution would be to test as many populations as possible and to extrapolate from the results obtained whenever a

recommendation is required for a field that has not been
examined. Although such an approach contains a certain
degree of risk, it appears to be the most feasible solution. Two examples of the possible use of this approach
are given below.

In a study of 25 single egg-mass populations from
Florida (Kirby *et al.*, 1975), only two reproduced on
strawberry and six on groundnut. Therefore, selection of
these two crops to include in rotations intended to reduce
root-knot nematode populations under Florida conditions
seems logical. However, use of single egg-mass cultures
should be avoided because such cultures do not represent
the physiologic variability of the populations from which
they were derived.

From similar studies on field populations and field
observations made in Sénégal, it may be concluded that
groundnut (Netscher, 1975) and strawberry (Netscher, 1970;
Taylor and Netscher, 1975) are rarely parasitized whereas resistant tomato cultivars are fairly often parasitized
(Netscher, 1977). Thus, groundnut and strawberry, from a
statistical viewpoint, can be recommended in rotations,
whereas resistant tomato should be used with caution. Although the great physiologic variability of root-knot
nematodes interferes with development of crop rotations
intended to control these parasites, unlike air-borne
pathogens the dissemination of nematodes is limited thus
making it possible to utilize crop management to prevent
the spread of specialized populations and the introduction
of root-knot nematodes into non-infested fields.

In discussing the complications of physiologic variability on control of *Meloidogyne*, Netscher (1978) stated,
"On the basis of considerable experience in the tropics
studying *Meloidogyne* problems the use of non-hosts and
resistant varieties should be recommended on slightly infested or non-infested land:
1) In most cases reduction of the nematode population is

accomplished by the trapping effect of the roots of the
plants involved, accompanied by necrosis of the invaded
root tissues. If the *Meloidogyne* populations are too
large, plants are badly damaged and a serious weed population, generally comprising several hosts of *Meloidogyne*,
will develop with a resultant multiplication of the root-
knot nematodes. In one extreme case, we observed a complete failure of a crop of groundnut grown with the in-
tention of reducing the root-knot population in a heavily
infested field: among the weeds that had become dominant
to the groundnuts, many were actively maintaining the
Meloidogyne populations based upon the degree of galling
of the roots. From inoculation experiments it was shown
that population densities of 32,000 juveniles per dm^3 of
soil or more, a level frequently encountered in the field,
severely damaged groundnut seedlings.

2) It is also important to avoid growing non-hosts in
heavily infested soils so as to reduce the possibility of
selecting and developing biological races from the exist-
ent *Meloidogyne* populations. Assuming that the frequency
of such biological races is independent of the number of
the nematodes present, it is logical to assume that in
soils containing few root-knot nematodes, these races will
be rare or absent.

   For these two important reasons it has been recommended
to use crop rotation and resistant varieties primarily as
preventive measures rather than as a cure for the *Meloid-*
ogyne problem (Netscher, 1974, 1975, 1977; Netscher and
Mauboussin, 1973). In cases of heavy infestations, *Meloid-*
ogyne populations should be reduced by either chemical or
other means (soil desiccation inundation, bare fallow)
before growing a non-host or a resistant variety."

   Until now no practical methods have been developed to
eradicate *Meloidogyne* populations from infested soils.
Therefore, it is necessary to live with this problem by
avoiding the introduction and establishment of these

nematodes in non-infested land and prevention of population increases in infested land. Unfortunately, no general formula can be given to achieve these goals. Although much essential information is still lacking, it is believed that in many cases intelligent use of available knowledge can develop integrated control systems in which non-hosts, resistant cultivars, cultural practices, and occasionally nematicides, will each have a place, always keeping in mind the physiologic variability within the genus *Meloidogyne*.

References

Barham, W.S. and Sasser, J.N. (1956). Root-knot nematode resistance in tomatoes. *Proc.Assn.Sth.Agric.Workers*, 53rd Ann.Conv., 150-151.

Berge, J.B., Dalmasso, A. and Ritter, M. (1974). Influence de la nature de l'hôte sur le développement et le déterminisme du sexe du nématode phytoparasite *Meloidogyne hapla*. *C.r.hebd.Seanc.Acad. Agri.Fr.*, **60**, 946-952.

Bessey, E.A. and Byars, L.P. (1915). The control of root-knot. *U.S. Dept.Agr.Farmers.Bul.*, **648**, 19.

Bird, A.F. (1974). Plant response to root-knot nematode. *Ann.Rev. Phytopathol.*, **12**, 69-85.

Chitwood, B.G. (1949). Root-knot nematodes. *Proc.helminth.Soc.Wash.*, **16**, 90-104.

Christie, J.R. and Albin, F.E. (1944). Host-parasite relationships of the root-knot nematode, *Heterodera marioni*. *Proc.helminth.Soc. Wash.*, **11**, 31-37.

Colbran, R.C. (1958). Studies of plant and soil nematodes. *Qd J.agric. Sci.*, **15**, 101-136.

Daulton, R.A.C. (1963). Controlling *Meloidogyne javanica* in Southern Rhodesia. *Rhod.agric.J.*, **60**, 150-152.

De Guiran, G. (1970). Le problème *Meloidogyne* sur tabac à Madagascar. *Cah.ORSTOM,Ser.Biol.*, **11**, 187-208.

Esser, R.P., Perry, V.G. and Taylor, A.L. (1976). A diagnostic compendium of the genus *Meloidogyne* (Nematoda, Heteroderidae). *Proc. helminth.Soc.Wash.*, **43**, 138-150.

Gilbert, J.C. and McGuire, D.C. (1956). Inheritance of resistance to severe root-knot from *Meloidogyne incognita* in commercial type tomatoes. *Proc.Amer.Soc.hort.Sci.*, **68**, 437-442.

Ibrahim, I.K.A. and El-Saedy, M.A. (1976). Development and pathogenesis of *Meloidogyne javanica* in peanut roots. *Nematol.medit.*, **4**, 231-234.

Kehr, A.E. (1966). *In* "Pest control by chemical, biological, genetic and physical means." 126-138. U.S. Dept.Agric. (A.R.S. 33 10 July 1966).

Kirby, M.F., Dickson, D.W. and Smart, G.C. Jr. (1975). Physiological variation within species of *Meloidogyne* occurring in Florida.

Plant.Dis.Reptr, **49**, 353-356.

Kuiper, K. (1977). Introduction and establishment of plant parasitic nematodes in newly reclaimed polders with special reference to *Trichodorus teres*. *Meded.Landbouwhogeschool Wageningen*, 77-84.

Linde, W.J. van der (1956). The *Meloidogyne* problem in South Africa. *Nematologica*, 1, 177-183.

McGlohon, N.E., Sasser, J.N. and Sherwood, R.T. (1961). Investigations of plant parasitic nematodes associated with forage crops in North Carolina. *N.C.agric.Exp.Stn Tech.Bull.*, **148**, 33.

Martin, G.C. (1956). The common root-knot nematode (*Meloidogyne javanica*). *Rhod.Fmr.*, **27**, 20.

Martin, G.C. (1962). *Meloidogyne javanica* infecting strawberry roots. *Nematologica* 7, 256.

Michell, R.E., Malek, R.B., Taylor, D.P. and Edwards, D.I.J. (1973). Races of the barley root-knot nematode, *Meloidogyne naasi*. *J.Nematol.*, **5**, 41-44.

Minton, N.A. (1963). Effects of two populations of *Meloidogyne arenaria* on peanut roots. *Phytopathology* 63, 79-81.

Minton, M.A., McGill, J.F. and Golden, A.M. (1969). *Meloidogyne javanica* attacks peanuts in Georgia. *Pl.Dis.Reptr*, **53**, 668.

Minz, G. (1958). *Meloidogyne javanica* in strawberry roots. *Pl.Prot. Bull.F.A.O.*, **6**, 92.

Netscher, C. (1970). Les nématodes parasites des cultures maraîchères au Sénégal. *Cah.ORSTOM,Ser.Biol.*, **11**, 209-229.

Netscher, C. (1974). L'arachide et le contrôle biologique des nématodes *Meloidogyne* spp. dans les cultures maraîchères du Sénégal. *C.r.hebd.Seanc.Acad.Agric.Fr.*, **60**, 1332-1339.

Netscher, C. (1975). Studies on the resistance of groundnut to *Meloidogyne* sp. in Sénégal. *Cah.ORSTOM.Ser.Biol.*, **10**, 227-232.

Netscher, C. (1977). Observations and preliminary studies on the occurrence of resistance breaking biotypes of *Meloidogyne* spp. on tomato. *Cah.ORSTOM.Ser.Biol.*, **11**, 1976, 173-178.

Netscher, C. (1978). Morphological and physiological variability of species of *Meloidogyne* in West Africa and implications for their control. *Meded.Landbouwhogeschool Wageningen*, 78-83, 46.

Netscher, C. and Mauboussin, J.C. (1973). Résultats d'un essai concernant l'efficacité comparée d'une variété de tomate résistante et de certains nématicides contre *Meloidogyne javanica*. *Cah.ORSTOM. Ser.Biol.*, **21**, 97-102.

Orton, W.A. (1903). Iron cowpea. *U.S.Bur.Plant Indus.Bull.*, **25**, 65-68.

Oteifa, B.A., Elgindi, D.M. and Moussa, F.F. (1970). Root-knot problem in recently reclaimed sandy areas of U.A.R. *Meded.Fak.Landb-Weten. Gent*, **35**, 1167-1176.

Perry, V.G. and Zeikus, J.A. (1972). Host variations among populations of the *Meloidogyne incognita* group. *J.Nematol.*, 4, 231-232.

Riffle, J.W. and Kuntz, J.E. (1967). Pathogenicity and host range of *Meloidogyne ovalis*. *Phytopathology* 57, 104-107.

Riggs, R.D. and Winstead, N.N. (1959). Studies on resistance in tomato to root-knot nematodes and on the occurrence of pathogenic biotypes. *Phytopathology* 49, 716-724.

Sasser, J.N. (1954). Identification and host-parasite relationships of certain root-knot nematodes (*Meloidogyne* spp.). *Bull.Md. agric.*

Exp.Stn., A-77, 31.
Sasser, J.N. (1966). Behaviour of *Meloidogyne* spp. from various geographical locations on ten host differentials. *Nematologica* 12, 97-98.
Sauer, M.R. and Giles, J.E. (1959). A field trial with a root-knot resistant tomato variety. C.S.I.R.O. *Austr.Irrig.Res.Sta.Tech Pap.*, 3, 1-10.
Sidhu, G. and Webster, J.M. (1974). Genetic control of resistance in tomato. *Nematologica* 19, 546-550.
Sidhu, G.S. and Webster, J.M. (1975). Linkage and allelic relationships among genes for resistance in tomato (*Lycopersicon esculentum*) against *Meloidogyne incognita*. *Can.J.Genet.Cytol.*, 17, 323-328.
Sivapalan, P. (1972). *In* "Economic Nematology." (Ed. J.M. Webster.) 285-311. Academic Press, London, New York.
Smith, P.G. (1944). Embryo culture of a tomato species hybrid. *Proc. Amer.Soc.hort.Sci.*, 44, 413-416.
Southards, C.J. and Priest, M.F. (1973). Variation in pathogenicity of seventeen isolates of *Meloidogyne incognita*. *J.Nematol.*, 5, 63-67.
Struble, F.B., Morrison, L.S. and Cordner, H.B. (1966). Inheritance of resistance to stem rot and to root-knot nematode in sweetpotato. *Phytopathology* 56, 1217-1219.
Taha, A.H.Y. and Yousif, G.M. (1976). Histology of peanut underground parts infected with *Meloidogyne incognita*. *Nematol.medit.*, 4, 175-181.
Taylor, D.P. (1975). Observations on a resistant and a susceptible variety of tomato in a field heavily infested with *Meloidogyne* in Sénégal. *Cah.ORSTOM.Sér.Biol.*, 10, 239-245.
Taylor, D.P. and Netscher, C. (1975). Occurrence in Sénégal of a biotype of *Meloidogyne javanica* parasitic on strawberry. *Cah.ORSTOM. Sér.Biol.*, 10, 247-249.
Toxopeus, H.J. (1956). Some remarks on the development of new biotypes in *Heterodera rostochiensis* that might attack resistant potato clones. *Nematologica* 1, 100-101.
Triantaphyllou, A.C. (1966). Polyploidy and reproductive patterns in the root-knot nematode *Meloidogyne hapla*. *J.Morph.*, 118, 403-413.
Triantaphyllou, A.C. and Sasser, J.N. (1960). Variation in perineal patterns and host specifity of *Meloidogyne incognita*. *Phytopathology* 50, 727-735.

Discussion

Olowe suggested that continuous cropping of strawberry for example, could lead to the selection of strains of root-knot nematodes to which the crop would be susceptible. Netscher agreed that continuous cropping is a bad policy. However he cited an example of a field in which strawberries were grown continuously for many years but in which a resistance breaking race of *M. javanica* did

not develop even though this species was present on the
property. On the contrary in experimental plots only 2 km
from this field a resistance breaking race of *M. javanica*
developed after growing strawberries for only one year.
Sasser remarked that regardless of how the crops react,
it is imperative that the species involved should be
known and that this was useful in designing rotations. He
also emphasized that "resistance is never absolute" and
it is quite common to have populations which break resistance; thus he could not agree with Netscher's view that
species identification is not important. Netscher replied
that although for a scientist it might be necessary to
identify the species of the populations with which he is
working he did not agree with Sasser that species identification was necessary to develop crop rotations for
existing populations, especially since specialists frequently disagree on the identity of certain populations.
To stress his point Netscher cited a case in which a
mixed population of species of *Meloidogyne* infected mil-
let successfully but declined when groundnut was grown;
the importance of this observation was such that the
identification of the species had no importance.

16 COWPEA, LIMA BEAN, CASSAVA, YAMS AND *MELOIDOGYNE* SPP. IN NIGERIA

F.E. Caveness

International Institute of Tropical Agriculture, Ibadan, Nigeria

Introduction

In Nigeria, *Meloidogyne* spp. were first recorded in the 1959/60 Annual Report of the Ministry of Agriculture (Anon., 1961) although in 1958 A.L. Taylor (personal communication) identified *M. incognita*, *M. acrita*, *M. arenaria* and *M. javanica* from cowpea (*Vigna unguiculata* (L.) Walp.) roots submitted to the U.S. Department of Agriculture. In the 1960's Wilson (1962a, 1962b) and Caveness (1962,1967a,1967b,1968,1971 and 1976) reported on the distribution and host range of root-knot nematodes in Nigeria. Life history studies on *M. acrita*, *M. arenaria* and *M. javanica* demonstrated that they had essentially the same life histories as reported in other warm climates (Caveness, 1967b).

Root-knot nematodes occur most frequently in the moist southern region of the country and are found in many home and market gardens in the northern region where soils are irrigated throughout the 9 month dry season (Fig. 1) Root-knot nematodes quickly became a limiting factor in large-scale irrigation projects in the north of Nigeria (Bos, 1977; Keay, 1967; Smit and Bos, 1976).

Root-knot nematodes are the most serious nematode pests of numerous food crops in the tropics. The objectives of the International Institute of Tropical

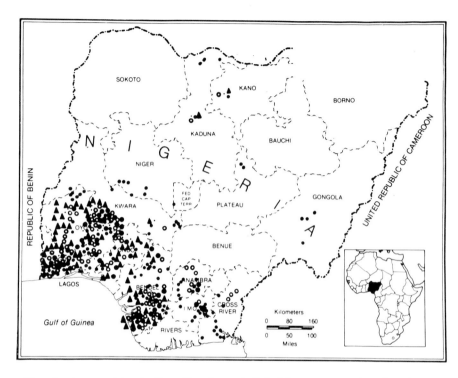

Fig. 1. Geographical distribution of *Meloidogyne* spp. in Nigeria.
• *Meloidogyne* spp. juveniles, ● *M. acrita,* ▲ *M. arenaria,* ○ *M. incognita,* · *M. javanica.*

Agriculture (IITA) nematology subprogramme are to increase and stabilize crop yields using minimum farmer input methods such as nematode resistant crop cultivars and crop rotations to suppress nematode populations, including plant mulches, intercropping and modification of agronomic practices. Resistant cultivars increase and stabilize yields, but at no extra cost to the farmer and with the additional advantage of mitigating nematode attack on the following crop. Thus, nematode resistant food crop cultivars in combination with improved cropping sequences could contribute greatly toward achieving optimum economic and long-term land use for growers within traditional farming systems as well as in more sophisticated and technical farming systems.

Cowpeas

Cowpeas (*Vigna sinensis* (Torner) Savi) in Nigeria are attacked by *M. incognita, M. acrita, M. arenaria* and *M. javanica* in all regions where root-knot nematodes have been reported (Amosu, 1974; Bos, 1977; Caveness, 1962, 1967a, 1967b, 1968, 1971, 1976; Odihirin, 1976; Ogunfowora, 1976; Olowe, 1976; Smit and Bos, 1976; Wilson, 1962a, 1962b); *M. incognita* and *M. arenaria* occur most commonly (Fig. 1). In one study in southern Nigeria (Ogunfowora, 1976) grain yields of cowpea were decreased by 59%. Systematic screening programmes at the University of Ife and IITA have identified cowpea lines resistant to *M. incognita* (Amosu, 1974, 1976; Caveness, 1976; Singh *et al.*, 1975). Amosu (1974, 1976) identified the cultivar Mississippi Silver as highly resistant to *M. incognita* in greenhouse and field tests and other lines with moderate resistance. At IITA four lines have been identified as resistant and 28 lines as moderately resistant to *M. incognita* attack, out of the 421 lines screened in over 4,000 greenhouse pot tests and confirmation tests. None has been identified as highly resistant. Only highly resistant and resistant germ plasm are deemed suitable as sources of resistance in breeding for crop improvement. The IITA cowpea cultivar Vita 3 is resistant to leafhoppers and many virus diseases as well as being resistant to *M. incognita* (Singh *et al.*, 1975).

Lima beans

The lima bean, *Phaseolus lunatus* (Benth.) Van Ess. is an important food, though of limited local interest in Nigeria. The IITA has included lima bean in its grain legume improvement programme because of its high yield potential. However, yellow mosaic disease, some insect pests and root-knot nematodes are a problem. In screening the lima bean collection to *M. incognita* 10 lines showed moderate resistance while the remaining 277 lines

were all susceptible or highly susceptible (Caveness and
Baudoin, 1976).

Cassava

Bulked cassava seed, *Manihot esculenta* Crantz, from
IITA was pot tested in the greenhouse to determine genet-
ic variation in relation to root-knot nematode attack
(Caveness and Hahn, 1976). All 190 seedlings were shown

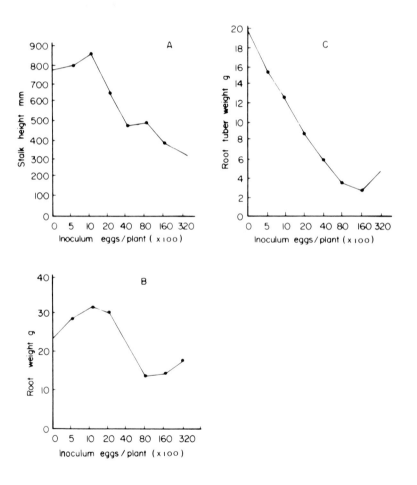

Fig. 2. The effect of *M. incognita* on seedling cassava, *Manihot esculenta* Crantz. A: stalk height, B: root weight, C: root tuber weight

to be highly susceptible to *M. incognita* attack.

Plants receiving 500-1000 eggs per plant increased in height slightly, compared with nematode-free controls (Fig. 2A); mean root weight increased with inoculations from 500-2000 eggs (Fig. 2B); root tuber weight was decreased at all levels of inoculum used (Fig. 2C). Higher inoculum levels caused considerable decreases of height and root and tuber weights (Fig. 2A,B,C).

Yams

The yams (*Dioscorea* spp.) are an important and preferred food in Nigeria. The yam nematode (*Scutellonema bradys*) and the root-knot nematodes are important field and post-harvest pests of yam tubers reducing market value and being involved in a tuber decay complex (Adesiyan and Odihirin, 1975; Adesiyan *et al.*, 1975a, 1975b; Bridge, 1972, 1973; Odihirin, 1976).

Two-month-old seedlings of *D. dumetorum* (Kunth) Pax, *D. praehensilis* Benth. and *D. rotundata* Poir. were screened for resistance to *M. incognita* (Caveness and Wilson, 1976). Most *D. dumetorum* seedlings tested were highly resistant to *M. incognita* attack. Two of the 48 tested plants showed light infections on the roots and 28 eggs were recovered from one plant. No gall development was seen on the tubers. *D. praehensilis* was highly susceptible to attack by *M. incognita* in both roots and tubers in a test of 48 seedlings. Seedlings of two breeding families of *D. rotundata* were highly susceptible. The roots and tubers of the 96 seedlings were all heavily galled and numerous eggs were recovered.

References

Anon. (1961). Report of the Department of Agricultural Research for the year 1959/60. Lagos: Federal Printing Division.

Bos, W.S. (1977). Root-knot nematodes in the Nigerian savanna zones with special reference to root-knot problems of irrigated crops in the Kano river project at Kadawa. Dept. Crop Protection Rpt. Inst.Agric.Res., Zaria.

Bridge, J. (1972). Nematode problems with yams, *Dioscorea* spp., in Nigera. *PANS* **18**: 89-91.

Bridge, J. (1973). Nematodes as pests of yams in Nigeria. *Meded.Fac. Landbouwwet.Rijksuniv.Gent*, **38**, 841-852.

Caveness, F.E. (1962). End of tour report of the nematology project. *Min.Agric.Nat.Resources*, Ibadan, 68.

Caveness, F.E. (1967a). Shadehouse host ranges of some Nigerian nematodes. *Pl.Dis.Rptr.* **51**, 33-37.

Caveness, F.E. (1967b). End of tour progress report on the nematology project. *Min.Agric.Nat.Resources*, *Westn.Regn.*, Nigeria and USA Agency for International Development. Lagos. 135.

Caveness, F.E. (1968). A survey of plant-parasitic nematodes in Nigeria. USDA, Shafter, California. 276.

Caveness, F.E. (1971). The distribution of plant-parasitic nematodes in Nigeria. *J.Assoc.Adv.Agric.Sc. in Africa*. **1**, 35-40.

Caveness, F.E. (1976). *In* "Proceedings of the Research Planning Conference on Root-knot Nematodes, *Meloidogyne* spp." (Ed. Anon.). 24137, I.I.T.A., Ibadan.

Caveness, F.E. and Baudoin, J.P. (1976). *In* "Farming Systems Program, Crop Protection Subprogram, Nematology internal report." (Ed. F.E. Caveness). 1-9, I.I.T.A., Ibadan.

Caveness, F.E. and Hahn, S.K. (1976). *In* "Farming Systems Program, Crop Protection Subprogram, Nematology internal report." (Ed. F.E. Caveness). 23-26, I.I.T.A., Ibadan.

Caveness, F.E. and Wilson, J.E. (1976). *In* "Farming Systems Program, Crop Protection Subprogram, Nematology internal report." (Ed. F.E. Caveness), 61-65, I.I.T.A., Ibadan.

Keay, M.A. (1967). Irrigation in northern Nigeria. Circular. Inst.Agric.Res., Zaria.

Odihirin, R.A. (1976). *In* "Proceedings of the Research Planning Conference on Root-knot Nematodes, *Meloidogyne* spp." (Ed. Anon.). 20-23, I.I.T.A., Ibadan.

Ogunfowora, A.O. (1976). *In* "Proceedings of the Research Planning Conference on Root-knot Nematodes, *Meloidogyne* spp." (Ed. Anon.). 9-14, I.I.T.A., Ibadan.

Olowe, T. (1976). *In* "Proceedings of the Research Planning Conference on Root-knot Nematodes, *Meloidogyne* spp." (Ed. Anon.). 15-19, I.I.T.A., Ibadan.

Singh, S.E., Williams, R.J., Rachie, K.O., Rawal, K., Nangju, D., Wien, H.C. and Luse, R.A. (1975). VITA-3 Cowpea. *Tropical Grain Legume Bull.* **1**, 18.

Smit, J.J. and Bos, W.S. (1976). *In* "Proceedings of the Research Planning Conference on Root-knot Nematodes, *Meloidogyne* spp." (Ed. Anon.). 41-43, I.I.T.A., Ibadan.

Wilson, W.R. (1962a). Root-knot nematodes (*Meloidogyne* spp.) in Northern Nigeria. Proc. 1st Int. Inter-African Pl. Nematol. Conf., Kikuyu, 1960, CCTA Publ. 16-17.

Wilson, W.R. (1962b). Root-knot eelworms. Techn. Rep. 24, Min. Agric. North. Region, Nigeria. 11.

R.N. Inserra and V. Vovlas

Laboratorio di Nematologia Agraria,
C.N.R., Bari, Italy

The root-knot nematode, *Meloidogyne arenaria* (Neal) Chitwood, was the first plant parasitic nematode recorded on
citrus. Neal (1889) reported infestation of the roots of
"bitter sweet orange" (*Citrus vulgaris* Risso) and *C. aurantium* L. in Florida. Although there are no further reports of its occurrence on citrus in Florida the association of *Meloidogyne* spp. with citrus has been recorded
in more than seventy papers (O'Bannon *et al.*, 1975).
Evidence of invasion or reproduction in citrus roots is
reported only in a few instances and many reports of
Meloidogyne juveniles from soil samples collected in citrus orchards were probably from the many weed hosts present. Usually root-knot nematode infection on citrus is
rare and of doubtful economic importance.

Five *Meloidogyne* spp. have been reported, on a worldwide base, to infest citrus roots: the Asiatic pyroid
citrus nema; *M. exigua* Goeldi, *M. incognita* (Kofoid &
White) Chitwood; *M. indica* Whithead; and *M. javanica*
(Treub) Chitwood.

The Asiatic pyroid citrus nema was first found in
Taiwan in 1959, and later that year in India (Chitwood
& Toung, 1960a). Host-parasite interactions and the
formations of egg masses were observed on naturally infected roots of *C. reticulata* Blanco var. *austera* Swing.

in Taiwan, and *C. sinensis* (L.) Osb. in India (Chitwood
& Toung, 1960b). In the glasshouse, several *Citrus* species
were injected together with piper sudan grass (*Sorghum
vulgare* Pers.), sweet potato (*Ipomea batata* Lam.) and
maize (*Zea mays* L.) (Chitwood & Toung, 1960b). Second-
stage juveniles were found to penetrate the roots, and
those becoming females established permanent feeding
sites. In this position, they developed into sexually
mature swollen females and deposited egg masses on the
root surface, while those becoming males migrated into
the soil. Gall formation, tip bloating and abnormal
branching were the symptoms observed on infected citrus
roots. Chitwood and Toung (1960a) suggested that the
species resembles *M. africana* Whitehead but differs in
not having the lateral striae interrrupted on the female
perineum, and in the position of juvenile and male phasmids; they did not name the species. The description of
M. indica from citrus roots collected in India (Whitehead,
1968) suggests the two species may be identical, which
could be confirmed by a comparison of the original material studied by Chitwood and Toung with paratypes of
M. indica.

M. *exigua* was reported from citrus in Surinam (Den
Ouden, 1965) and Guadeloupe (Scotto la Massese, 1969).
It is common on coffee (*Coffea arabica* L.) and is able
to reproduce on citrus (bitter orange) roots on which it
forms small, separate galls (Scotto la Massese, 1969).
Infection on citrus appears to be more serious in soils
previously planted with coffee.

In Australia, *M. incognita* caused galls and produced
egg masses on sweet orange (citronelle) (*C. sinensis*)
roots (Colbran, 1958); in glasshouse inoculation tests it
caused tip swelling on feeder roots of *Poncirus trifoliata*
(L.) Raf. but with no evidence of nematode reproduction.
In Italy, galls containing adult females but no eggs were
observed on the roots of sour orange growing in sandy

soils (Accorti & Ambrogioni, 1976).

M. indica is specific to citrus and has been described only from *C. aurantifolia* (Christm.) Swing. and *C. sinensis* in India (Whitehead, 1968). Infective second-stage juveniles invaded roots and gave rise to sedentary swollen females or active vermiform males. Infection resulted in gall formation, and protruding egg masses, similar to other *Meloidogyne* species.

M. javanica has been reported more frequently than other species on citrus. Galls were found on Cleopatra mandarin (*C. reticulata*), sour orange (*C. aurantium*), and Troyer citrange (*C. sinensis* X *P. trifoliata*) in California (Gill, 1971; Van Gundy *et al.*, 1959), on *C. aurantifolia* var. *dulcis* (Minz, 1956), and *C. reticulata* var. Wilking (Orion & Cohn, 1975) in Israel, and on *P. trifoliata* (Fig. 1), sour orange and Troyer citrange in Italy (Accorti & Ambrogioni, 1976; Inserra *et al.*, 1978;

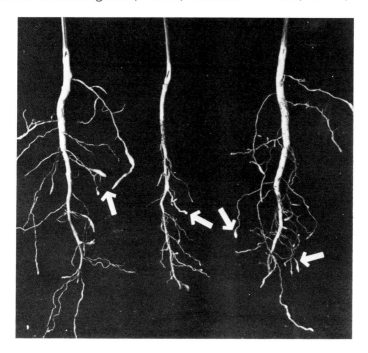

Fig. 1. Galls on *Poncirus trifoliata* caused by *Meloidogyne javanica*

Inserra & Vovlas, 1977). According to Orion & Cohn (1975) *M. javanica* fails to reproduce on citrus because no syncytia form after juvenile penetration, but Gill (1971) reported reproduction on citrus in California. Infection by *M. javanica* juveniles causes a general disorder in stelar tissues on *P. trifoliata* and Troyer citrange infected roots and necrosis of vascular and cortical cells. Syncytial formation in the primary tracheal elements and in secondary vascular tissues were detected in *P. trifoliata* infected roots (Inserra *et al.*, 1978). Mixed populations of *M. javanica* and *M. incognita* were observed on sour orange roots in citrus orchards in southern Italy (Accorti & Ambrogioni, 1976).

The Asiatic pyroid citrus nema and *M. indica* are specific to citrus but other reported *Meloidogyne* spp. appear unable to complete their life cycle on citrus roots. In glasshouse tests, *M. javanica*-infected citrus seedling grown with infected tomato plants recovered quickly when removed from the inoculum and transplanted into non-infected soil; no further gall formation was seen on the root systems because of the failure of the nematode to reproduce on citrus (Fig. 1.; Inserra *et al.*, 1978). A consequence of growing nematode hosts with non-host is the possibility of selecting, as a result of population pressure, a biotype able to infect and reproduce on citrus. The development of a citrus-infecting root-knot nematode strain would be a serious menace and therefore it is important to keep plantations and nurseries of *Citrus* or related genera free of *Meloidogyne* susceptible hosts.

References

Accorti, M. and Ambrogioni, L. (1976). Infestazioni da nematodi del gen. *Meloidogyne* su *Citrus* spp. e su *Myoporum* sp. in Italia. *Redia* **59**, 323-330.
Chitwood, B.G. and Toung, M.C. (1960a). *Meloidogyne* from Taiwan and New Delhi. *Phytopathology* **50**, 631-632.

Chitwood, B.F. and Toung, M.C. (1960b). Host-parasite interaction of the Asiatic pyroid citrus nema. *Plant.Dis.Rep.* 44, 848-854.

Colbran, R.C. (1958). Studies of plant and soil nematodes. *Queensl. J.Agric.Sci.* 15, 101-135.

Den Ouden, H. (1965). Een aantasting van citrus in Suriname veroorzaakt door *Meloidogyne exigua*. *Surinaamse Landbouw* 13, 34.

Gill, H.S. (1971). Occurrence and reproduction of *Meloidogyne javanica* on three species of *Citrus* in California. *Plant Dis.Rep.* 55, 607-608.

Inserra, R.N., Perrotta, G., Vovlas, N. and Catara, A. (1978). Reaction of citrus rootstocks to *Meloidogyne javanica*. *J.Nematol.* 10, 181-184.

Inserra, R.N. and Vovlas, N. (1977). Nematodes other than *Tylenchulus semipenetrans* pathogenic to citrus. *Int. Citrus Congress, Orlando, Florida (U.S.A.)*, 8.

Minz, G. (1956). The root-knot nematode, *Meloidogyne* spp. in Israel. *Plant Dis.Rep.* 40, 798-801.

Neal, J.C. (1899). Root-knot disease of peach, orange and other plants in Florida, due to the work of *Anguillula*. *U.S. Dep.Agric. Bur.Entomol.Bull.* 20, 1-31.

O'Bannon, J.H., Esser, R.P. and Inserra, R.N. (1975). Bibliography of nematodes of citrus. *ARS-S-68, U.S. Dep.Agric.* 41.

Orion, D. and Cohn, E. (1975). A resistant response of *Citrus* roots to the root-knot nematode *Meloidogyne javanica*. *Marcellia* 38, 327-328.

Scotto La Massese, C. (1969). *In* "Nematodes of tropical crops" (Ed. J.E. Peachey). *Commonw.Bur.Helminthol.Tch.Commun.* 40, 168-169.

Van Gundy, S.D., Thomason, I.J. and Rackham, R.L. (1959). The reaction of three *Citrus* spp. to three *Meloidogyne* spp. *Plant Dis. Rep.* 43, 970-971.

Whitehead, A.G. (1968). Taxonomy of *Meloidogyne* (Nematodea:Heteroderidae) with descriptions of four new species. *Trans.Zool.Soc. Lond.* 31, 263-401.

18

PATHOGENICITY VARIATIONS OF THREE
SPECIES OF ROOT-KNOT NEMATODES

D. Stoyanov

*Plant Protection Institute,
Kostinbrod, Bulgaria*

Introduction

Three species of root-knot nematodes, *Meloidogyne arenaria*, *M. hapla* and *M. incognita*, are widespread and of
economic importance in Bulgaria; *M. thamesi* and *M. javanica* are also present but have a limited geographical distribution. Crop rotation and the use of resistant cultivars are employed as methods of control in the field, and
chemical nematicides for soil disinfection are economical
only in glasshouses. In either situation, successful control depends on identification of the *Meloidogyne* species
and knowledge of their interrelations with the host
plants.

Recently, reliable criteria have been sought to identify species and to distinguish inter-species (Dalmasso
and Bergé, 1975). Variations in the perineal pattern are
taxonomically unreliable and attention has been directed
to the use of host ranges as a means of separating root-
knot nematode races. Dropkin (1959) used cultivars of
maize, oat and soya bean to distinguish races within four
Meloidogyne spp. Southards & Priest (1973) separated six
physiological races of *M. incognita* using six host
species. Ogbuji and Jansen (1972, 1974) established five
biotypes of *M. hapla* out of 14 populations examined, and
Phipps *et al* (1972) demonstrated that a race of

M. incognita was capable of infesting the resistant tobacco cv. NC95.

In Bulgaria, resistant tomato cultivars have been infested with root-knot nematodes, suggesting the existence
of races within *Meloidogyne* spp.   This prompted the investigations reported here.

Materials and methods

Cultures of *M. arenaria*, *M. hapla* and *M. incognita*
were established from single egg masses obtained from
several different populations originating in the field
or in the glasshouse. *M. arenaria* and *M. incognita* populations were from tomato, with the exception of one
population of *M. arenaria* from potato. Crops from which
M. hapla populations were obtained included tomato, pepper,
lavender and hop in the field, and tomato, pepper and
cucumber grown under glass. *M. arenaria* and *M. incognita*
were cultured on tomato cv. Triumph and *M. hapla* on
lettuce.

Populations of each of the three species were tested
by infecting seedling plants at the 3-leaf stage with 10
egg masses (1500 freshly hatched larvae). There were
five pots of each treatment. Experiments were done in a
glasshouse at 24-28°C; soil moisture in the pots was
maintained at 75% field capacity. After 50 or 65 days
samples of roots were stained with lactophenol and cotton
blue (Goodey, 1963) to ascertain the presence of larvae;
roots were visually examined for the presence of galling
and mature females with eggs; and the viability of eggs
was determined by incubating 5 egg masses from each infected root in tap water for 2 weeks. The degree of attack was then classified according to the extent of infection and galling. For *M. arenaria* and *M. incognita*
a 6-point scale ranged from (0) no larval penetration of
the roots, to (2) 1-9 galls on the roots with mature
females, and to (5) more than 40 galls on the roots. For

M. hapla a 4-point scale ranged from (0) no larvae (1) galls but no mature females, (2) 1-6 galls, (3) many mature females with egg masses.

Results and discussion

M. incognita populations were tested on six tomato cultivars and two grafts, considered to be resistant. Cv. Rossol was infected to a varying extent by five populations and was immune only to the populations from Pazardjik. *M. arenaria* populations were exposed to tomato cv. Rossol and *Arachis hypogaea*. Of the 11 populations tested four infected cv. Rossol and two infected *A. hypogaea*.

TABLE I

Infestation of test plants by nine
Bulgarian *Meloidogyne hapla* populations

Origin of *M. hapla* populations[1]

Test plant	Liubimets	Berkovitsa	Knezha	Samokov	Pazardjik	Lom	Kazanluk	Botevgrad	Blagoevgrad
Cucumis melo	0/1	0/1	0/1	1?	0/1	0/1	0/1	1	0/1
Hibiscus esculentum	0	0	0/1	2	1?	2	2	1?	2
Gossypium hirsutum	0	0	0	0/1	1?	0/1	0	2	0/1
Medicago sativa	2	2	2	2	2	2	2	2	2
Tagetes erecta	0	0	0	2	2	2	2	0	2
Zea mays	0	0	0	2	1?	0	0	0	2
Avena sativa	0	0	0	0	0	0	2	0	0
Lactuca sativa	2	2	2	2	2	2	2	2	2
Lycopersicon esculentum	2	2	2	2	2	2	2	2	2

[1]Numbers indicate degree of infestation (see text)

All of the nine populations of *M. hapla* developed successfully on tomato and lettuce, but varied in their degree of infestation of other test plants (Table I). Populations from Berkovitsa, Knezha and Liubimets infected lucerne only and are identified as a separate race (No. 1);

populations from Blagoevgrad and Samokov are separated as
another race (No. 2) on the basis of their infection of
H. esculentus, *M. sativa*, *T. erecta* and *Z. mays*; populations from Botevgard, Kazanluk, Lom and Pazardjik from
a further four races (3, 4, 5 & 6).

The observed differences in pathogenicity demonstrate
the feasibility of grouping root-knot nematode populations
into races (Dropkin, 1959; Ogubji and Jensen, 1974). They
also demonstrate the need to precede any work involving
resistant varieties by investigations on the race composition of the populations of root-knot nematodes being
used.

References

Dalmasso, A. and Bergé, J.B. (1975). Variabilité génétique chez
    Meloidogyne et plus particuliérement chez *M. hapla*. *Cah.ORSTOM*.
    ser. Biol. **10**, 233-238.
Dropkin, V.H. (1959). Varietal response of soybeans to *Meloidogyne*.
    A biossay system for separating races of root-knot nematodes.
    Phytopathology **49**, 18-23.
Goodey, J.B. (1963). Laboratory methods for work with plant and
    soil nematodes. *Tech.Bull.* M.A.A.F. **2**, 72.
Ogbuji, R.O. and Jensen, H.J. (1972). Pacific Northwest biotypes of
    M. hapla. *Plant Dis.Reptr.* **56**, 520-523.
Ogbuji, R.O. and Jensen, H.J. (1974). Two Pacific Northest biotypes
    of *M. hapla* reproduce on corn and oat. *Plant Dis.Reptr.* **57**, 128-129.
Phipps, P.M., Stripes, R.S. and Miller, L.J. (1972). A race of *M.
    incognita* from *Albizzia julibrissin* parasitizing *Nicotiana tab-
    acum* NC 95. *J.Nematol.* **4**, 32.
Southards, C.J. and Priest, M.F. (1973). Variation in pathogenicity
    of seventeen isolates of *M. incognita*. *J.Nematol.* **5**, 63-67.

19

A SUGGESTED MODEL TO VISUALIZE
VARIABILITY OF *MELOIDOGYNE* POPULATIONS

D.P. Taylor and C. Netscher

Laboratoire de Nematologie, O.R.S.T.O.M.
Dakar, Senegal

Introduction

It has been stated that "the *Meloidogyne*, or root-knot
nematode, problem is beyond doubt the most important nematode problem in tropical Africa" (Taylor, 1976). Thus,
considerable research is done on this group of pathogens.
A fundamental problem faced by the research worker is the
specific identification of field populations. The most
commonly identified species in this area are *M. incognita*,
M. javanica and *M. arenaria* - all characterized by being
polyphagous, polyploid and parthenogenetic species. Regarding the implication of this form of reproduction on
the definition of species, Triantaphyllou & Hussey (1973)
wrote, "Since there is no definite species concept for
obligatorily parthenogenetic organisms, the species concept applied to all these forms is, by necessity, subjective.". Further they wrote, "In reality, each one of these
parthenogenetic species consists of a large number of
field populations which proceed in evolution independently of each other, but at their present state of evolution, they share some common characteristics of taxonomic value.".

It is well recognised that within these parthenogenetic
species of *Meloidogyne*, populations exist that vary from
the "norm" in various characteristics (Netscher, 1978).

Perineal pattern configuration is the most widely used
morphological character to separate species in this genus,
and yet Baldwin (1976) has stated that "this character is
variable within species, and some of its features overlap
among the large number of species."

The subject of variability in host-parasite relations
in the genus *Meloidogyne* has recently been reviewed (Netscher, 1978). Although numerous examples could be cited,
it is sufficient to note that Southards and Priest (1971),
employing only six differential hosts, were able to dis-
cern six physiological races within *M. incognita* among
only seventeen isolated from a relatively limited area in
the United States.

Hackney (1977) reported successful separation of species
of *Meloidogyne* based upon chromosome number in a limited
area containing a mixed population of four species. Yet
Triantaphyllou (1976) has cited considerable variation in
chromosome numbers within these tropical species, i.e.
M. incognita, 2n = 36-44; *M. javanica*, 2n = 43-48; and
M. arenaria, 2n = 36 and 51-54. Thus, on the basis of
these figures an overlap occurs between *M. incognita* and
M. javanica (at 2n = 43 and 44) and between *M. incognita*
and *M. arenaria* (at 2n = 36). This overlap in chromosome
numbers and the generally low frequency of metaphase figures having countable chromosome configurations limits
the usefulness of this technique in identification of
species.

Nematologists and agronomists who regard the root-knot
problem from the practical or production point of view
thus are confronted with a baffling situation viz. in
light of the admitted variability of the characters mentioned, how can such species be easily identified with a
high degree of confidence? In an attempt to visualize
this situation a model has been developed which is a
schematic, three-dimensional drawing showing the relationships between certain hypothetical populations of

Meloidogyne.

The model

The horizontal axis represents time, starting at the left with a single "species", or ancestral progenitor, and progressing to the right where the present time is depicted in cross-section (Fig. 1). For simplification, only three major diversions are shown giving rise to four main branches. For the sake of argument, the three branches at the top represent the three commonly-occurring, polyploid, parthenogenetic, polyphagous, tropical species under discussion i.e. *M. incognita, M. javanica* and *M. arenaria.* It is intentional that these three taxa are illustrated as being physically close to each other. The fourth branch represents another species not as closely related e.g. *M. hapla.*

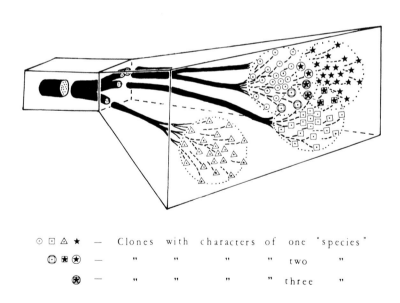

⊙ ⊡ △ ★ — Clones with characters of one ˙species˙

⊚ ✳ ✪ —   "   "   "   "   two   "

✆ —   "   "   "   "   three   "

Fig. 1. A model representing variability in hypothetical *Meloidogyne* populations (explanation in the text).

In the cross-section of the figure representing the present time, a large number of populations of the four "species" are represented by individual points at various co-ordinates. Their exact positions are determined by an evaluation of <u>all</u> their biological characteristics, e.g. morphology, physiology and cytology, in such a way that the more closely the characteristics of two populations coincide the closer these populations appear in the diagram. Thus, each "species" is represented by a cloud of points, compact towards the centre ("typical" populations), more diffuse towards the periphery ("abnormal" populations). It should be noted that an overlap of three of the "species" is shown in the diagram. This is also intentional, because a point in the area of overlap represents a population of one "species" which, in fact, behaves as another "species" depending upon criteria considered, e.g. a population of M. *arenaria* which does not attack peanut may be represented as a point situated in the cloud of M. *arenaria*, on the basis of morphological similarity, and also in that of M. *javanica*, on the basis of host specificity.

Although the complexity of this situation cannot be explained by any verified data at the present time, logic dictates that this variability is the result of genotypic variation within the "species". Such variation could result from the accumulation of numerous recurrent and/or non-recurrent mutations and chromosome additions or deletions giving rise to various degrees of aneuploidy. Triantaphyllou (1971) has suggested that additional chromosomes may have been incorporated into certain clones by occasional fertilization of unreduced eggs by reduced or unreduced sperm. However, it should be noted that in the thousands of eggs observed by various workers, no such fertilization has been reported. Triantaphyllou (1971) has also suggested that the polyploid forms "must have been derived at different occasions from a variety of ancestral types" which may have given rise to "the wide diversity of forms present in each parthenogenetic species".

Although these species reproduce by mitotic parthenogenesis, the possibility cannot be completely excluded that meiotic parthenogenesis has occurred occasionally which would have given great opportunity for genetic exchange. These genetic changes, regardless of their cause or causes, are represented by branchings in Fig. 1. If such changes had selective advantages, they may have become stabilized in a clone derived from the original individual possessing that change. If this hypothesis is accepted, it becomes obvious that what are called "species" are really mixed assemblages of clones which are interrelated to a greater or lesser extent dependent upon the degree of genotypic relationship.

From a taxonomic point of view, those clones located near the centre of a "species" cloud can be identified with little difficulty. However, what is to be done with a clone in an area of overlap? Is it a clone of M. *javanica* in the M. *arenaria* cloud, or is it M. *arenaria* in the M. *javanica* cloud? Sometimes on a morphological basis such a specific determination can be made; sometimes not!

It is our view that, to a large degree, it does not really matter. To the practical nematologist the fundamental importance of a clone of *Meloidogyne* is its biological activity (host range, for example) and not its nomenclatorial status. For example, assume that on the basis of morphology a clone is identified as M. *javanica*. However, from a practical agricultural viewpoint, what information does this convey? Is it a clone that attacks strawberry, or not? Does it attack peanut? Does it break resistance in tomato? Populations of M. *javanica* having each of these characteristics have been reported! Morphology cannot answer these questions! The clone in question must be tested against specific potential hosts.

It is for this reason that we have de-emphasised the importance of specific determinations in our work. We prefer to consider each *Meloidogyne* population as an entity which, after study, has been determined to possess certain biologically significant characters. We consider this approach to be the only practical method of understanding and combatting the *Meloidogyne* problem in tropical Africa.

References

Balwin, J.G. (1976). Morphology of *Meloidogyne* spp.-summary and prospects for further investigation. Proc. Research Planning Conf. 12-16 January 1976, North Carolina State Univ., Raleigh, 25-28.

Hackney, R.W. (1978). Identifications of field populations of *Meloidogyne* spp. by chromosome number. *J.Nematol.* **9**, 248-249.

Netscher, C. (1978). Morphological and physiological variability of species of *Meloidogyne* in West Africa and implications for their control. *Meded.LandbHogesch.Wageningen.* **78-3**, 46.

Southards, C.J. and Priest, M.F. (1971). Physiologic variations of seventeen isolates of *Meloidogyne incognita*. *J.Nematol.* **3**, 330.

Taylor, D.P. (1976). Plant nematology problems in tropical Africa. *Helminthol. Abstr. Ser. B* **45**, 269-284.

Triantaphyllou, A.C. (1971). *In* "Plant parasitic nematodes" (Eds. B.M. Zuckerman, W.F. Mai and R.A. Rohde) **II**, 1-34, Academic Press, New York.

Triantaphyllou, A.C. (1976). Cytogenetics of root-knot nematodes. Proc. Research Planning Conf. 12-16 January 1976, North Carolina State University, Raleigh, 39-42.

Triantaphyllou, A.C. and Hussey, R.S. (1973). Modern approaches in the study of relationships in the genus *Meloidogyne*. *OEPP/EPPO Bull.* **9**, 61-66.

W.A. Coolen

*Rijksstation voor Nematologie en
Entomologie, Merelbeke, Belgium*

Introduction

The methods described here are based on those of Coolen
and D'Herde (1972) and Gooris and D'Herde (1972) but with
slight modification to simplify them and to ensure uniformity for root and soil examination. Incubation and
maceration are the basic extraction methods for root-knot
and other endoparasitic nematodes.

Incubation

This method is based on the fact that when the nematodes
are kept in a moist and well aerated environment they
actively leave the plant material. Various methods described essentially are all modifications of the Baermann
funnel technique (Baermann, 1917; Young, 1954; Mountain
and Patrick, 1959; Seinhorst, 1950; Oostenbrink, 1960;
Chapman, 1957). Stemerding's (1964) mixer-cottonwool-
filter method combines incubation and previous crushing
of the roots.

With reference to the extraction of *Meloidogyne*, it
appears that the result is completely dependent on the
stage of development of the parasite in the roots. Only
second stage juveniles (J_2s) which are free in the root
tissue or which have hatched during incubation can be
collected.

Maceration

In this method plant tissue is comminuted in a mixer to
release the nematodes (Taylor and Loegering, 1953; Fenwick,
1963; Dropkin *et al.*, 1960; Zacheo and Lamberti, 1974).
It has the advantage of releasing both the swollen stages
and the eggs of *Meloidogyne*. However, if sieves are used
for separation, then many juveniles (J₂s) are lost. Caveness and Jensen (1955) developed a method in which no
sieves are used by modifying the centrifugal-flotation
technique of Faust *et al.* (1938). The suspension of soil
or macerated plant material in water is centrifuged and
the sediment is then resuspended in a concentrated sugar
solution and again centrifuged. The supernatant, containing the nematodes, is then poured into a relatively large
volume of water and after allowing sufficient time for the
nematodes to settle, the excess of water is decanted.

The method described in this paper specifically for
Meloidogyne extraction is a combination of maceration and
centrifugal-flotation. In developing it, the following
were considered:
- optimal maceration period for each stage of development
- type of roots
- modification of the centrifugal-flotation technique.
Further investigations have shown that this method is also
suitable for the extraction of the swollen juvenile stages
of *Heterodera* from roots and of migratory endoparasitic
nematodes from roots and other plant tissues.

The examination of soil samples for plant parasitic
nematodes can be carried out by different techniques,
such as direct microscopic observation of the soil, the
Baermann-funnel and its modification, or the centrifugal-
flotation technique.

A quantitative extraction method requires the recovery
of a high percentage of nematodes present in the soil and
it should also be reproducible and practicable. The
current techniques, summarized by Oostenbrink (1960,1970),

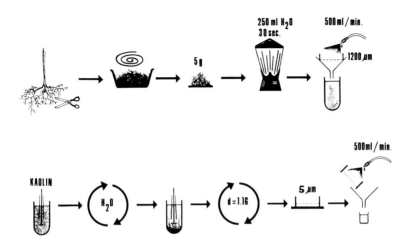

Fig. 1. Schematic representation of the method for the extraction of nematodes from plant tissue.

allow an almost quantitative extraction of active and free moving stages of nematodes from soil, but this is not the case with less active nematodes and eggs. The latter, however, are very important in soil sampling for *Meloidogyne* spp., which, during certain periods occur as eggs in the soil, mainly in egg masses attached to or embedded in root remains. The extraction technique described below answers the requirements, as both eggs and J_2-stages of *Meloidogyne* spp. and other nematode species and their eggs are extracted from soil quantitatively in an exact, reproducible and quick way.

Description of the method

Root processing (Fig. 1)

The root material is carefully washed and cut into pieces about 0.5 cm long, homogenized by thoroughly stirring in a large volume of water, and then collected on a sieve. Nematodes are then released from the root pieces

in a mixer (Waring-Blendor) which has two running speeds,
12,600 and 20,500 rpm unloaded. The 0.5 litre mixer jar
containing a 5g sample of root pieces is filled with 250
ml of water and mixed for 30 seconds at the lower speed.
The optimal combination for obtaining the maximum of all
developmental stages in one operation is based on a study
involving *Meloidogyne* and *Pratylenchus* and described by
Coolen and D'Herde (1972).

The nematodes are separated by pouring the suspension
through a 1200 μm sieve, placed on top of a 0.5 litre
centrifuge tube. The residue on the sieve is rigorously
washed with a fine, powerful fan-shaped water jet, produced by a low volume fog spray nozzle[1] (500 ml per minute).
The final residue on the 1200 μm sieve is discarded.

Before centrifugation, about 5 ml of kaolin powder are
added to the tube before thoroughly mixing the contents
of the tube by means of a mechanical stirrer[2], which is
carefully cleaned after each operation. Because kaolin
particles are small and flat (2-3 μm) they sink more
slowly than the nematodes, even though the kaolin specific
gravity of 2.6 is greater than that of the nematodes, with
the result that the kaolin spreads out like a tough skin
over the loose sediment and seals it off when decanting
occurs. When the sediment is resuspended in the separation
liquid, the kaolin layer must be thoroughly disintegrated
with the stirrer. An additional advantage of kaolin is
that it is also precipitated during the second centrifugation; this prevents the remixing of the sedimented deb-
ris during the decanting of the solution. In this way a
suspension of the nematodes in clear water is obtained.

In the first centrifugation[3] step the mixture is spun
for 4 minutes at about 1500 G when loaded. The supernatant

---

[1]spray nozzle Bray 00.
[2]Vibro mixer, laboratory model E1 with a vertical vibration, vibrating frequency 50 t/min, amplitude 0-3 mm.
[3]centrifuge U J III S

water is then poured off and the residue resuspended in the separation liquid, e.g. solutions of sucrose, $ZnSO_4$ or $MgSO_4$ with a density of 1.16, by means of the mechanical stirrer for at least 30 seconds and centrifuged for 4 min at 1500 G.

The nematodes are finally released from the suspension in a special way, because even with the most careful pouring of a suspension of eggs and J_2-stages through the finest sieve (5 μm sieve), an important fraction of the J_2-stages is carried through the meshes with the liquid or is caught by the sieve. In the latter case the J_2-stages constitute a possible source of contamination when the sieve is used for the following sample examination. Such a loss does not occur with the eggs or swollen stages because of their larger diameter. In order to prevent the loss of J_2-stages it is sufficient to avoid a direct flow through the meshes. This can be obtained by retaining the liquid in the sieve, until the J_2-stages are gathered on the surface of the fluid. When the liquid is allowed to flow gently through the sieve the nematodes descend slowly and settle on the meshes in a horizontal position, without any loss in numbers.

The procedure is as follows: the 5 μm sieve is stuck between two rings, an upper one sufficiently large to contain the 500 ml suspension, and a lower one with a smooth edge. When the sieve is placed on a smooth surface, e.g. a glass plate, an air cushion is formed under the sieve and no liquid can pass through. After some time, the sieve can be moved gently over the edge of the glass plate, so that the air escapes, followed by the liquid. A spray is used to wash the nematodes from the sieve into about 100 ml water. A more sophisticated version of this device is shown in Fig. 3.

Soil processing (Fig. 2)

A soil sample of about 1 litre is thoroughly but gently

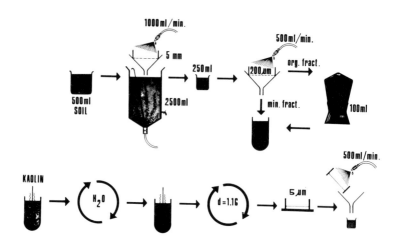

Fig. 2. Schematic representation of the method for the extraction of nematodes from soil.

mixed in a wide plastic bag. A 75 ml beaker is filled with the soil, firmly pressed into place (this is equivalent to 100 ml of soil of normal field structure). A quantity of soil equal to 5 times the weight of 75 ml of pressed soil is placed in a 1 litre plastic beaker and water added. After a short soaking period the soil is mixed with a stirring-vibrator and the suspension then poured onto a 0.5 cm sieve and washed through a funnel into the homo-genizer (3.5 litre capacity) using a jet[1] of water (1000 ml/min) until the apparatus contains 2.5 litre suspension (Figs. 2,4). The suspension is stirred for a few minutes by compressed air injected at the bottom of the apparatus. Finally 250 ml of the suspension is drawn off into a measuring beaker. This volume of suspension corresponds to 0.1 of the original 500 ml soil sample (i.e. 5 x 75 ml compressed soil samples = 500 ml field soil).

---

[1]Spray nozzle Bray 0

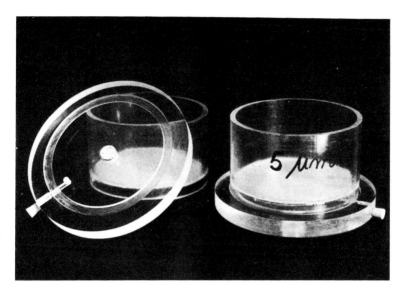

Fig. 3. Arrangement of the 5 μm sieve between two rings as used for the release of nematodes from the separation liquid.

The 250 ml sample is separated into an organic and a mineral fraction. To do this a 200 μm sieve is placed in a funnel on top of a 500 ml centrifuge tube. The suspension is washed through the sieve into the tube using a jet of water (500 ml/min), until the tube is filled to about 350 ml. Root remains and some larger mineral particles are retained by the sieve whereas almost the total mineral fraction enters the tube.

The organic fraction of the sample collected on the sieve is washed into a 250 ml beaker with a small amount of water (max. 100 ml) and mixed for 30 seconds at low speed in the Waring Blendor. The suspension is then transferred to the centrifuge tube which already contains the mineral fraction. Water is added up to 500 ml and centrifugation completed as for the root sample.

Longidorus and *Xiphinema* are extremely intolerant to osmotic pressure and collapse as soon as they come into contact with the separation liquids mentioned previously. In this case colloidal silica (Ludox[R], Du Pont) can be

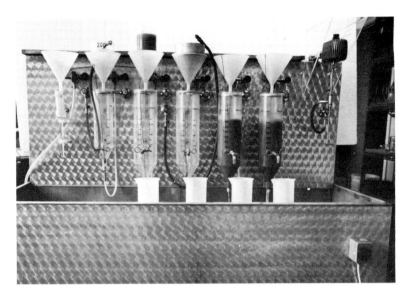

Fig. 4. General view of the arrangement of the homogenizers.

used. It has no osmotic effect, a low viscosity and is
inexpensive (Coolen and D'Herde, 1977).

Experiments indicate that more than 80% of the nematodes may be collected after a single centrifugal treatment, both for roots and soil (Coolen and D'Herde, 1972;
Gooris and D'Herde, 1972).

Description of equipment

The homogenizer is a transparent methylmethacrylate
(Perspex) jar of about 4 litres capacity (Fig. 4). The
cylindrical body is 360 mm height and 110 mm diameter with
a funnel-shaped base. A rubber tube with clamp leads from
the basal outlet. Compressed air is supplied to the homogenizer via a small tube inserted through the rubber
stopper in the base of the funnel. Within the homogenizer
the neck of the funnel can be closed by means of a rubber
stopper attached to a metal rod. The separate pieces comprising the homogenizer can be glued together with a poly-
merizing liquid (Acrifix 92-liquid) which hardens in 3-6

hr in sunlight or UV light.

The 5 μm sieve consists of a sieve base placed between
two rings. The rings are made from a methylmethacrylate
tube with an inner diameter of 110 mm, the upper ring
being 60 mm high and the lower one 10 mm. The sieve mesh
is of woven nylon thread, commercially obtainable as
"Monodur". To make the sieve a piece of the nylon mesh
is tightened over a glass plate, while some chloroform is
poured onto another piece of glass, in which the upper
ring is rubbed with a turning motion for about a minute
until the edge is soft and sticky. The ring is then immediately placed on the mesh and pressed down with a weight
of 5-10 kg. After about 15 min the sticky surface is sufficiently hardened and the excess mesh is cut away. The
lower ring is treated in the same way and pressed against
the mesh with the same weight. When making sieves with
larger meshes, such as the 1200 μm sieve, it is not sufficient to soak the edges of the rings in chloroform and
instead they should be coated with a pasty solution of
methylmethacrylate in chloroform.

Advantages of the method

It is simple, using standard and relatively inexpensive
apparatus such as a centrifuge, a mixer and a vibrating
stirrer. The homogenizing jar, the sieves and the spray
are easily made from materials which readily are obtainable commercially. Also, the apparatus is easy to clean,
because of the transparency or the mat white colour of
the materials.

It is rapid. The total time needed to extract the nematode from four root or soil samples, is less than 1 hr,
or about 15 min per sample.

It is reliable because it gives accurate and reproducible results. All stages, including the immobile ones,
of *Meloidogyne* are extracted from the roots, hence quantitative assessments can be made at each stage of

TABLE I

Numbers of nematodes extracted from roots with the mixer-centrifuge
method compared with three techniques currently used

Methods	Mean number of nematodes per 5 g of root					
	M. hapla in rose Eggs + J_2s		*P. penetrans* in *Acer* mobile stages		*Ditylenchus dipsaci* in fodder beet mobile stages	
	$\overline{X}$	C	$\overline{X}$	C	$\overline{X}$	C
Mixer-centrifuge	21700	11	12160[1]	10	5175[2]	8
Funnel spray	1300	25	3950	15	1510	8
Mixer-cottonwool filter	2500	20	3890	14	–	–

$\overline{X}$ = mean no. nematodes; C = Coefficient of variation

[1] in addition 8210 eggs

[2] in addition 4770 eggs

TABLE II

Numbers of nematodes extracted from soil with the mixer-centrifuge
method compared with other techniques

Nematode species	Mixer-centrifuge		Elutriation cottonwool-filter		Elutriation sieve	
	$\overline{X}$	C	$\overline{X}$	C	$\overline{X}$	C
Meloidogyne naasi[1]	25530	8	2771	6	–	–
Pratylenchus penetrans[1]	845	8	111	15	–	–
Rotylenchus robustus[2]	1583	11	625	23	–	–
Trichodorus primitivus[2]	250	19	71	36	–	–
Xiph. diversicaudatum[2,3]	277	8	15	65	129	8
Long. caespiticola[2,3]	30	15	1	46	13	17

$\overline{X}$ = mean no. nematodes; C = Coefficient of variation

[1] mineral and organic fraction

[2] mineral fraction only

[3] the separation liquid was colloidal silica

development. Eggs as well as nematodes can be extracted
from soil samples.

The mixer-centrifuge method has been compared, both for plant parts and roots and for several types of nematodes, with the following methods: the funnel spray (Oostenbrink, 1960) and the mixer-cottonwool filter (Stemerding, 1964) for roots; the elutriation-cottonwool filter (Oostenbrink, 1954; De Maeseneer and D'Herde, 1963), and the elutriation-sieve method (D'Herde and Van Den Brande, 1964) for soil. The results are summarized in Tables I and II.

From these comparisons it appears that the mixer-centrifuge method for roots and soil always gives a higher yield and a lower coefficient of variation. If these data are accepted as a measure of the accuracy and reproducibility of a method, then it may be concluded that the mixer-centrifuge method is more reliable than the others.

Applications of the method

The method provides an accurate and quantitative extraction of nematodes, especially of *Meloidogyne* spp., from soil. It can be applied at any time of the year, even when eggs occur in egg-masses that are still attached to, or embedded in, root remains in the soil. The separate extraction from mineral and organic fractions allows an accurate estimation of the eggs and J_2-stages in the soil as a proportion of the total population. The estimation of the egg and J_2-stage ratio in *Meloidogyne* spp. may be of help in studying the biology of the species, e.g. diapause and ecological influences on hatching.

The fact that endoparasitic nematodes can be collected in a pure suspension in large numbers, and in all stages of development, means that this method can be useful in specialized nematological work, e.g. systematics, cytogenetics and biochemical research. Also, because of the great accuracy and reproducibility of the method, it is useful in relating the development of an endoparasitic nematode to the influence of ecological factors such as temperature, light, host plants, chemicals.

The method has also been usefully applied for accurate
and speedy diagnosis in advisory work and phytosanitary
inspections.

References

Baermann, G. (1971). Eine einfache methode zur Auffindung von anchy-
lostomum-(nematoden)-larven in erdproben. *Geneesk.Tijdschr.Ned.Ind.*
57, 131-137.
Caveness, R.F. & Jensen, H.J. (1955). Modification of the centrifugal-
flotation technique for isolation and concentration of nematodes
and their eggs from soil and plant tissue. *Proc.Helminthol.Soc.
Wash.*, 22, 87-89.
Chapman, R.A. (1957). The effects of aeration and temperature on the
emergence of species of *Pratylenchus* from roots. *Plant Dis.Rep.*,
41, 836-841.
Coolen, W.A. & D'Herde, C.J. (1972). A method for the quantitative
extraction of nematodes from plant tissue. *Publication of the State
Nematology and Entomology Research Station, Merelbeke, Belgium.* 77.
Coolen, W.A. & D'Herde, C.J. (1977). Extraction de *Longidorus* et
Xiphinema spp. du sol par centrifugation en utilisant du silice
colloïdal. *Nematol.medit.*, 5, 195-206.
De Maeseneer, J. & D'Herde, C.J. (1963). Methodes utilisées pour
l'étude des anguillules libres du sol. *Revue Agric.*, 3, 441-447.
D'Herde, C.J. & Van Den Brande, J. (1964). Distribution of *Xiphinema*
and *Longidorus* spp. in strawberry fields in Belgium and a method
for their quantitative extraction. *Nematologica* 10, 454-458.
Faust, E.C., d'Antoni, J.S., Odom, V., *et al.* (1938). A critical
study of clinical laboratory techniques for the diagnosis of
protozoan cysts and helminth eggs in feces. *Amer.J.Trop.Med.*,
18, 169.
Fenwick, D.W. (1963). Recovery of *Rhadinaphelenchus cocophilus* from
coconut tissues. *J.Helminthol.*, 37, 11-14.
Gooris, J. & D'Herde, C.J. (1972). A method for the quantitative extraction of eggs and second stage juveniles of *Meloidogyne* spp.
from soil. *Publication of the State Nematology and Entomology Research Station, Merelbeke, Belgium* 36.
Gooris, J. & D'Herde, C.J. (1977). Study of the biology of *Meloidogyne
naasi* Franklin 1965. *Publication of the State Nematology and Entomology Research Station, Merelbeke, Belgium* 115.
Mountain, W.A. & Patrick, Z.A. (1959). The peach replant problem in
Ontario. *Can.J.Bot.*, 37, 459-470.
Oostenbrink, M. (1954). Een doelmatige methode voor het toesten von
aaltjesbestrijdingsmeddelen in grond met *Hoplolaimus uniformis* als
proefdier. *Meded.Lanbdouwhogesch.Opzoekingstat.Gent* 19, 377-408.
Oostenbrink, M. (1960). *In* "Nematology: Fundamentals and Recent Ad-
vances." (Eds. J.N. Sasser, W.R. Jenkins.) 85-102. Univ. of North
Carolina Press, Chapel Hill.
Oostenbrink, M. (1970). *In* "Methods of Study in Soil Ecology." (Ed.
J. Philipson.) 249-255. Proc. Paris Symp. UNESCO.
Seinhorst, J.W. (1950). De betekenis van de toestand van de grond voor
het optreden van een aantasting door het stengelaaltje *Ditylenchus*

dipsaci (Kuehn) Filipjev. *Tijdschr.Plantenziekten* **56**, 291-349.
Stemerding, S. (1964). Een mixer-wattenfilter methode om vrijbeweeg-
lijke endoparasitaire nematoden uit wortels te verzamelen. *Versl.
Meded.plziektenk.Dienst Wageningen (Jaarboek 1963)*, 170-175.
Taylor, A.L. and Loegering, W.Q.· (1953). Nematodes associated with
root lesions in Acaba. *Turrialba* **3**, 8-13.
Young, T.W. (1954). An incubation method for collecting migratory
endo-parasitic nematodes. *Plant Dis.Rep.*, **38**, 794-795.
Zacheo, G. and Lamberti, F. (1974). Un metodo rapido per l'estrazione
di uova di nematodi dal suolo e dai tessuti vegetali. *Nematol.medit*
2, 55-59.

Discussion

Netscher suggested that it was essential to use a
'Waring Blendor', and not any other make, if the method
was to be successfully followed. Coolen said he could give
no absolute figure for the number of nematodes that might
be killed by blending, but investigations had been made
on blending times and speeds to give an optimal number of
nematodes recovered, which corresponded with the smallest
proportion of nematodes killed. Asked whether it was
necessary to modify the sugar solution for different nematodes, Coolen replied that he did not do this but that
some species were sensitive to osmotic pressure, e.g. *Xiphinema, Longidorus*. For these nematodes "Ludox" (colloidal silica) could replace sucrose. In the method described,
Meloidogyne larvae were found to be active and infective
after centrifugation and this also applied to many other
species.

Coolen explained that four root samples could be processed in 40 min, or four soil samples in 1 hr.

Mary T. Franklin

*Commonwealth Institute of Helminthology,
St. Albans, England*

Introduction

Temperate climates are taken to include Europe from the
northern shores of the Mediterranean northwards, central
and northern parts of the USSR, northern USA and Canada,
southern South America, Japan, the southern parts of Australia, New Zealand and isolated areas at high altitudes
in tropical or sub-tropical zones, where crops may be
grown that belong essentially to temperate climates. In
the cooler parts of these areas crops of high market value are often produced in glasshouses where species of *Meloidogyne* that could not survive out of doors find ideal
conditions for multiplication. In glasshouses in northern
climates, tomatoes, cucumbers, lettuce and many ornamental
plants are often heavily attacked by *M. incognita* (Kofoid
& White) Chitwood, *M. javanica* (Treub.) Chitwood, and *M.
arenaria* (Neal) Chitwood. Out-of-doors in northern regions
the most important and widespread species are *M. hapla*
Chitwood and *M. naasi* Franklin, with *M. artiellia* Franklin
occurring sporadically. In the more southern and sheltered
areas, *M. incognita* and *M. arenaria* are present out of
doors and occasionally also *M. javanica*. Other cold cli
mate species not yet known to be important pests of crop
plants are *M. mali* Ito, Ohshima & Ichinohe, recorded from
Malus prunifolia L; *M. ovalis* Riffle on species of *Acer*,

Betula, Ulmus and *Fraxinus; M. carolinensis* Fox on *Vacci-
nium* and *Rhododendron; M. litoralis* Elmiligy on *Ligustrum;
M. deconincki* Elmiligy on *Fraxinus, Rosa* and *Solanam nig-
rum* L.; *M. graminis* (Sledge & Golden) Whitehead on *Ammo-
phila arenaria* (L.) Link, cereals and grasses; *M. arden-
sis* Santos on *Vinca, Ligustrum, Fraxinus* and *Sambucus;
M. ottersoni* (Thorne) Franklin on *Phalaris* and *Hordeum;* and
M. microtyla Mulvey, Townshend & Potter on *Festuca rubra*
L., *Phleum pratense* L., cereals and *Trifolium repens* L.

Over such a wide and diverse area as that under consideration the crops affected are necessarily varied in
kind and in relative value and economic losses similarly
depend not only on the severity of the nematode attack
but also on the cash value of the crop.

Geographical distribution

A review of some examples of root-knot damage to crops
in temperate regions follows, starting in the north of
the temperate regions and moving southwards. In Iceland
(Siggeirsson & van Riel, 1975) root-knot damage to crops
such as tomatoes and cucumbers growing in glasshouses,
sometimes heated from natural sources, can be severe. *M.
incognita, M. arenaria* and *M. javanica* have been recorded.
Occasionally infestations may be present in warm soils
around hot springs. These nematodes must have been introduced in the past and have found ideal conditions for
multiplication. Although a good degree of nematode control in glasshouses is relatively easy, complete eradication is often extremely difficult due to the great
depths to which some nematodes penetrate. The small numbers of individuals that escape nematicidal treatments
soon multiply in the favourable sheltered conditions and
are a continuing problem.

Meloidogyne hapla

In Sweden (Anderson, 1970) and Denmark (Lindhart &

Bagger, 1967) severe attacks of *M. hapla* have been recorded on field-grown carrots. If young seedlings are
heavily infested they may be killed, while lighter infestations can result in losses because the unsightly appearance of the galled roots makes them unmarketable. Damage
to carrots by *M. hapla* has been reported from Germany
(Decker, 1961), Poland (Berbeć, 1972), the Low Countries,
France (Ritter, 1972) and Britain. In England a field
infestation on carrots, parsnips and some weed species
was recorded by Triffitt as long ago as 1931. Sugar beet
is another crop to suffer from root-knot disease, 20%
losses having been reported in Yugoslavia (Grujicić &
Paunović, 1971) and heavy infestations in France (Christmann, 1971) and in Japan (Ichinohe, 1955); potatoes have
also been badly affected in Japan. *M. hapla* is probably
indigenous in north-western Europe where it also occurs
on other field crops, though often without causing not-
iceable damage, and on weeds which serve as reserve hosts.
Among such crops are lettuce, celery, brassicas, legumes
such as lucerne (*Medicago sativa* (L.)), .sainfoin and
clovers, ornamental plants of many species both perennial
shrubs and annuals, fruit trees and strawberry. *M. hapla*
was recorded in 26% of strawberry plantations in a survey
of Poland (Szczygiel, 1974). Damage to *Trifolium repens*
in New Zealand is associated with combined infestations
of *M. hapla* and *Heterodera trifolii* Goffart (Healy *et
al.*, 1973). One of the more unusual instances of damage
due to *M. hapla* occurs in Kenya where *Pyrethrum*, growing
close to the equator at an altitude of 2,000 m, has suffered losses of 50% in flower yield and a decrease in
pyrethrin content (Parlevliet, 1971).

Although data on the economic losses due to *M. hapla*
in practice are scarce and not necessarily relevant out
of their own context, the seriousness with which this
nematode is regarded can be judged from the amount of
research that has been carried out, particularly in

Canada, into the losses caused by infestations on various
vegetables when grown either as summer or autumn crops.
Studies have been made to determine the infestation levels
causing yield reductions in such plants as cabbage, potato
onion, lettuce and clovers (Olthof & Potter, 1972). Other
investigations are reported from Belgium on resistance to
M. hapla in rose rootstocks cultivated under glass (Cool-
en & Hendrickx, 1972), and in France populations of *M. hap
la* from 9 localities were tested on 6 grapevine (Dalmasso
& Cuany, 1976) rootstocks. Such investigations indicate
the importance of the nematode on various cultures of
plants as well as on food crops.

Meloidogyne naasi

The other species of *Meloidogyne* which is widespread in
field crops in the northern temperate regions is *M. naasi*.
This species has been reported causing damage to cereals
and sugar beet in USA, Germany (Sturhan, 1973), Holland,
Belgium, Britain and northwestern France where 50 infested
sites are known (Caubel *et al.*, 1971). The most frequently
recorded host is barley but losses or patchiness in crops
of wheat and oats and occasional crop failure of seedling
sugar beet and onions have been noted in Belgium (Gooris
& D'Herde, 1969).

In Wales damage to barley by *M. naasi* is often associated with wet, poorly drained or compacted soil. Information supplied by P.A. York of the Welsh Plant Breeding
Station shows a marked increase during the past 30 years
in the proportion of host to non-host crops being grown
in Wales: barley has largely replaced oats, and the host
plant ryegrass is grown in preference to several non-host
crops. Experiments in Wales have shown that infestation
of perennial ryegrass with *M. naasi* adversely affects
seedling establishment and vigour and reduced yields. In
field and pot experiments infestations of 9 or 10 larvae
per g soil have resulted in loss of yield from barley cv.

Sabarlis. As measured by a gall index arrived at from the
number and size of galls, wheat is more susceptible than
barley and barley than oats but severity of galling is
variable within cereal species (see Welsh Plant Breeding
Sta., Report for 1976, p. 71-72).

M. *naasi* is apparently indigenous in both northwest
Europe and northwestern USA. In Europe it has been found
in Holland in new polders, having apparently spread in
from naturally infected reeds (*Phragmites communis* Trin.)
growing nearby (Kuiper, 1977). The reed beds are frequented
by large numbers of ducks and geese and it is likely that
they may be instrumental in transporting the nematodes.
In Yugoslavia *M. naasi* has been recorded on wheat, barley,
beet and mangolds (Grujičić, 1967) and it is present in
Italy, Malta and Gozo (Lamberti, 1976) and in other European countries. In the USA barley has been severly affected
in Oregon (Jensen *et al.*, 1968), particularly in low-lying,
poorly drained areas and here it occurs also on oats and
grasses. The same is true in Illinois (Golden & Taylor,
1967) where sugar beet is also infested. In northern
California, in the Tulelake area (Allen *et al.*, 1970),
barley losses of from 50 to 75% have been recorded. It
appears to be a species that thrives best in somewhat anaerobic soil. In southern Chile (Kilpatrick *et al.*, 1976)
wheat has been found severly damaged by *M. naasi* with a
recorded loss of more than 70%. Oats were also affected,
but less severely.

Meloidogyne artiellia

Another species apparently widely scattered from north
to south of the European temperate zone is *M. artiellia*.
First described from eastern England it has since been
recorded in France, Spain and Greece. This is a polyphagous
species that has been found feeding on brassicas, clovers
and cereals but not in numbers sufficient to cause appre-
ciable damage. In Spain, in non-irrigated areas around

Granada, it is widespread on wheat and barley and to a
lesser extent on oats, chickpea (*Cicer arietinum* L.) and
vetch (Tobar Jimenez, 1973).

Meloidogyne spp.

In warmer parts of the temperate zones, along the north
ern shores of the Mediterranean, the southern parts of
Australia and parts of the USSR, *M. incognita*, *M. arenaria*
and *M. javanica* are the prevalent species. Because of thei
wide host range these species are liable to cause losses
wherever there is intensive cultivation of susceptible
crops and precautions against population build up are not
taken. *M. incognita* is the most widespread of the three
species in the warmer temperate regions and also in glasshouses further north where it would not survive out-of-
doors. It is found in parts of France, in Italy, Spain,
Portugal, Yugoslavia, Turkey, Greece and on the Black Sea
coast. In many of these countries perennial crops such as
grapevine, peach and other fruit trees are infested. In
southern France and Italy *M. arenaria* is often commonly
encountered and is a particularly serious pest in flower
nurseries in the Mediterranean coastal areas (Ritter,
1972). This species has also appeared in glasshouses in
a new polder in Holland (Kuiper, 1977). *M. javanica* tends
to occur in the warmer parts of the Mediterranean region.
In Australia, peach is attacked by *M. javanica* in New
South Wales and grapevines are infested by several *Meloidogyne* species in South Australia. Unnamed species have
been reported on many vegetable crops, on strawberries
and on grapevines from all the Australian States.

Economic losses

Compared to the situation in the tropics, damage and
crop losses due to root-knot nematodes in cool temperate
climates are insignificant, but very real to the farmer.
In isolated instances field crops may be completely lost

and, where conditions favour the nematode such as on sandy
soils or in glasshouses, serious economic losses can and
do occur. Losses may be direct or secondary. Direct finan-
cial losses may be due to decreased yields, reduced stands
or unmarketable produce, as when root vegetables such as
carrots or potatoes are disfigured by galls. Indirect
losses may result from the rejection of plants for export
due to light infestation or contamination, or from increased production costs incurred by the need to use costly
control measures, to grow less productive resistant crops
on infested land or, where appropriate, to use nematicides.
It is not possible to estimate losses in terms of cash
because market values are too variable, but their significance is demonstrated by the amount of research devoted
to overcoming or forestalling them.

References

Allen, M.W., Hart, W.H. & Baghott, K. (1970). Crop rotation controls
barley root-knot nematode at Tulelake. *Calif.Agric.*, 24, 4-5.
Andersson, S. (1970). Rotgallnematoder och deras förekomst i skånska
odlingar. *Vaxtskyddsnotiser* 34, 22-28.
Berbeć, E. (1972). Investigations on the appearance and damage caused
by northern root-knot nematode *Meloidogyne hapla* Chitwood on
carrots. *Prace Wydzialu Nauk Przyrodniczych Bydgoskiego Towarzystwa
Naukowego Ser. B.* 15, 3-32.
Caubel, G., Ritter, M. & Rivoal, R. (1971). Observations relatives à
des attaques du nématode *Meloidogyne naasi* Franklin sur céréales
et graminées fourragères dans L'Ouest de la France en 1970. *C.r.
hebd.Séanc.Acad.Agric.Fr.*, 57, 351-256.
Christmann, J.C. (1971). *In* "Les nématodes des cultures." 257-272.
Paris, France, ACTA.
Coolen, W.A. & Hendrickx, G.J. (1972). Investigations on the resistance
of rose root-stocks to *Meloidogyne hapla* and *Pratylenchus penetrans*.
Nematologica 18, 155-158.
Dalmasso, A. & Cuany, A. (1976). Résistance des porte-greffes de vignes
à différentes populations du nématode *Meloidogyne hapla*. *Progrès
agric.vitic.*, 93, 800-807.
Decker, H. (1961). Der wurzelgallennematode *Meloidogyne hapla* Chitwood
and sein freilandauftreten im Norden der DDR. *Wiss.Z.Univ.Rostock.*
10, 57-70.
Golden, A.M. & Taylor, D.P. (1967). The barley root-knot nematode in
Illinois. *Pl.Dis.Reptr,* 51, 974-975.
Gooris, J. & D'Herde, C.J. (1969). De l'infestation et des dégats
causés par le nematode des racines de graminées *Meloidogyne naasi*
Franklin sur quelques cultures. *Revue Agric.Brux.*, 22, 581-590.

Grujičić, G. (1967). Korenova nematoda (*Meloidogyne naasi* Franklin) u Sribji. *Zast.Bilja.*, 18, 193-197.

Grujičić, G. & Paunović, M. (1971). A contribution to the study of the root-knot nematode (*Meloidogyne hapla* Chitwood). *Zaštita Bilja.*, 22, 147-152.

Healy, W.B., Widdowson, J.P. & Yeates, G.W. (1973). Effect of root nematodes on growth of seedling and established white clover on a yellow-brown loam. *N.Z.Jl agric.Res.*, 16, 70-76.

Ichinohe, M. (1955). Two species of the root-knot nematodes in Japan. *Oyo-Dobutsugaku-Zasshi.*, 20, 75-82.

Jensen, H.J., Hopper, W.E.R. & Loring, L.B. (1968). Barley root-knot nematode discovered in western Oregon. *Pl.Dis.Reptr*, 52, 169

Kilpatrick, R.A., Gilchrist, L. & Golden, A.M. (1976). Root-knot on wheat in Chile. *Pl.Dis.Reptr*, 60, 135.

Kuiper, K. (1977). Introductie en vestiging van planteparasitaire aaltjes in nieuwe polders, in het bijzonder van *Trichodorus teres*. *Meded.LandbHoogesch.Wageningen* 77, 140.

Lamberti, F. (1976). Laboratorio di nematologia agraria applicata ai vegetali, Bari. Attività scientifica svolta nel 1975. *Ricerca scient.*, 46, 500-512.

Lindhart, K. & Bagger, O. (1967). Plantesygdomme i Danmark 1966. *Tidsskr.PlAvl.*, 71, 285-337.

Olthof, T.H.A. & Potter, J.W. (1972). Relationship between population densities of *Meloidogyne hapla* and crop losses in summer-maturing vegetables in Ontario. *Phytopathology* 62, 981-986.

Parlevliet, J.E. (1971). Root-knot nematodes, their influence on the yield components of pyrethrum and their control. *Acta.Hort.*, 21, 201-205.

Ritter, M. (1972). Rôle économique et importance des *Meloidogyne* en Europe et dans le bassin Méditerranéen. *Bull.OEPP.*, 6, 17-22.

Siggeirsson, E.I. & Riel, H.R. van (1975). Plant-parasitic nematodes in Iceland. *Rannsóknastofnunin Nedri Ás, Hveragerdi, Island Skýrsla* 20, 32.

Sturhan, D. (1973). *Meloidogyne naasi* - ein fur Duetschland neuer Getreideparasit. *NachrBl.dt.PflSchutzdienst., Braunschweig.*, 25, 102-103.

Szczygiel, A. (1974). Plant parasitic nematodes associated with straw berry plantations in Poland. *Zesz,Problem.Postep.Bauk Roln.*, 154, 9-132.

Tobar, Jimenez, A. (1973). Nematodes de los "secanos" de la comarca de Alhama. I Niveles de población y cultivos hospedadores. *Revta iber.Parasit.*, 33, 525-556.

Discussion

Asked why it took so long for *M. naasi* to be identified as a problem, Mary Franklin explained that it was first found in Wales on rye grass growing near a glasshouse, from which it was thought to have escaped. A watch was kept for symptoms of infestation on rye grass and barley, and it was found repeatedly, particularly more recently

with the increase in barley production. Lamberti observed
that galls produced by *M. naasi* are extremely small and
require microscopic examination, suggesting that it could
be overlooked.

Ferris stated that in the USA the principal host of
M. naasi is wheat and occasionally it is found on oats.
Coolen suggested that this may be partly an effect of environmental temperature and explained that in experiments
undertaken in Belgium, oats were resistant to infestation
by *M. naasi* at 15°C but were infested at 25°C.

Netscher commented that *M. naasi* appeared to consist of
several different races as a study with a number of populations from various regions of the world showed that
they varied in their ability to reproduce on sorghum
(*Sorghum bicolor*) and creeping bentgrass (*Agrostis palu-
tris*) (Mitchell, R.E. *et al.*, *J. Nematol.*, 5, 41 (1973)).

F. Lamberti

*Laboratorio di Nematologia Agraria
del C.N.R., Bari, Italy*

Introduction

Seven species of *Meloidogyne* Goeldi hav been found in association with crops of agricultural importance in regions
with subtropical and mediterranean climate. They are, in
order of importance:

> *M. incognita* (Kofoid and White) Chitwood
>
> *M. javanica* (Treub) Chitwood
>
> *M. arenaria* (Neal) Chitwood
>
> *M. hapla* Chitwood
>
> *M. naasi* Franklin
>
> *M. graminis* (Sledge and Golden) Whitehead
>
> *M. artiellia* Franklin

Only records from the last 20-25 years are referred to
subsequently as prior to Chitwood's revision of the genus
(1949) all the nematodes causing root-knot were reported
as *Anguillula radicicola* Müller or *A. marioni* Cornu, or
Heterodera marioni (Cornu) Marcinowski.

Root-knot species

Meloidogyne incognita

This is undoubtedly the commonest and most widespread
root-knot nematode causing considerable losses of many
crops wherever the soil structure (light and sandy soil)

is favourable to its attack, outdoor or under glass. Many
vegetable crops suffer the greatest damage; they are included in the botanical families Solanaceae, Cucurbitaceae,
Leguminosae, Liliaceae, Chenopodiaceae, Compositae, Umbel-
liferae, Cruciferae, Hypericaceae and Malvaceae. *M. incognita* is also highly pathogenic to many other industrial
ornamental and fruit crops.

Field experiments done in southern Italy (Lamberti and
Cirulli, 1970; Lamberti, 1971 and 1975) have demonstrated
that severe attacks of *M. incognita* on canning tomato
(*Lycopersicon esculentum* Mill.) may cause, under the climatic conditions of a mediterranean summer, yield losses
of about 50%. The same figure can be considered valid for
table tomato in autumn-winter production in unheated glasshouses in Malta (Lamberti *et al.*, 1976), or in Algeria
where the nematode is a major pest of this crop around
Alger (Lamberti *et al.*, 1975).

These data have been confirmed in outdoor experiments
undertaken in North Carolina (Barker *et al.*, 1976) which
showed that *M. incognita* can suppress yields of tomato by
up to 85% in coastal plains and 20-30% in mountain cultivations. Apparent differences in pathogenicity were considered to be due possibly to the different climatic conditions and this hypothesis is corroborated by the results
of two experiments done in unheated glasshouses in Malta.
In the first, tomato cv. Fountain Cross planted in October
1974 was cropped until June 1975, after a warm autumn and
a mild winter; plants severely attacked by *M. incognita*
yielded 56% less than plants only moderately galled (Lamberti *et al.*, 1976). The second experiment was conducted
in the same glasshouse in soil containing an initial
population of the nematode similar to the first experiment.
The plants were transplanted in mid-October 1975 with the
mean soil temperature at 14-16°C during the 2½ months following transplanting and 12°C in January to March 1976.
At the end of the experiment examination of the roots of

TABLE I

Effect of soil type on root invasion by *Meloidogyne incognita*

Type of soil			Number of juveniles/root system Days after inoculation			
			7	10	15	18
Sandy	Sand Silt Clay pH	85.6% 10.2% 4.2% 8.6	3.4	7.8	10.2	10.6
Volcanic	Sand Silt Clay pH	71.5% 11.2% 16.3% 7.0	0.9	0.9	0.9	3.6
Sandy with gravel	Sand Silt Clay pH	84.7% 8.1% 7.2% 7.5	0.2	0.7	2.2	8.0
Sandy loam	Sand Silt Clay pH	62.5% 28.5% 9.0% 8.0	0	0.2	1.4	1.2

the plants showed only a light attack overall by the nematode. Nevertheless the attack caused yield losses of 34-
37% of marketable fruits (Dandria *et al.*, 1977).

Soil structure has an important influence on the host-
parasite relationship involving *M. incognita*. Invasion of
tomato roots by this nematode is greatly affected by the
type of soil in which the plants are grown. (Table I;
Lamberti, unpublished).

The age of the plant when nematode attack is initiated
influences the extent of root galling and the rate of
reproduction of *M. incognita*. Tomato seedlings of different ages, artificially inoculated with *M. incognita* juveniles at sowing or at intervals up to 60 days were examined for root galling at either 30 or 60 days after inoculation; the greatest amount of galling was seen in 30
day old plants inoculated at sowing (Volvas, unpublished;

TABLE II

Degree of galling and rate of reproduction of M. *incognita*
in tomato plants of different ages

Age of plant at inoculation	No. galls/g root		No.♀♀/g root 30 days after inoculation
	30 days after inoculation	60 days after inoculation	
0	52.5	7.4	47.3
3	17.2	6.5	11.5
8	17.8	42.3	10.9
18	12.5	6.0	9.3
24	9.8	1.2	0.7
30	9.9	2.2	15.4
40	5.3	1.9	3.4
60	1.8	0.5	0.4

Table II).

All these factors combined might be important in the
field when a delayed attack by the nematode will allow the
plant to develop initially in the absence of the nematode
or, more generally, in the establishment of a population
of the nematode in a particular environment (Sasser, 1977)
In fact, in southern Italy it has been observed that M.
incognita very seldom starts to invade root tissues when
the soil temperature is below 18°C (Lamberti *et al.*, unpublished). However, in assessing crop losses caused by
M. *incognita* it should be remembered that there are pathotypes (Riggs and Winstead, 1959) or strains of the nematode with different degrees of pathogenicity (Di Vito and
Lamberti, 1976; Table III). The behavioural diversity of
populations of M. *incognita* and the importance of the environmental condition on the host-parasite relationships
moreover are demonstrated by tests on differential hosts
(Greco and Di Vito, unpublished; Table IV) and by the different threshold densities reported by various authors
(Barker and Olthof, 1976).

The eggplant (*Solanum melongena* L.) is another solan-

TABLE III

Degree of galling on the roots of tomato plants cv. Roma VF,
grown in soil infested with different
Italian populations of *M. incognita*

Origin of population	Degree of galling (scale 0-5)	Weight of plant tops (g)
Control	0.0	16.9
Altomonte, Cosenza (Tomato)	1.1	22.9
Monopoli, Bari (Lettuce)	1.3	14.3
Vico Equense, Napoli (Squash)	2.6	16.1
Lecce (Tobacco)	2.8	22.8
Torino (Celery)	2.9	12.6
Margherita di S., Foggia (Tomato)	2.9	13.5
Bari (Tomato)	3.0	13.0
Fondi, Latina (Egg plant)	3.3	8.8
Ragusa (Tomato)	3.4	11.9
Castellamare di S., Napoli (Tomato)	4.1	7.8
Scafati, Salerno (Anemone)	4.5	8.4

aceous crop frequently attacked by *M. incognita* resulting
in yield losses of 30-60% in southern Italy (Lamberti,
1975) and economically important damage in the Sudan
(Yassin, 1974). Pepper (*Capsicum frutescens* L.) is usually less affected, at least in the Mediterranean Region
where it is transplanted very early (end of winter) and
thus escapes attack by the nematode during the growth of
the plants. Late attack by *M. incognita* can reduce crop
yield by up to 15% (Lamberti, 1975).

Among the Cucurbitaceae, watermelon (*Citrullus vulgaris*
Schrad), cantaloupe (*Cucumis melo* L.), cucumber (*C. sati-
vus* L.), squash (*Cucurbita maxima* Dcn.), and vegetable
marrow (*C. pepo* L.) are severely damaged by *M. incognita*
wherever they are grown. In southern Italy the nematode
has reduced the yields of cantaloupe and watermelon by
50% (Di Vito, unpublished) and occasionally crops of cucumber have been completely destroyed, especially under

TABLE IV

Response of plant species to attack by *M. incognita*

Origin of Population	HOSTS							
	Water-melon	Peanut	Sweet Potato	Cotton	Straw-berry	Corn	Pepper	Tobacco NC 95
Bari, Italy	+	–	+	–	–	+	+	–
Taranto, Italy	+	–	+	–	–	+	+	–
Foggia, Italy	+	–	+	–	–	+	+	–
Lecce, Italy	+	–	+	–	–	+	+	–
Torino, Italy	+	–	+	–	–	+	+	–
Salerno, Italy	+	–	+	–	–	+	+	+
Latina, Italy	+	–	+	–	–	+	+	–
Messina, Italy	+	–	+	+	–	+	+	–
Ragusa, Italy	+	–	+	–	–	+	+	–
Cosenza, Italy	+	–	+	–	–	+	+	–
Bulgaria	+	–	–	–	–	+	+	+
Rumania	+	–	+	–	–	+	+	–
Algeria	+	–	+	+	–	+	+	–
Peru	+	–	+	–	–	+	+	–
Venezuela	+	–	+	+	–	+	+	–
California	+	–	?	–	–	+	+	–
Florida	+	–	+	–	–	+	+	–

+ = gall formation with production of eggs; – = no root invasion by infective stages.

glass (Lamberti, unpublished).

Yield decreases caused by *M. incognita* are well known on okra (*Hibiscus esculentus* L.), sweet potato (*Ipomea batatas* (L.) Lam.) and celery (*Apium graveolens* L.). The nematode is particularly destructive to carrot (*Daucus carota* L.) as it not only reduces the production of tap-roots but also makes them unmarketable by altering their shape (Lamberti, 1971a; Fig. 1).

M. incognita is less pathogenic and not so frequently encountered on some vegetable crops belonging to the

Fig. 1. Galling caused by *Meloidogyne incognita* on roots of carrot (left and tobacco (right).

families Leguminosae, Liliaceae, Compositae, Chenopodiaceae and Cruciferae, although damage has often been observed
on pea (*Pisum sativum* L.), French bean (*Phaseolus* spp.),
onion (*Allium cepa* L.), lettuce (*Lactuca sativa* L.), artichoke (*Cynara scolimus* L.) (Furia and Mancini, 1973; Di
Vito and Botta, 1976), garden beet (*Beta vulgaris* L.) and
cauliflower (*Brassica oleracea* L.).

M. *incognita* is a major pest of tobacco (*Nicotiana tabacum* L.) wherever it is cultivated (Fig. 1). In many regions of the United States the crop has become uneconomi-
cal unless resistant cultivars are used. High yield losses
are caused by the nematode in the Mediterranean Region
(Greece, Jugoslavia, Turkey, Middle East and North Africa).
In southern Italy massive attacks may reduce the yield of
the light Levantine cultivars to nil (Lamberti, 1969 and
1971b) and in years with dry summers the crop reduction
in non-irrigated fields ranges from 30-75% (Lamberti,

1972a; 1975). Damage is less (10-30%) in irrigated crops,
although the roots are often severely galled (Lamberti,
unpublished); the higher soil and air humidity reducing
plant transpiration and root functioning result in reduced damage to the plants.

Sugarbeet (*Beta vulgaris* L.) is another industrial
crop considerably affected by this root-knot nematode,
especially when cultivated in light or sandy soils as in
central California or in the Mediterranean region. In
southern Italy, in heavily infested soil, root yields were
reduced by 60% in 1976 and 80% in 1977 and sugar yield
70-80% (Di Vito and Lamberti, 1977; Di Vito *et al.*, 1977).
M. *incognita* also causes reduction in the yield of sugar
from sugarcane in the United States and in North Africa.

M. *incognita* is also the causal agent of the most
important nematode disease of cotton (*Gossypium hirsutum*
L.) in the American Cotton Belt and its attacks cause
considerable damage on *G. barbadense* L. in Egypt and other
North African Countries.

On potato (*Solanum tuberosum* L.) M. *incognita* is not
usually an important parasite as in Mediterranean or subtropical regions the crop is planted in winter when the
soil temperature is too low for the nematode attacks. However, the nematode may cause considerable damage to crops
which are planted late (March instead of January or Feb-
ruary, as normally done) or to summer plantings as done
in the Mediterranean region for the production of early
potatoes.

Soybean (*Glycines max* (L.) Merr.) and peanut (*Arachis
hypogea* L.) are two other row crops severely damaged by
M. *incognita* in the United States, South America and North
Africa. Particularly in the United States, its importance
on soybean is exceeded only by *Heterodera glycines* Ichinohe

Reports from different countries indicate that M. *incognita* is an important pathogen on many field crops, of
which the most important are: rice (*Oryza sativa* L.) in

Egypt (Ibrahim *et al.*, 1973), sorghum (*Sorghum vulgare* Pers.) in Texas (Orr, 1967), ginger (*Zingiber officinale* Roscoe) in Australia (Pegg *et al.*, 1974), and leguminous forage plants such as alfalfa (*Medicago sativa* (L.) L.) in the south-west of United States and South America (Eriksson, 1972), red clover (*Trifolium pratense* L.) in North and South America and Australia (Eriksson, 1972), and white clover (*T. repens* L.) in southern Italy (Lamberti, 1970). Some pasture legumes and turf grasses are also susceptible to this root-knot nematode (Eriksson, 1972).

Carnations (*Dianthus caryophyllus* L.) are also severely damaged by *M. incognita*, especially when cultivated repeatedly in the same soil. Severe attack reduces the quantity of flowers and quality (shortening of the stem) may be affected to such an extent to make them unmarketable (Lamberti, 1975). Other ornmanetal plants affected by *M. incognita* are: anemone (*Anemone coronaria* L.) in southern Italy, especially when planted in late summer or early autumn (Vovlas *et al.*, 1973), begonia (*Begonia rex-cultorum* Bailey) in France, various cacti and other less important flower and ornamental plants grown under glass or outdoors in different countries (Picard, 1971; Hague, 1972; Kolicpanos and Vovlas, 1977).

Among fruit crops, the most susceptible to *M. incognita* is undoubtedly peach (*Prunus persica* Stock.) which is damaged by this nematode in most countries where the crop is grown (Sharpe *et al.*, 1969). In the Mediterranean region heavy infestations are often observed on the roots of almond (*P. amygdalus* Batsch) in the field (Scotto La Massese, 1971) or in nurseries (Lamberti and Di Vito, 1972), while apricot (*P. armeniaca* L.) seems to be less susceptible (Siniscalco *et al.*, 1976).

Root galls induced by *M. incognita* have been observed on olive (*Olea europaea* L.) in California (Lamberti and Baines, 1969; Lamberti, 1969a) and in the Mediterranean

region (Minz *et al.*, 1963; Lamberti and Di Vito, 1972;
Inserra *et al.*, 1976), and glasshouse experiments have
demonstrated the pathogenicity of this nematode to different olive cultivars (Lamberti and Baines, 1969).

Grapevine (*Vitis vinifera* L.) is infested by *M. incognita* in California (Raski and Lear, 1962), Israel (Minz
et al., 1963) and Bulgaria (Katalan-Gateva and Choleva
Abadjieva, 1977), but not very frequently in Italy (Lamberti, unpublished). The nematode is sometimes found on
fig (*Ficus carica* L.) planted in sandy soils (Lamberti,
unpublished; Koliopanos and Vovlas, 1977) and on forest
trees (Wang *et al.*, 1975). Also Lamberti *et al.* (1977)
have demonstrated experimentally that the nematode can
be pathogenic to date palm (*Phoenix dactylfera* L.)

Meloidogyne javanica

This is the second most widespread root-knot species
in areas with subtropical and mediterranean climate and
in some regions of the Sudan (Yassin, 1974), Cyprus (Philis and Siddiqi, 1976; Philis and Vakis, 1977), Greece
(Koliopanos, personal communication) and Egypt (Ibrahim,
personal communication) is predominant. In California
and Italy *M. javanica* frequently occurs outdoors in the
southern regions, or on glasshouse crops.

Again, the most affected plants are various vegetables
such as tomato, eggplant, beans, squash, watermelon, and
sometimes okra, carrot, celery and yam (*Dioscorea* spp.).
Also industrial and row crops, such as potato, sugarbeet,
sugarcane (*Saccharum officinarum* L.), rice, tobacco and
some forage legumes suffer varying levels of damage. On
ornamentals *M. javanica* is a serious pest of carnation,
especially under glass in many subtropical countries, and
of poinsettia (*Euphorbia pulcherrima* Willd.) in California (Siddiqui *et al.*, 1973) and of tuberose (*Polianthes
tuberosa* L.) in Italy (Melis, 1959).

Peach and almond are subject to heavy attacks by *M.*

javanica in the southern United States (California, Florida, Louisiana, Georgia) (Sharpe *et al.*, 1969) and in the
Mediterranean region (Minz *et al.*, 1963). Grapevine may be
heavily attacked in California (Raski and Lear, 1962), Australia (Meagher, 1969), Israel (Minz *et al.*, 1963), Bulgaria (Katalan-Gateva and Choleva-Abadjeva, 1977) and
Egypt (Lamberti, unpublished), and date palm in California
(Carpenter, 1964) and Algeria (Lamberti *et al.*, 1977).

The pathogenicity of *M. javanica* to olive trees has
been shown by Diab and El-Eraki (1968) in Egypt and by
Lamberti and Lownsbery (1968) and Lamberti and Baines
(1969) in California, while Gill (1971) in California and
Inserra *et al.* (1978) in Italy (Sicily) report the occurrence and damage, but without reproduction, on *Citrus*
spp., *Poncirus trifoliata* (L.) Raf. and "Troyer" citrange
(*Citrus sinensis* (L.) Osb. x *P. trifoliata*). In Sicily
(Inserra and Cartia, 1977) and other subtropical countries
M. javanica has been found on the roots of papaya (*Carica
papaya* L.) and pistachio (*Pistacia* spp.) (Lamberti, unpublished; Thomason, personal communication).

Meloidogyne arenaria

This is another polyphagous species pathogenic to various
ous crops. It has a worldwide distribution but its local
occurrence is less frequent than *M. incognita* or *M. javanica*. *Meloidogyne arenaria* is mainly a problem on vegetables such as tomato, pepper, watermelon, celery, beet,
carrot and onion, but the crops most susceptible to its
attack are peanut, tobacco and sugarbeet, especially in
the southern United States. However, in all the Mediterranean region as well as in other American countries, in
Australia, in South Africa or Asia it is found on potato,
forage legumes, sugarcane, carnation and other ornamentals,
peach, grapevine and fig. In pathogenicity tests carried
out in California by Lamberti and Baines (1969) the nematode failed to invade olive roots.

Meloidogyne hapla

Although considered a nematode of cooler climates, *M. hapla* is common also in subtropical or mediterranean countries where it attacks a wide range of plants during the winter months or when temperatures are relatively low.

In plot experiments in North Carolina yields of tomato were decreased by 50% (Barker *et al.*, 1976) and in Canada by 40% (Olthof and Potter, 1977). Considerable yield losses are also reported on alfalfa (36%), peanut (70%), carrot (50%), sugarbeet (20%), potato (46%), onion (64%) (Williams, 1974), various *Trifolium* spp. (Eriksson, 1972) and lettuce (Olthof and Potter, 1972; Potter and Olthof, 1974). In some areas *M. hapla* may be a major pest of strawberry (*Fragaria chiloensis* Duchsne) (McElroy, 1972), raspberry (*Rubus idaeus* L.) (Griffin *et al.*, 1968), actinidia (*Actinidia chinensis* Planch.) (Scotto La Massese, 1973; Vovlas and Roca, 1976) and hop (*Humulus lupulus* L.) (Maggenti and Hart, 1963).

Some fruit trees e.g. peach, grapevine (McElroy, 1972) and olive (Minz, 1961) are reported to be hosts for *M. hapla* but it does not seem to cause economic damage to them. Experimental inoculations of the nematode on rooted olive plants failed (Lamberti and Baines, 1969).

Meloidogyne naasi

Meloidogyne naasi occurs mostly in temperate regions where it is found principally on cereals and sugarbeet.

It is fairly common in California, especially on turf-grass (Radewald *et al.*, 1970) and barley (*Hordeum vulgare* L.) (French, 1965; Allen *et al.*, 1970) on which losses of 50-75% have been recorded (Franklin, 1973). A survey on the nematofauna of durum wheat (*Triticum durum* Desf.) in southern Italy has shown that 17% of the fields sampled had poorly growing plants, with roots infested by *M. naasi* (Inserra and Vovlas, unpublished). In the Mediterranean region *M. naasi* has also been found in the Maltese islands on barley (Inserra *et al.*, 1975) and on sweet vetch (*Hedy sarum coronarium* L.) (Lamberti, unpublished).

Other crops on which the nematode might cause some damage are onion, sugarbeet, cotton, alfalfa and sorghum, and various grasses (Eriksson, 1972; Kort, 1972).

Meloidogyne graminis

This species is rather rare. Its distribution is limited to the southern United States where it is associated with cereals and more often with turfgrasses such as bermuda grass (*Cynodon dactylon* (L.) Pers.) and St. Augustine grass (*Stenotaphrum secondatum* (Wolt.) Kuntze) (Eriksson, 1972). It may also attack alfalfa and *Trifolium* spp.

Meloidogyne artiellia

This species has been found in France on the roots of cabbage (*Brassica oleracea* L.) and turnip (*B. rapa* L.), but it may also attack legumes and cereals (Ritter, 1972). In Greece *M. artiellia* has been found associated with roots of declining wheat plants (Kyrou, 1969).

References

Allen, M.W., Hart, W.H. and Baghott, K. (1970). Crop rotation controls barley root-knot nematode at Tulelake. *Calif.Agric.*, **24**, 4-5.

Barker, K.R. and Olthof, Th. H.A. (1976). Relationships between nematode population densities and crop responses. *Ann.Rev.Phytopath.*, **14**, 327-353.

Barker, K.R., Shoemaker, P.B. and Nelson, L.A. (1976). Relationships of initial population densities of *Meloidogyne incognita* and *M. hapla* to yield of tomato. *J.Nematol.*, **8**, 232-239.

Carpenter, J.B. (1964). Root-knot nematode damage to date palm seedlings in relation to germination and stage of development. *Date Growers' Inst.Ann.Rep.*, **41**, 10-14.

Chitwood, B.G. (1949). Root-knot nematodes. *Proc.helminthol.Soc.Wash.*, **16**, 90-104.

Dandria, D., Lamberti, F. and Vovlas, N. (1977). La lotta chimica contro i nematodi galligeni del Pomodoro a Malta. *Nematol.medit.*, **5**, 127-131.

Diab, K.A. and El-Eraki, S. (1968). Plant parasitic nematodes associated with olive decline in the United Arab Republic. *Pl.Dis. Reptr.*, **52**, 150-154.

Di Vito, M. and Botta, S. (1976). Infestazioni di *Meloidogyne incognita* su carciofo in Puglia. *Nematol.medit.*, 4, 237-239.

Di Vito, M. and Lamberti, F. (1976). Reazione di varietà di Pomodoro
a popolazioni di *Meloidogyne* spp. in serra. *Nematol.medit.*, 4,
211-215.

Di Vito, M. and Lamberti, F. (1977). Prove di lotta chimica contro i
nematodi galligeni su Barbabietola da zucchero. *Nematol.medit.*,
5, 31-38.

Di Vito, M., Lamberti, F. and Carella, A. (1977). Ulteriori risultati di prove di lotta chimica contro *Meloidogyne incognita* su Barbabietola da zucchero. *Nematol.medit.*, 5, 339-344.

Eriksson, K.B. (1972). *In* "Economic Nematology". (Ed. J.M. Webster).
66-96. Academic Press, London and New York.

Franklin, M.T. (1973). *Meloidogyne naasi*. *C.I.H. Descrip.Pl.par.Nem.*,
19, 4.

French, A. (1965). Air and ground surveys in Tulelake basin for new
root-knot species of barley. *Nema News*, 2, 1-2.

Furia, A. and Mancini, G. (1973). Infestazione da *Meloidogyne incognita* (Kofoid *et* White, 1919) Chitwood, 1949 su *Cynara scolimus*
L. e modificazioni istologiche associate. *Riv.Pat.Veg.*, 9, 63-73.

Gill, H.S. (1971). Occurrence and reproduction of *Meloidogyne javanica*
on three species of Citrus in California. *Pl.Dis.Reptr.*, 55, 607-
608.

Griffin, G.D., Anderson, J.L. and Jorgenson, E.C. (1968). Interaction
of *Meloidogyne hapla* and *Agrobacterium tumefaciens* in relation to
raspberry cultivars. *Pl.Dis.Reptr.*, 52, 492-493.

Hague, N.G.M. (1972). *In* "Economic Nematology." (Ed. J.M. Webster.)
409-434. Academic Press, London and New York.

Ibrahim, I.A., Ibrahim, I.K.A. and Rezk, M.A. (1973). Host-parasite
relationship of *Meloidogyne incognita* (Kofoid *et* White) Chitw. on
rice. *Nematol.medit.*, 1, 8-14.

Inserra, R.M. and Cartia, G. (1977). *Meloidogyne javanica* su Papaya
in Sicilia. *Nematol.medit.*, 5, 137-139.

Inserra, R.N., Lamberti, F., Vovlas, N. and Dandria, D. (1975). *Meloidogyne naasi* nell'Italia Meridionale e a Malta. *Nematol.medit.*,
3, 163-166.

Inserra, R.N., Perrotta, G., Vovlas, N. and Catara, A. (1978). Reaction of citrus rootstocks to *Meloidogyne javanica* infestations.
J.Nematol., 10, 181-184.

Inserra, R.N., Vovlas, N., Lamberti, F. and Bleve-Zacheo, T. (1976).
Plant parasitic nematodes associated with declining olive trees
in Southern Italy. *Proc.IV.Cong.Un.fitopat.medit.*, 419-424.

Katalan-Gateva, S. and Choleva-Abadjeva, B. (1977). Gall-forming
nematodes (genus *Meloidogyne* Goeldi, 1887) on the vine in the
district of Blagoevgrad. *Acta Zool.Bulg.*, 6, 35-39.

Koliopanos, C.N. and Vovlas, N. (1977). Records of some plant parasitic nematodes in Greece with morphometrical descriptions. *Nema-
tol.medit.*, 5, 207-216.

Kort, J. (1972). *In* "Economic Nematology." (Ed. J.M. Webster). 97-126.
Academic Press, London and New York.

Kyrou, N.C. (1969). First record of occurrence of *Meloidogyne artiellia* on wheat in Greece. *Nematologica* 3, 432-433.

Lamberti, F. (1969). Prospettive di lotta nematocida sul tabacco in

provincia di Lecce. Nota preliminare. *Il Tabacco* 733, 8-10.
Lamberti, F. (1969a). Nematodi parassiti dell'Olivo. *Inf.tore Fito-patol.*, 19, 405-408.
Lamberti, F. (1970). Danni da nematodi in prati di una zona residen-ziale di Bari. *Inf.tore Fitopatol.*, 20, 8-9.
Lamberti, F. (1971). Prove di lotta chimica contro i nematodi galli-geni del Pomodoro in Puglia. *Phytiatr.Phytopharm.circum-médit.*, 3, 140-146.
Lamberti, F. (1971a). Nematode-induced abnormalities of carrot in Southern Italy. *Pl.Dis.Reptr.*, 55, 111-113.
Lamberti, F. (1971b). Primi risultati di prove di lotta nematocida su tabacchi levantini in provincia di Lecce. *Il Tabacco* 738, 5-10.
Lamberti, F. (1972). Chemical control of root-knot nematodes on tob-acco in Apulia. *Meded.Fakul.Landbouw.*, *Gent*, 37, 790-797.
Lamberti, F. (1975). Fumiganti e nematocidi sistemici nella lotta con-tro i fitoelminti ipogei. Report presented at the Round Table of S.I.F., Cagliari, Italy.
Lamberti, F. and Baines, R.C. (1969). Pathogenicity of four species of *Meloidogyne* on three varieties of olive trees. *J.Nematol.*, 1, 111-115.
Lamberti, F. and Cirulli, M. (1970). Prove preliminari di lotta a nematodi del gen. *Meloidogyne* su Pomodoro con due nematocidi sper-imentali. *Italia agric.*, 107, 721-723.
Lamberti, F. and Di Vito, M. (1972). Sanitation of root-knot nematode infected olive-stocks. *Proc.III.Cong.Un.fitopat.medit.*, 401-411.
Lamberti, F. and Lownsbery, B.F. (1968). Olive varieties differ in reaction to the root-knot nematode *Meloidogyne javanica* (Treub) Chitw. *Phytopath.medit.*, 7, 48-50.
Lamberti, F., Greco, N. and Vovlas, N. (1977). Patogenicità di due specie di *Meloidogyne* nei confronti di quattro varietà di Palma da dattero. *Nematol.medit.*, 5, 159-172.
Lamberti, F., Greco, N. and Zaouchi, H. (1975). A nematological survey of date palms and other major crops in Algeria. *FAO Pl.Prot.Bull.*, 23, 156-160.
Lamberti, F., Dandria, D., Vovlas, N. and Aquilina, J. (1976). Prove di lotta contro i nematodi galligeni del Pomodoro da mensa in serra a Malta. *Colture Protette* 5, 27-30.
Maggenti, A.R. and Hart, W.H. (1963). Control of root-knot nematode on hops. *Pl.Dis.Reptr.*, 47, 883-885.
McElroy, F.D. (1972). *In* "Economic Nematology." (Ed. J.M. Webster.) 335-376. Academic Press, London and New York.
Meagher, J.W. (1969). Nematodes and their control in vineyards in Victoria, Australia. *Int.Pest.Control* 11, 14-18.
Melis, G. (1959). *Meloidogyne javanica* (Treub, 1885) Chitwood, 1949 su Tuberosa (Nematoda, Heteroderidae). *Redia* 64, 51-54.
Minz, G. (1961). Additional hosts of the root-knot nematode, *Meloid-ogyne* spp., recorded in Israel during 1958-1959. *Isr.J.Agric.Res.*, 11, 69-70.
Minz, G., Strick-Harari, D. and Cohn, E. (1963). *In* "Plant parasitic nematodes in Israel and their control." 84. Sifriat Hassadeh Publishing House, Tel Aviv.
Olthof, Th. H.A. and Potter, J.W. (1972). Relationship between popul-ation densities of *Meloidogyne hapla* and crop losses in summer-maturing vegetables in Ontario. *Phytopathology* 62, 981-986.

Olthof, Th. H.A. and Potter, J.W. (1977). Effects of population densities of *Meloidogyne hapla* on growth and yield of tomato. *J.Nematol.*, 9, 296-300.

Orr, C.C. (1967). Observations on cotton root-knot nematode in grain sorghum in West Texas. *Pl.Dis.Reptr.*, 51, 29.

Orton Williams, K.J. (1974). *Meloidogyne hapla. C.I.H.Descrip.Pl.par. Nem.*, 31, 4.

Pegg, K.G., Moffett, M.L. and Colbran, R.C. (1974). Diseases of ginger in Queensland. *Queen.Agric.Jour.*, 100, 611-618.

Philis, J. and Siddiqi, M.R. (1976). A list of plant parasitic nematodes in Cyprus. *Nematol.medit.*, 4, 171-174.

Philis, J. and Vakis, N. (1977). Resistance of tomato varieties to the root-knot nematode *Meloidogyne javanica* in Cyprus. *Nematol. medit.*, 5, 39-44.

Picard, M. (1971). *In* "Les Nématodes des Cultures." 401-426. A.C.T.A. Paris.

Potter, J.W. and Olthof, Th. H.A. (1974). Yield losses in fall-maturing vegetables relative to population densities of *Pratylenchus penetrans* and *Meloidogyne hapla*. *Phytopathology* 64, 1072-1075.

Radewald, J.D., Pyeatt, L.E., Shibuya, F. and Humphrey, W. (1970). *Meloidogyne naasi* a parasite of turfgrass in Southern California. *Pl.Dis.Reptr.*, 54, 940-942.

Raski, D.J. and Lear, B. (1962). Influence of rotation and fumigation on root-knot nematode populations on grape replants. *Nematologica* 8, 143-151.

Riggs, R.D. and Winstead, N.N. (1959). Studies on resistance in tomato to root-knot nematodes and on the occurrence of pathogenic biotypes. *Phytopathology* 49, 716-724.

Ritter, M. (1972). Rôle economique et importance des *Meloidogyne* en Europe et dans le bassin Méditerraneen. *OEPP/EPPO Bull.*, 2, 17-22.

Sasser, J.M. (1977). Worldwide dissemination and importance of root-knot nematodes *Meloidogyne* spp. *J.Nematol.*, 9, 26-29.

Scotto La Massese, C. (1971). *In* "Les Nématodes des cultures." 377-400. A.C.T.A., Paris.

Scotto La Massese, C. (1973). Nouvel hôte Europeen de *Meloidogyne hapla* et *Rotylenchus robustus : Actinidia chinensis. Nematol. medit.*, 1, 57-59.

Sharpe, R.H., Hesse, C.O., Lownsbery, B.F., Perry, V.G. and Hansen, C.J. (1969). Breeding peaches for root-knot nematode resistance. *J.Am.Soc.hort.Sci.*, 94, 209-212.

Siddiqui, I.A., Sher, S.A. and French, A.M. (1973). Distribution of plant parasitic nematodes in California. *Dept. of Food and Agric. Div. of Plant Ind. Sacramento.* 324.

Siniscalco, A., Lamberti, F. and Inserra, R.N. (1976). Reazione di portainnesti del Pesco a popolazioni italiane di due specie di nematodi galligeni (*Meloidogyne* Goeldi). *Nematol.medit.*, 4, 79-84.

Vovlas, N., Avgelis, A. and D'Urso, M.P. (1973). Nematodi e virus, agenti di gravi danni a colture di Anemone nel Mezzogiorno. *Inf. tore Fitopatol.*, 23, 19-22.

Vovlas, N. and Roca, F. (1976). *Meloidogyne hapla* su *Actinidia chinensis* in Italia. *Nematol.medit.*, 4, 115-116.

Wang, K.C., Bergeson, G.B. and Green, R.J. (1975). Effect of

Meloidogyne incognita on selected forest tree species. *J.Nematol.*,
7, 140-149.
Yassin, A.M. (1974). Root-knot nematodes in the Sudan and their chemical control. *Nematol.medit.*, 2, 103-112.

Discussion

Asked about root-knot problems on peach and almond,
Lamberti said that they were infrequent in field situations
and heavy galling had been found only in older plantations.
Orion said that in Israel he had seen almond and peach
trees killed by root-knot nematodes when irrigation was
applied, but Lamberti suggested that death may have been
caused by the release of HCN from amygdatin, which is
abundant in peach and almond. Lamberti considered that
root-knot nematodes caused little damage to *Vitis vinifera*
in Italy. Asked whether older olive trees would be more
resistant than young ones, he explained that it is difficult to evaluate nematode damage because of the extensive
root system of olives which can compensate for nematode
attack; damage can be seen more readily on young trees.

In discussing the niche occupied by *M. javanica* and
M. incognita in mixed infections, Lamberti suggested that
the species may be favoured or selected by the host or
by environmental conditions. Bird observed that *M. hapla*
occurs in some hot locations in the Australian desert indicating an extreme degree of adaptation, but Sasser
thought that the species is introduced to an area and
survives for only a few years because it is unable to
adapt to the environmental conditions. Ritter added that
in Southern France *M. javanica* is uncommon but has been
introduced *via* glasshouses in some areas, and that *Vitis
vinifera* growing in soils with less than 3% clay were
destroyed by *M. arenaria*.

J.N. Sasser

Dept. of Plant Pathology, North Carolina State University, Raleigh, North Carolina, U.S.A.

Introduction

Root-knot nematodes (*Meloidogyne* spp.) are worldwide in distribution attacking many economically important crops. Serious losses resulting from yield reductions occur wherever these pests are found. Yield reductions are often related to the biological stress placed on the crop plant following infection regardless of the particular *Meloidogyne* species involved. Some of these stresses are brought about by man's practices while others are uncontrollable environmental influences. Crop plants reflect this stress with suppressed root and shoot growth, discoloration of foliage, small size and quality of product produced, and yield.

We are just beginning to get information on crop losses caused by nematodes in the tropics[1]. The lack of such information is due to the fact that nematology is a young science compared with entomology and plant pathology, and only recently have agricultural universities and government agencies of countries in the tropics added nematologists to their staffs. Many institutions are still without such trained personnel. Nevertheless, an attempt is made in this chapter to synthesize information in the

---

[1] Tropics – conveniently defined as the area between the Tropic of Cancer and the Tropic of Capricorn.

literature, personal experience, and recent data obtained
from nematologists working in the tropics. Some statistics
are given concerning losses and distribution of *Meloidogyne*
species in the tropics. Included also is a brief discussion
of climatic and edaphic factors, agronomic practices, and
other unique conditions, associated with the tropics, which
influence development of *Meloidogyne* spp. Specifically,
consideration is given to

(i) *Meloidogyne* species known to occur in the tropics,
their relative frequency and importance;

(ii) estimated crop losses caused by *Meloidogyne* spp. in
major geographical regions of the tropics; and

(iii) favourable and unfavourable factors for survival,
infection and pathogenicity of these nematodes in the
tropics.

Meloidogyne species known to occur in the tropics, their
relative frequency and importance

Reports of distribution, pathogenicity, and relative
importance of *Meloidogyne* species in the tropics are recorded in more than a hundred publications. Most of these
are cited in the Proceedings of several recent Research
Planning Conferences on *Meloidogyne* spp. (Anonymous, 1976a,
b,c,d,e.). Additional unpublished information has been obtained during the past 3 years through activities of the
International *Meloidogyne* Project (IMP).* This information
is available only because of the excellent cooperation of
more than 75 cooperators participating in the Project. The
known distribution of *Meloidogyne* species in the tropics,
as indicated in published reports and as determined through
additional studies of the IMP, is shown in Table I.

Frequency of *Meloidogyne* species identified from populations studied from 54 countries on six continents

*Funded by the United States Agency for International Development
(USAID) through a contract with North Carolina State University
(Contract No. AID-ta-C-1234).

TABLE I

Known distribution of *Meloidogyne* species by major geographical
regions in the tropics

	Geographical regions		
Central and South America		Africa	Asia and Australia
1st Order	*M. incognita*	*M. incognita*	*M. incognita*
	M. javanica	*M. javanica*	*M. javanica*
2nd Order	*M. arenaria*	*M. arenaria*	*M. arenaria*
	M. hapla	*M. hapla*	*M. hapla*
	M. exigua	*M. acronea*	*M. graminicola*
3rd Order	*M. coffeicola*	*M. africana*	*M. brevicauda*
	M. lordelloi	*M. kikuyensis*	*M. indica*
	M. inornata	*M. ethiopica*	*M. lucknowica*
	M. bauruensis	*M. decalineata*	
		M. megadora	
		M. oteifae	
		M. propora	

[1]Arbitrarily ranked according to their probable importance in agriculture.

(including temperate, subtropical, and tropical regions)
and the occurrence of races (Sasser and Triantaphyllou,
1977) are shown in Table II. Of a total of 400 populations
identified thus far, 97% were identified as belonging to
only four species. When only populations from tropical
countries are considered, the relative frequencies of
species are similar except for fewer *M. hapla* populations
(Table III). *M. incognita* is more dominant, accounting for
approximately 64% of the populations occurring in tropical
countries. Second in importance is *M. javanica*, 28%. *M.
arenaria*, *M. hapla* and *M. exigua* account for about 8%. *M.
exigua* is important primarily on coffee in Central and
South America. *M. hapla* occurs infrequently in tropical
countries and only at high elevations. Thus far it has
been identified from Costa Rica, El Salvador, Taiwan,
Kenya and Colombia. In each case, the population came from

TABLE II

Frequency of *Meloidogyne* species identified from populations studied[1]

Identification	No. Populations	% of Total Populations	
Species and Races			
M. incognita[2]	222	55	
Race 1	146		37
Race 2	43		11
Race 3	24		6
Race 4	9		2
M. javanica	107	27	
M. hapla	37	9	
M. arenaria[3]	22	6	
Race 1	8		2
m Race 2	14		4
M. exigua	7	2	
M. graminicola	2		
M. megatyla	1	1	
M. microtyla	1		
M. naasi	1		
Total 400		100%	

[1]Populations included are from all areas of the world. Mixed populations and some not yet identified are not included.

[2]There are at least 4 host races of *M. incognita* as determined by their parasitism on tobacco (*Nicotiana tabacum*) cv. NC-95 and cotton (*Gossypium hirsutum*) cv. Deltapine 16. Race 1 does not reproduce on root-knot resistant tobacco cv. NC-95 or cotton cv. Deltapine 16. Race 2 reproduces on cv. NC-95 tobacco. Race 3 reproduces on cv. Deltapine 16 cotton. Race 4 reproduces on both the cotton and tobacco cultivars.

[3]There are 2 host races of *M. arenaria*. Race 1 reproduces on peanut (*Arachis hypogaea*) cv. Florrunner. Race 2 does not reproduce on peanut.

an altitude of 350 m or higher. *M. arenaria* is found infrequently also.

Little is known concerning the occurrence and importance of other species shown in Table I other than that reported in the original descriptions (Coetzee, 1956;

TABLE III

Frequency of *Meloidogyne* species identified from the tropics[1]

Identification	No. Populations	% of Total Populations	
Species and Races			
M. *incognita*	166	64	
Race 1		115	44
Race 2		34	13
Race 3		11	4
Race 4		6	2
M. *javanica*	73	28	
M. *hapla*	5	2	
M. *arenaria*	9	3	
Race 1		1	>1
Race 2		8	3
M. *exigua*	7	3	
Total	260	100%	

[1] Populations are from the following tropical countries: Mexico, Panama, Costa Rica, Jamaica, Puerto Rico, Guadeloupe, Trinidad, El Salvador, Bermuda, Dominican Republic, Suriname, Colombia, Ecuador, Bolivia, Peru, Venezuela, Brazil, Senegal, Liberia, Ivory Coast, Ghana, Nigeria, Kenya, Malawi, Tanzania, Rhodesia, Philippines, Taiwan, Thailand, Sri Lanka, Malaysia, Indonesia, Fiji Islands, Australia, Sudan, Yemen.

De Grisse, 1960; Elmiligy, 1968; Goeldi, 1887; Loos, 1953; Lordello, 1956a,b; Lordello and Zamith, 1958, 1960: Ponte, 1969; Singh, 1969; Spaull, 1977; Whitehead, 1959), and in subsequent reviews (Esser, *et al.*, 1976; Sasser, 1977a,b; Taylor and Sasser, 1978; Whitehead, 1968). *Meloidogyne acronea* has recently been reported as causing severe damage to cotton in Malawi (Bridge *et al.*, 1976), but has not been reported as occurring elsewhere. *Meloidogyne graminicola* has established as a pest or rice in nurseries and uplands in India (Rao *et al.*, 1969; Roy, 1975), and is probably causing considerable damage in other rice producing countries.

Percentage occurrence of the various species is based
on populations sent to North Carolina State University for
identification and further study. Identifications were
based on morphology, host response, and cytological characteristics. Although far from complete, these studies
represent the most comprehensive effort to date to gain
information on the distribution and relative importance
of the various species.

Estimated crop losses caused by *Meloidogyne* spp. in
major geographical regions of the tropics

Only a few estimates of crop losses due to *Meloidogyne*
species in the tropics were available 6 years ago when
the U.S. Agency for International Development decided to
conduct a series of studies on crop pests in developing
countries. Fortunately, it is relatively easy for anyone
trained in nematology to recognize root-knot infected
plants and to correlate knotted root systems with suppressed growth and low yields in farm fields. These symptoms
were so common in the tropics that there was little doubt
that crop losses were very large; thus USAID agreed to
finance the International *Meloidogyne* Project.

The IMP enlisted the help of about 75 scientists in
developing countries to gather additional information.
Each scientist was asked to estimate losses from root-
knot nematodes on the principal food and fibre crops of
the part of his country with which he was most familiar.
These estimates (Table IV) are the best available information on root-knot losses.

Except for Region IV, West Africa, where the estimated
average loss was 25%, the average losses for the other
regions were similar with losses in Region I, II, III and
VI being 15, 15, 13 and 11% respectively. No estimates
were available for Region V (Kenya, Uganda and Tanzania).

Of the several objectives of the International *Meloidogyne* Project, one has been to make a study of the influ-

TABLE IV

Estimated crop losses due to *Meloidogyne* spp. in major geographical regions of the tropics

Region I — Mexico, Central America and the Caribbean		Region II — South America		Region III — Brazil		Region IV — West Africa		Region VI — Southeast Asia	
Crop	% Loss	Crop	% Loss	Crop	% Loss	Crop	% Loss	Crop	% Loss
banana	7	banana	10	banana	8	banana	7	bean (common)	18
bean (common)	16	bean (common)	24	bean (common)	11	bean (common)	22	black pepper	16
cassava	5	cassava	10	cassava	3	carrot	38	cassava	5
chayote	38	coffee	13	citrus	5	cocoyam	23	Chinese Pechay	16
citrus	7	cucumber	33	coffee	24	cowpea	43	citrus	4
coffee	10	eggplant	20	cotton	17	eggplant	29	coconut	2
cowpea	10	grain legumes	19	grapes	6	lima bean	27	eggplant	17
guava	35	grapes	5	papaya	15	lowland rice	5	maize	6
maize	8	papaya	18	rice	4	maize	14	melon	18
pigeon pea	8	peach	6	soybean	23	melon	33	mung bean	15
pineapple	11	pepper	22	sugarcane	9	okra	42	papaya	15
plantain	11	pineapple	5	tomato	25	papaya	12	peanut	15
pumpkin	22	plantain	10	yam	15	peanut	15	pineapple	10
rice	10	potato	3			pigeon pea	35	rice	2
soybean	8	sugarcane	6			pineapple	2	sorghum	4
sugarcane	7	sweetpotato	15			plantain	3	soybean	10
tomato	38	tomato	27			soybean	4	sugarcane	8
yam	16	watermelon	23			sweetpotato	24	sweetpotato	6
						tomato	46	tea	13
						upland rice	4	tomato	24
						yam	23	yam	8
Mean % loss (all crops)	**15**		**15**	Average loss for all crops – 12.69%			**25**		**11**

ence of various ecological factors - soil, climate, previous cropping history, and other factors - on nematode survival and pathogenicity. In these studies, the various cooperators were asked to send to project headquarters cultures of *Meloidogyne* populations and soil samples collected from the rhizosphere of the infected plant. At the time
of collection of the nematode (egg masses) and soil samples,
the cooperators recorded data pertaining to each sample.
One question asked was the degree of infection of host
plants and, where economic crops were involved, estimates
of the percent loss in yield due to the nematode. The
scale used to estimate percent loss was: very light,
(0-5); light, (>5-10); moderate, (>10-20); severe, (>20-
50); and very severe, (>50).

Estimated percent losses for selected crops in the
tropics as reported herein indicate the probable crop
losses caused by *Meloidogyne* species in tropical countries
(Table V). Ultimately data will be analyzed to determine
whether significant correlations exist between root-knot
indices and/or percent loss, and cropping sequence, soil
type, climatic conditions, or other local factors.

From personal observations by the author during recent
visits to most of these countries, it appears that the
estimates given are conservative. It is clear that the
direct losses each year are hundreds or thousands of
times greater than the amounts which will be expended
during the foreseeable future for research on means for
alleviation of damage caused by nematodes.

Favourable and unfavourable factors for survival,
infection and pathogenicity of *Meloidogyne* in the tropics

Favourable factors

Important requirements for *Meloidogyne* species to survive, infect, and cause disease, whether in a tropical or
temperate region, are a suitable host plant, adequate

moisture, and a favourable temperature for the host plant and development of the nematode. In the tropics, survival of a species essentially is dependent upon a continuously available host. Population densities of pathogenic nematodes remain high in the presence of suitable hosts and damage can be severe because conditions are generally optimum for the nematode. Interaction with other soil-borne microorganisms may further enhance root destruction. The population density of *Meloidogyne* will decline rapidly in

TABLE V

Estimated percent loss due to *Meloidogyne* spp. for selected crops in the tropics[1]

Crop[2]	No. Estimates	0-5	>5-10	>10-20	>20-50	>50	Mean % Loss
Tomato	42	1	3	15	11	12	29
Eggplant	10	1	0	3	1	5	23
Okra	10	1	2	3	3	1	22
Beans	7	1	1	1	2	2	28
Pepper (all types)	6	1	1	3	1	0	15
Cabbage	6	1	0	0	3	2	26
Soybean (Brazil only)	5	0	0	3	1	1	26
Papaya	4	0	2	1	0	1	20
Potato	4	1	1	0	1	1	24
Yams	3	0	0	2	1	0	22
Coffee	3	0	1	1	1	0	19
Banana	2	1	1	0	0	0	5
Rice	2	0	0	1	1	0	25

[1] Data from Mexico, Panama, Costa Rica, Jamaica, Puerto Rico, Guadeloupe, Trinidad, El Savador, Bermuda, Dominican Republic, Suriname, Colombia, Ecuador, Bolivia, Peru, Venezuela, Brazil, Senegal, Liberia, Ivory Coast, Ghana, Nigeria, Kenya, Malawi, Tanzania, Rhodesia, Philippines, Taiwan, Thailand, Sri Lanka, Malaysia, Indonesia, Fiji Islands, Australia, Sudan, Yemen.

[2] Crops not included in the table are carnation, 10-20%; corn, 20-50%; chrysanthemum, 10-20%; cucumber, 10-20%; beet, 10-20%; plantain, 10-20%; cotton, 20-50%; carrot, >50%; sugarcane, 20-50%; jute, 20-50%; spinach, 10-20%.

the absence of a host, especially during the dry seasons. Under conditions of high temperature, active nematode larvae rapidly deplete their food reserves and die from starvartion or become easy prey to predaceous organisms in the soil.

The situation is different in the temperate regions. Sufficient eggs and larvae survive in the soil over the winter months, even under fallow conditions, to cause some injection during the next growing season, causing considerable damage. The magnitude of damage is influenced by population density, soil type, and other factors which place the plant under stress. If fallowing is practised during the hot summer months, the population density is greatly reduced.

The potential for damage caused by *Meloidogyne* species is ever present in the tropics. Susceptible crops such as tomato, bean, cowpea, and other vegetables can be grown throughout the year, and if this continuous cropping is practiced, the nematodes build up to very high populations. Other tropical crops - yams, cassava, pineapple, sugar cane - require longer periods of time to mature, thus providing an even greater opportunity for *Meloidogyne* species to increase to damaging levels. Often in addition to multi-cropping, there is inter-cropping with as many as six different crops growing in the same field. This intensive type of agriculture is conducive to the survival and maintenance of high populations of *Meloidogyne* species. Furthermore, during periods when economic crops are not growing on the land, weeds often are allowed to grow and many of these serve as suitable hosts for *Meloidogyne* species.

Another favourable factor for root-knot nematode development in many tropical countries, is the low level of agricultural technology practised. Often there is no awareness of the problem because of the lack of nematologists. Low yields and poor quality are accepted as normal

or thought to be due to other causes. Seed or planting
stock of such crops as potato, yams and citrus are often
infected with nematodes. This practice often introduces
the nematode into new areas. The use of infected planting
stock favours the nematode since it is in the root tissue
at the time of planting. Resistant varieties are not
widely used or are not available, and rotations designed
to reduce the nematode density are not economically practical. These technical disadvantages, when combined with
unfavourable growing conditions such as drought, which may
occur, place infected plants under extreme stress, resulting in lower yields, poorer quality, and sometimes complete crop failure.

Unfavourable factors

As pointed out in the previous section, survival of
Meloidogyne species in the tropics is dependent upon an
almost continuously available host. Because of high temperatures, excess moisture (in some areas), and presence
of other microorganisms (many that can attack nematodes),
larvae of *Meloidogyne* do not survive long in the absence
of host. Larvae expend their reserve food supply and/or
become subject to attack by predaceous fungi and nematodes
if suitable hosts are not available. Thus, one method of
control would be to utilize complete fallow for a few
months. In areas where there are dry seasons (4 or 5
months) and wet seasons, growers could use the dry season
to reduce nematode density by practising fallow. Occasional ploughing to expose as much of the soil as possible to
the drying effects of the sun would improve the effect-
iveness of this practice. In some countries, this method
of control may not be practical because of labour costs and
the difficulty of effecting complete fallow conditions.

In some areas, Southeast Asia for example, flooding of
fields is a normal farming practice in the production of
rice. This practice probably results in a reduction of the

population of the two most important species (*M. incognita* and *M. javanica*) which do not attack rice. Thus, there are some natural conditions and certain farming practices unique to the tropics, which, if properly used, can impede survival and reproduction of *Meloidogyne* populations. Utilization of such practices will be governed by many local constraints such as availability of equipment, cost of control practices, cropping systems normally used, degree of nematode infestation, and other factors.

Discussion and conclusions

Estimates of disease losses, regardless of the disease, are usually subjective because of the lack of information based on experimentation. Ideally, one of the best ways to gain information on losses caused by nematodes would be to compare yields of crops grown in nematode-infested and nematode-free soils. A large number of such tests over a wide geographic area would indicate fairly closely the percent loss caused by nematode for those or similar fields. It would not be a reflection of losses generally because fields, whether for a small area within a country, the entire country, or the whole region, vary widely in their infestation levels and, consequently, would vary in their response to control measures. Furthermore, there are so many other factors such as climate, soil type, and agronomic practices which would influence response to treatment for nematode control. It is for these and other reasons that we do not have more precise estimates of losses due to *Meloidogyne*.

Nevertheless, scientists recognizing these limitations can obtain the best estimates possible. Nematologists, after some experience with chemical soil treatments and other control measures, can predict probable losses resulting from given root-knot nematode population densities. The important conclusion, based on what we now know, is that the amount of loss is substantial.

Data on distribution and frequency of occurrence of the various species of *Meloidogyne* in the tropics are indeed limited. The number of populations examined, compared to the number present in the various habitats, is admittedly too small. A much longer and more intensive study is needed before any definite conclusions can be drawn. From the small number of populations studied, it appears that *M. incognita* and *M. javanica* are the most prevalent and that other species are very rare. It must be remembered, however, that most populations have come from cultivated fields and that more species will be found when natural vegetations are examined. However, the importance of *Meloidogyne* species other than *incognita* and *javanica* appears to be negligible at the present time. The relative importance of other described and undescribed species will be largely dependent on future agricultural practices.

The race problem within the species *Meloidogyne incognita* will have to be dealt with. Use of resistant cultivars or development of new ones through breeding programmes will have to take these biological differences into account, if maximum usefulness is to be obtained.

The tropics are inherently different from other regions with respect to soils, climates, and type of agriculture. Some of these are favourable to the root-knot nematodes - some are detrimental. Growers, with the help of agriculturists, must learn to use these phenomena to their advantage in minimizing damage caused by *Meloidogyne* species.

Acknowledgement

This is paper number 5714 of the Journal Series of the North Carolina Agricultural Experiment Station, Raleigh, North Carolina 27650.

References

Anonymous, (1976a). Proceedings of the research planning conference on root-knot nematodes, *Meloidogyne* spp. *North Carolina State Univ. Raleigh, N.C. 27650.*

Anonymous, (1976b). Proceedings Asian regional planning conference on root-knot nematode research program. *Univ. of the Philippines College, Laguna, Philippines.*

Anonymous, (1976c). Memorias de la conferencia de trabajo sobre el Projecto Internacional *Meloidogyne. Centro Internacional de Agric. Tropical, Palmira, Colombia.*

Anonymous, (1976d). Proceedings of the regional planning conference of the International *Meloidogyne* Project. *Instituto de Investigaciones Agropecuarias de Panama, Facultad de Agronomia, Universidad de Panama.*

Anonymous, (1976e). Proceedings of the research planning conference on root-knot nematodes, *Meloidogyne* spp. *International Institute of Tropical Agriculture, Ibadan, Nigeria.*

Baldwin, J.G. and Sasser, J.N. (1979). *Meloidogyne megatyla* n. sp., (Heteroderidae) a root-knot nematode from Loblolly Pine, *J.Nematol.* (in press).

Bridge, J.G., Jones, E., and Page, L.J., (1976). *Meloidogyne acronea* associated with reduced growth of cotton in Malawi. *Pl.Dis.Reptr.* **60,** 5-7.

Coetzee, V. (1956). *Meloidogyne acronea,* a new species of root-knot nematode. *Nature* **4515,** 899-900.

De Grisse, A. (1960). *Meloidogyne kikuyensis* n. Sp., a parasite of KiKuYu grass (*Pennisetum cladestinum*) in Kenya. *Nematologica* **5,** 303-308.

Elmiligy, I.A. (1968). Three new species of the genus *Meloidogyne* Goeldi 1887. (Nematoda: Heteroderidae). *Nematologica* **14,** 577-590.

Esser, R.P., Perry, V.G., and Taylor, A.L. (1976). A diagnostic compendium of the genus *Meloidogyne* (Nematoda: Heteroderidae). *Proc. Helminthol.Soc.Wash.,* 43, 138-150.

Goeldi, E.A. (1887). Relatoria sobre a molestia do cafeeiro na provincia do Rio de Janeiro. *Arch.Mus.Nac.Rio de Janeiro* 8, 7-121.

Loos, C.A. (1953). *Meloidogyne brevicauda* n. sp. - cause of root-knot in mature tea in Ceylon. *Proc.Helminthol.Soc.Wash.,* 20, 83-121.

Lordello, L.G.E., (1956a). Nematoides que parasitam a soja na regaio de Bauru. *Bragantia* **15,** 55-64.

Lordello, L.G.E. (1956b). *Meloidogyne inornata* sp. n., a serious pest of soybean in the state of Sao Paulo, Brazil (Nematoda: Heteroderidae). *Rev.Brazil Biol.,* **16,** 65-70.

Lordello, L.G.E. and Zamith, A.P. (1958). The morphology of the coffee root-knot nematode *Meloidogyne exigua* Goeldi, 1887. *Proc.Helminthol.Soc.Wash.,* 25, 133-137.

Lordello, L.G.E. and Zamith, A.P. (1960). *Meloidogyne coffeicola* sp. n., a pest of coffee tress in the state of Parana, Brazil. *Rev. Brasil Biol.,* **20,** 375-379.

Ponte, J.J. da., (1969). *Meloidogyne lordelloi* sp. n., a nematode parasite of *Cereus macrogonus* Salm-Dick. *Bolm.Soc.Cearense Agron.,* 10, 59-63.

Roa, V.N., Rao, Y.S. and Israel, P. (1969). Control of parasitic nematodes in rice. *All India Nematol.Symp.* New Delhi, 46.

Roy, A.K. (1975). Studies on resistance of rice to the attack of *Meloidogyne graminicola. Z.PflKrankh.PflSchutz.,* **82,** 384-387.

Sasser, J.N. (1977a). Worldwide dissemination and importance of the root-knot nematodes, *Meloidogyne* spp. *J.Nematol.*, **9**, 26-29.

Sasser, J.N. (1977b). The distribution and ecology of the root-knot nematodes in the potato-production areas of the world. *Organization of Tropical American Nematologists (OTAN), Annual Meeting, Peru,* 16.

Sasser, J.N. and Triantaphyllou, A.C. (1977). Identification of *Meloidogyne* species and races. *J.Nematol.*, **9**, 283.

Singh, S.P. (1969). A new plant parasitic nematode (*Meloidogyne lucknowica* n. sp.) from the root galls of *Luffa cylindrica* (sponge gourd) in India. *Zool.Anz.*, **182**, 259-270.

Spaull, V.W., (1977). *Meloidogyne propora* n. sp. (Nematoda: Meloidogynidae) from Aldabra Atoll, Western Indian Ocean, with a note on *M. javanica* (Treub). *Nematologica* **23**, 177-186.

Taylor, A.L. and Sasser, J.N. (1978). Biology, identification and control of root-knot nematodes (*Meloidogyne* species). *Dept.of Plant Path.N.C.State Univ.Raleigh.*, 111.

Whitehead, A.G. (1959). The root-knot nematodes of East Africa. *Nematologica* **4**, 272-278.

Whitehead, A.G. (1968). Taxonomy of *Meloidogyne* (Nematoda: Heteroderidae) with descriptions of four new species. *Trans Zool.Soc. London* **31**, 263-401.

Discussion

Several participants remarked on the occurrence and distribution of root-knot nematodes in relation to crop losses. Mankau suggested that long-season crops should be added to those listed by Sasser, as the more rapid increase of nematode populations in tropical areas may cause late-season damage, which would not be encountered in temperate areas. Netscher referred to the 100-200% increases in yield of bananas in West Africa obtained with the necessary twice-yearly treatment with Nemagon. Olowe remarked that on government farms in Nigeria, monocropping is practised and yield losses of 60% may occur. Some farms burn the crop remains after harvest and use shifting cultivations so that losses are not so large. *Meloidogyne* occurs in nearly all fields in southern Nigeria; in northern Nigeria, which is mainly savanna, *M. javanica* is more prevalent than other *Meloidogyne* spp. Netscher observed that root-knot nematodes are not a real problem in the wet tropics where any nematode damage is masked by leaf diseases and insect problems. In savanna the

situation is similar to temperate regions as the long dry
season replaces the winter. In Dakar, a rapid increase of
root-knot nematode populations occurs within a few months
but decreases to about 4% in winter, building up again in
the following year. Netscher added that with good crop
rotations few problems are encountered; problems arise
when fields are irrigated and short rotations are practised. He felt that the crop losses for West Africa presented by Sasser are likely to be overestimates in view
of the number of small farms and the practice of shifting
cultivation. Sasser pointed out that in the areas referred to, growers are tending to run out of land where they
can practise "burning" and long rotations. Evans referred
to the need for careful land use in the Carribean region
and explained that where seedbeds are fumigated with DBCP
the transplants from them give economically acceptable
yields in the field.

Bird asked if the control of fungal pathogens in the
tropics would result in a greater expression of root-knot
nematode problems. Sasser replied that this would be so
as the nematodes increase in numbers more rapidly on
healthy plants.

Orion observed that in Israel, *Meloidogyne* is not a
problem on primitive farms and becomes one only where
agriculture is improved. Sasser responded by asking if it
is ever profitable to do anything inefficiently?

C.E. Taylor

*Scottish Horticultural Research Institute,
Invergowrie, Dundee, Scotland*

Introduction

Plant parasitic nematodes are often regarded as pathogens in their own right, capable of producing a single, recognizable disease but some 46 years ago Fawcett (1931) recognised that "nature does not work with pure cultures" and that many plant diseases are influenced by associated organisms. In the soil environment plants are constantly exposed to many and various microorganisms which are likely to influence one another, as they occupy the same habitat. Infection by one pathogen also may alter the host response to subsequent infection by another.

Much of the literature on interactions between microorganisms in a plant disease situation is concerned with interrelationships or interactions between nematodes and fungi. As early as 1892 Atkinson observed that *Fusarium* wilt of cotton was more severe in the presence of root-knot nematode infection. Since this early observation the *Meloidogyne-Fusarium* interaction has been described on several different hosts (Powell, 1963).

Field observations in the 1940s (Smith, 1948) showed that cotton wilt disease, as well as nematodes, was controlled by soil fumigation with ethylene dibromide,which was conceded generally to have little if any fungicidal action. Later experiments in Louisiana (Newsom & Martin,

1953) also showed that fumigation of soils infested with
the cotton wilt organism resulted in striking control of
the disease; in these experiments a wide range of plant
parasitic nematodes were present in the unfumigated plots.

The type of experiments undertaken by Martin *et al.*,
(1956) are illustrative of those which have attempted to
clearly identify an interrelationship between *Meloidogyne*
and *Fusarium* wilt. They studied development of *Fusarium*
wilt disease of cotton (*F. oxysporum* f.sp. *vasinfectum*)
in steam sterilized soil to which were added either nematodes alone, the fungus alone, or combinations of nematode
plus fungus. *M. incognita* and *M. incognita acrita*, but not
Trichodorus sp., *Tylenchorhynchus* sp. or *Helicotylenchus*
sp., significantly increased the incidence of wilt in two
cotton cultivars, Deltapine 15 (wilt susceptible) and
Coker 100 Wilt (wilt resistant).

In Uganda, Perry (1963) showed that DBCP fumigation
of the soil decreased the root-knot galling index from 64
to 18 and this was associated with a decrease in wilt incidence from 14 to 2%; in inoculation experiments the development of wilt was much more extensive in plants growing in soil inoculated with *M. incognita* plus *F. oxysporum*
f.sp. *vasinfectum* compared with plants in soil with the
fungus alone, indicating an interaction between the fungal
pathogen and the nematode.

Anatomical studies of cotton seedling roots led Perry
(1963) to conclude that the entry of *M. incognita* did not
facilitate infection by *F. oxysporum*; also, he found no
evidence of attraction of the fungus to entry points of
nematodes. Other workers studying the interaction of *Meloidogyne* and *Fusarium* in cotton and other crops have reported that mechanical wounding has increased the susceptibility of the host *less* than nematode attack (Jenkins
& Coursen, 1957; Thomason *et al.*, 1959) and they suggest,
as Perry also concludes, that the interaction stems from
the effect of the root-knot nematodes on the physiology

of the host.

Verticillium causes a wilt disease similar to that of *Fusarium* and also appears to be influenced by the presence of plant parasitic nematodes but with *Pratylenchus* species mainly involved in interactions. Investigations by McLellan *et al.* (1955) showed that the incidence of *Verticillium* wilt in cotton was unaffected by the presence of root-knot nematodes. On the contrary, Overman & Jones (1970) showed that the incidence of *Verticillium* wilt in tomato was increased by *Tylenchorynchus capitatus* and by *M. incognita*; however, the optimum temperature for wilt development was 23°C with *T. capitatus* and 29°C with *M. incognita*.

The black shank disease of tobacco caused by *Phytophthora parasitica* var. *nicotianae* is the only *Phytophthora* species that has so far been shown to be enhanced by root-knot infection. Nusbaum & Chaplain (1952) found the disease was greatly reduced when nematodes were controlled by soil fumigation, and since the fumigants at the rates used had little effect on fungi they supposed that this provided strong circumstantial evidence that parasitic nematodes were contributing to the increased damage caused by black shank in certain fields. Glasshouse experiments were undertaken by Sasser *et al.* (1955) to examine this supposition. They grew two cultivars resistant to black shank in sterilized soil to which *P. parasitica* var. *nicotianae* and *Meloidogyne* spp. were added alone and in combination. Those plants inoculated with the fungus plus root-knot nematodes developed black-shank symptoms earlier and to a greater extent than plant grown in soil inoculated only with the fungus. Allowing the plants to become severely infested with root-knot nematodes before adding the fungus produced a level of disease similar to the addition of both organisms together. The experiments showed that root-knot nematodes increased the severity of black shank, but that factors other than mechanical injury by the nematodes were involved in the breakdown of resistance, possibly

biochemical alterations of the host cells caused by the
nematodes making them a more congenial substrate for the
fungus.

Rhizoctonia solani causes damping-off in seedlings of
many different crop plants. *M. incognita acrita* increased
the incidence of the disease in cotton (Reynolds & Hansen,
1957) and the adverse effect of the fungus on the emergence of soybean seedlings was greatly increased by *M. haplc*
and *M. javanica*. In a series of laboratory experiments
Batten & Powell (1971) found that *R. solani* was unable to
colonize healthy tobacco roots but when the plants had
been exposed to *M. incognita* for at least 10 days, the
fungus was then capable of penetrating the roots and promoting necrosis. As also noted with fungi other than *R.
solani* (Porter & Powell, 1967) predisposition of plants
to attack by some microorganisms is most pronounced after
the nematodes have been in the roots for some time, supp-
orting the suggestion made by many authors that physiological changes in galled roots are the key to the interactions between fungi and root-knot nematodes.

Associations of root-knot nematodes with other fungi
are less extensively reported. Agarwal & Goswani (1973)
reported a significant synergistic effect in soybean when
M. incognita preceded *Macrophomina phaseoli* by three weeks,
and considered that the root-knot nematode predisposes
plants to fungal infection, making them more susceptible
than when the fungus is present alone. Fungal infection
of jute seedlings did not appear to be enhanced by root-
knot infections (Majumdar & Goswani, 1974).

In experiments designed to study the role of root-knot
nematodes in the production of root necrosis of celery,
M. hapla caused more damage in the presence of a natural
soil microflora than in autoclaved soil (Starr & Mai,
1976); of the numerous organisms isolated from necrotic
root-knot galls, *Pythium polymorphon* appeared to be the
most important cause of root necrosis.

The earliest published work of an association between
nematodes and a bacterial disease appears to be that of
Hunger (1901) who showed that tomatoes were readily attacked by the wilt bacterium, *Pseudomonas solanacearum*,
in root-knot infested soil but remained healthy in nematode free soil. Lucas *et al.* (1955) demonstrated an interrelationship between Granville wilt or southern bacterial wilt, caused by *P. solanacearum*, and root-knot
nematodes in tobacco. They showed that tobacco plants
grown in soil with the bacterium and *M. incognita acrita*
exhibited wilt symptoms earlier and to a greater extent
than plants grown in soil infested only with the bacterium. The development of wilt symptoms in plants growing
in soil infested with bacteria and nematodes was comparable to that in plants inoculated by pouring a bacterial
suspension on freshly cut roots, and hence the authors
suggestion that the role of the nematodes was principally
that of providing wounds in the roots through which the
bacteria may enter. Similarly, Stewart & Schindler (1956)
also concluded that root-knot and other nematodes provided
wounds for the entry of the bacterial wilt organism, *Pseudomonas caryophyli*, in carnations. However, the investigations of Fukudome & Sakusegawa (1972) showed that the
symptoms of Granville wilt on tobacco were more severe
when *M. incognita* and *P. solanacearum* were inoculated
than when a bacterial inoculum was added to articifially
wounded roots, suggesting a synergistic effect between
bacteria and nematodes. Also, *M. incognita* was found to
increase the severity of bacterial canker (*Corynebacterium
michiganense*) on both susceptible and resistant cultivars
of tomato when nematode was inoculated before the bacterium, but not when the pathogens were concomitantly inoculated or the bacterium inoculated 10 days before the
nematode (Moura *et al.*, 1975); again suggesting a more
complex interaction than nematodes being involved simply
as wounding agents.

Interaction between plant and fungus

Root-attacking fungi typically invade only living roots and in the absence of the host generally exist in the soil as resting spores or sclerotia. Interaction between the plant host and fungal pathogens may be divided into three consecutive phases which although interdependent are physically separated by the host surface -

(i) growth of the pathogen prior to penetration,

(ii) penetration,

(iii) growth within the host.

Fungal spores apparently remain dormant in the soil until stimulated to germinate by changes in the micro-environment promoted by the presence of host roots. There is abundant evidence that plant exudates influence the germination and growth of many soil microorganisms. Hiltner (1904) introduced the term "rhizosphere" to describe the abundant association of microorganisms with plant roots and speculated whether the microflora might be stimulated by soluble organic substances, but Knudson (1920) provided the first conclusive evidence of exudation from roots by identifying carbohydrates, amino acids, organic acids, vitamins, nucleotides, flavonones and enzymes. These substances represent the contents of the cytoplasm which are released to the soil in sloughed-off root cap cells, by injured cells and root hairs, or through autolysis of epidermal cells. Exudation is affected by many factors. For example, temporary wilting increases the release of amino acids; high light and temperature increases exudation, especially during the first few weeks of plant growth. Patterns of exudation may be altered by the presence of microorganisms - they may alter the permeability of root cells, modify the metabolism of the roots, or may modify some of the material released from the roots. The microflora of the rhizoplane and rhizosphere may be regarded as a selective sieve, absorbing some exudate constituents and in turn releasing exudates from their own cells.

Spore germination in the rhizosphere occurs as the result of exudation of carbon and nitrogen compounds, principally amino acids and simple sugars. The influence of root exudates on germination is believed to be largely unselective, but there are many references (Schroth & Hildebrand, 1964) to possible toxic components of root exudates which may specifically affect fungal pathogens. For example, in laboratory experiments Buxton (1957) showed that exudates from each of three pea cultivars, differentially susceptible to three races of *Fusarium oxysporum* f.sp. *pisi* specifically depressed conidium germination of the races to which the cultivar appeared to be resistant. Mycelial growth of the three races of the fungus was inhibited by exudates from resistant plants and stimulated by those from susceptible cultivars.

Following germination, the fungus must reach its host by vegetative growth involving the further expenditure of energy which is obtained from root exudates. After reaching the host root some pathogens may penetrate with a minimum of ectotrophic growth, whereas other produce a mycelial mat or thallus on the host before penetrating. In *Rhizoctonia solani* the most frequent type of structure is the dome-shaped infection cushion formed by the aggregation of short, stubby side branches of the hyphae. Infection cushions are closely appressed to the host surface and penetration of the host cells is usually by means of fine infection pegs from the swollen, flattened cells at the base of the structure and in contact with the plant surface. *Fusarium* species form a thallus on the root surface from which they penetrate the host. The motile zoospores of *Phytophthora* species encyst on the root surface and then produce a hypha which grows into the epidermal cell. This appears to be a non-specific association in many species but further development in the root, beyond the first cell infection, is highly specific.

Bacteria can only enter plants through wounds or possibly through natural openings, or by vectors. The role of nematodes is that of wounding agents in bacterial associations as noted earlier, but modification of plant cells as caused by *Meloidogyne* and other nematodes may provide a substrate which enhances or detracts from bacterial infection. The act of wounding may also enhance infection by bacteria because of increased cell leakage, rather than by simply provided a method of entry; nutrients at the wound may be non-specific or may be inhibiting to the pathogen.

Mechanisms of interaction between nematodes and other plant pathogens

Experimental evidence is sufficient to indicate a biological interaction between *Meloidogyne* and other plant pathogens, particularly certain soil-borne fungi. However, it should be recognised that no experiments have been conducted under sterile conditions which can completely exclude all organisms other than the three being considered viz. host plant, nematode, pathogen, and hence it cannot be proven beyond all doubt that the disease observed is solely due to the interaction between nematode and pathogen. But assuming the simple situation of a one to one relationship, the question remains as to the nature of the interaction and in particular to the role of *Meloidogyne* in enhancing the development of the disease. The possible roles that nematodes might play in disease complexes have been the subject of several reviews (Powell, 1963; 1971a,b; Pitcher, 1965; Bergeson, 1972) and are outlined below.

Vectors of fungal pathogens

Some nematodes may carry fungal spores on their external surface and bacteria may be transported by ingestion. Nematode transport increases the mobility of the pathogens, but with a few exceptions the association is non-specific.

Viruses have been shown to be specifically transmitted by nematodes but the vectors are confined to the ectoparasitic genera *Xiphinema, Longidorus, Trichodorus* and *Paratrichodorus*. *X. index* has been implicated as a vector of Rickettsia-like organisms causing yellows disease of grapevines (Rumbos *et al.*, 1977). *Meloidogyne* species have not been implicated as vectors of any microorganisms.

Mechanical wounding agents

All plant parasitic nematodes produce a wound of some sort in the plant host during the process of feeding, either by a simple micro-puncture or by rupturing or separating cells. *Meloidogyne* larvae usually penetrate near the root cap with subsequent penetration into the meristematic zone of the root apex; evidence of entry can been seen by dot-like penetration marks but once within the plant the larvae move intercellularly without inducing necrosis in host cells surrounding the nematode body. Nevertheless, previously and generally nematodes were thought simply to provide avenues of entry for fungal or bacterial pathogens. This concept remained until it was observed that the severity of the fungal disease was greater when the nematode was present in the plant host several weeks prior to exposure to the fungus than when the host was exposed to both pathogens simultaneously. Whilst more complicated interactions underlie fungus-nematode complexes, it is likely that wounding by the nematode still accounts for interaction with bacteria.

Host modifiers

All plant parasitic nematodes induce changes in their host tissue as part of the feeding process and in order to derive nourishment from it. *Meloidogyne* nematodes induce syncytia, primarily in the stele, which are highly specialized structures in which there is an accumulation of certain chemicals resulting from the nematode-induced cellular modification. Owens & Specht (1966) measured increases in several organic and inorganic substances in the galled tissue of tomatoes. Goswani *et al.* (1976) measured increased of calcium and magnesium in tomatoes infested by *M. incognita* and increases in phosphorus have been noted by other investigators (Ishibashi & Shimizu, 1970; Oteifa & El-Gindi, 1962).

The mechanism whereby nutrients accumulate in *Meloidogyne* giant cells is still a matter for speculation and investigation. Owens and Specht (1965) observed that giant cells are sharply contrasted against the surrounding cells and suggest that this is indicative of a decrease in the permeability of the cell walls. Dropkin (1972) suggested a transformation in membrane-permeability which would allow some ions to enter the cell in larger quantities, but Wang *et al.* (1975) concluded from their experiments with *M. incognita* on tomato that there was no significant change in membrane permeability. Owens & Bottino (1966) found differences in the constituents of giant cell walls compared with those of healthy tissues; giant cell walls contained cellulose and pectin, but no lignin, suberin or starch and because of the absence of arabinoses they concluded that nematode infection had induced a significant change in pentose metabolism.

In general, it may be surmised that the biochemical changes induced by the nematode may be favourable to the fungal or bacterial pathogen, resulting from
a) a nutritionally improved substrate
b) the destruction of certain chemicals antagonistic to the pathogen; or obstruction of defence reactions by which the host would normally fend off invaders.

Bergeson (1972) considers that the influence of *Meloidogyne* on host tissues is to maintain them in a juvenile state, which is in effect a nutritional improvement and would enhance infection by many fungi. Brodie and Cooper (1964) found that cotton seedlings infected by root-knot

nematodes, and hence delayed in their maturation, remained
susceptible to damping-off by *Rhizoctonia solani* and *Pythium debaryanum* for much longer than did healthy seedlings.
Powell (1968) showed that the incitant effect of the nematode on the fungi occurred only when the nematode had infected the plant for 3 to 4 weeks and galled tissue had
been formed and considered that the galled tissue afforded
an improved substrate for the growth of the pathogens. In
a subsequent study, Batten & Powell (1971) observed that
R. solani was unable to colonise healthy tobacco roots
unless preceded by root-knot nematode infection and concluded that "the nematode is capable of elevating a normally minor pathogen to major status". Similarly, Mayol &
Bergeson (1970) found that whilst the general soil microflora could account for only a non-significant 18% reduction in the growth of tomato plants, when the plants
were infected with root-knot nematodes the growth reduction
was increased to 75%, compared with 37% attributed to nematodes alone.

Many authors have illustrated by experimental evidence
that host plants are predisposed to fungal infection only
when *Meloidogyne* infection precedes the addition of the
fungal inoculum by 2 to 4 weeks. It has also been noted
that the greatest metabolic activity and physiological
change occurs in the giant cell and galled tissue when
the nematode development is at the stage of moulting to
the adult and at the commencement of egg laying, which is
usually 3-4 weeks after initial infestation by the larvae.
At this time there is a considerable increase in proteins,
nucleic acids and DNA, together with increases in sugars,
and various inorganic ions. Which, if any, of the various
substances is of special importance in stimulating the
growth of fungal pathogens is unknown, although it has
been stated that amino acids have a positive influence
on the virulence and growth of *Fusarium* (Van Andel, 1956;
Woltz & Jones, 1970; Jones & Woltz, 1972).

The effect of nematode infection is not necessarily
confined to the locality of the galled tissue but may be
systemically translocated to other parts of the host. In
experiments in which the root system was divided between
two different soil environments (the split-root technique)
it has been shown that *Meloidogyne* can result in a systemic change in the host so that plant tissues remote from
the site of nematode infection are rendered more susceptible to fungal infection. Bowman & Bloom (1966) exposed
one part of the root system of a wilt-resistant variety
of tomato to *Fusarium oxysporum* f.sp. *lycopersici* and the
other to *M. incognita*, and demonstrated that the plant
became diseased even though the nematode and fungus were
on separate root systems. Melendez & Powell (1967) found
hyphae of *Fusarium* wilt fungus growing vigorously within
the vascular system well above the soil line and far from
the site of nematode infection. Wang & Bergeson (1974)
showed in experiments with tomatoes and *M. incognita* that
as the root-knot infection progressed, the volume of xylem
sap gradually decreased, probably as the result of a reduction in absorbing surface of galled roots (Bergeson,
1968) and injury to vascular tissue (Bloom & Burpee, 1973).
An increase in sugars in xylem sap from infected plants
was considered to be possibly due to intensified photo-
synthsis of infected plants resulting in excessive sugar
translocation to galled roots (Bodrova, 1961) and this
increased as the nematode inoculum increased at 2, 4 and
6 weeks after inoculation. Wang & Bergeson (1974) concluded that the increase in total sugars and decrease in
amino acids of xylem sap in nematode-infected plants and
the similar difference in root exudates are a probable
mechanism by which tomato plants are predisposed to *Fusar-*
ium wilt.

Rhizosphere modifiers

A wide range of organic compounds exude from intact

plant roots, and soil microbiologists have demonstrated the various effects they have on the soil microflora. The exudates may be attractants for the motile stage of the plant pathogens or may be a source of nutrients for the microflora (Rovira, 1965), or may be a stimulus for the germination of dormant spores. Nematodes may directly effect the release of exudates in a quantitative way by rupturing the root cell membranes during feeding, penetration and migration within the root. *Meloidogyne* species probably do little damage of this sort except when they break through the epidermis to extrude their eggs on the root surface. *Meloidogyne* nematodes probably influence the rhizosphere by indirectly inducing changes in the root exudates.

The activity of *Fusarium* spp. in the rhizosphere is strongly influenced by the host root exudates. For example, the chlamydospores of *F. oxysporum* f.sp. *pisi* germinate more readily near disease-susceptible pea seeds than in the region of disease-resistant seeds and young roots are more effective in this respect than mature roots (Schroth & Snyder, 1961; Cook & Snyder, 1965; Kraft, 1974). This activity has been correlated with increased amounts of amino acids and sugars released from susceptible varieties and from seedlings.

Golden & Van Gundy's (1972) study of the penetration response of *R. solani* to stimuli originating from *M. incognita*-infected roots and passing through a semi-permeable cellophane membrane suggested that root exudates were responsible for attracting the fungus to the galls and for initiating sclerotial formation. They reasoned that if the stimuli inducing *R. solani* attack of nematode-infected roots were root exudates, then removal of the exuded substances as they were produced should reduce the severity of root disease. In subsequent experiments (Van Gundy *et al.*, 1977) it was shown that when tomato plants were inoculated simultaneously with *M. incognita* and *R. solani* and

subjected to continuous leaching to remove the exudates,
no root decay occurred. In contrast when leachates were
collected from *M. incognita* infected roots and applied to
roots of tomatoes inoculated with *R. solani* alone, a severe rot developed in roots receiving the leachates; a
severe rot also developed on tomato roots inoculated with
M. incognita and *R. solani* and not subjected to leaching.
During the first 14 days following nematode infection,
carbohydrates were the major organic constituents in exudates leaking from nematode-infected roots but 14 days
after nematode infection, nitrogenous compounds became
the major constituents leaking from the roots. This shift
in the C/N ratio of root exudates was associated with the
development of *R. solani* sclerotia on the gall surfaces
14-21 days following nematode infection.

Van Gundy *et al.* (1977) found that the levels of
electrolytes leaking from nematode-infected roots increased progressively, following infection, from 7 to 42
days, with maximum leakage occurring after 21 days. Between days 28 and 35 after initial infection, the swollen
females ruptured the cortex and epidermis of the roots
and deposited their eggs in a gelatinous matrix on the
root surfaces. Minerals and amino acids were the major
electrolytes detected on root gall exudates. Wang *et al.*
(1975) on the contrary, found a lower percentage of electrolyte leakage from galled roots than from healthy ones
and suggested that an increase in nutrients released to
organisms in the rhizosphere by galled roots may result
from a higher internal concentration of these nutrients,
rather than from a significant change in membrane permeability. As Van Gundy *et al.* (1977) observe, the 60% increase in the concentration of electrolytes in galled
roots homogenates found by Wang & Bergeson (1974) confirms
that galls do act as nutrient sinks.

The concept of nematode-induced changes in root exudates relates to the effect on the pathogen in the

rhizosphere, and experimental evidence points to some en-
hancement such as the increase in *F. oxysporum* f.sp.*lycopersici* observed by Bergeson *et al.* (1970) in the rhizo-
sphre of tomato infected with *M. javanica*. But they also
noted that whereas *Fusarium* propagules increased, actino-
mycetes around galled roots significantly decreased. Act-
inomycetes are antagonists of *Fusarium* (Walker, 1965;
Mitchell & Alexander, 1962) and thus root-knot infection
may have the effect of stimulating the plant pathogen
while at the same time inhibiting their antagonists (Bergeson, 1972). Thus modification of the rhizosphere resulting from changes in root exudates has a general influence
on the rhizosphere microflora, with the possibility of a
chain of reactions in the nematode-fungus disease complex.

Resistance breakers

Many of the early observations on root-knot fungus interactions referred to the breakdown of resistant cultivars
which led to the necessity of breeding plants resistant to
both nematode and fungus. Experiments by Jenkins & Coursen
(1957) illustrate the sort of results that have been obtained in a variety of circumstances. They exposed three
cultivars of tomato (susceptible, moderately resistant
and resistant to *Fusarium*) with and without *M. incognita
acrita* and *M. hapla*. In the susceptible cv. Red Beefsteak,
100% wilt occurred 15 days after inoculation by the fungus
and this was unaffected by the presence of the nematodes.
With cv. Rutgers, with some resistance to *Fusarium* wilt,
the presence of the fungus alone resulted in 60% wilt
after 22 days, and this was increased to 100% in 17 days
when the nematodes were present. The highly resistant cv.
Chesapeake showed no wilt with the fungus alone, compared
with 100% wilt in 26 days with *M. incognita* and the fungus and 60% wilt after 26 days with *M. hapla*. The authors
considered that the nematodes cause some host response
that lowers the natural resistance of the tomato plants

to fungal infection, with differences between the two
species of *Meloidogyne*. Goode & McFuire (1967) suggested
that root-knot nematode infection sensitizes certain tomato cultivars to attack by races of *F. oxysporum* f.sp.
lycopersici that are ordinarily non-pathogenic in those
cultivars, possibly because the fungus mutates within the
nematode-infected host.

   Meloidogyne infection may not always result in the
breakdown of resistance to *Fusarium* wilt. Jones *et al.*
(1976) found that the resistance to races of *F. oxysporum*
f.sp. *lycopersici* in the tomato cultivars Manapal and
Florida MH-1 was unaffected by *M. incognita*. Fassuliotis
& Rau (1969) found that *M. incognita* did not reduce the
resistance of cabbage cultivars to *Fusarium* wilt and Johnson & Littrell (1969) reported that three different species
of *Meloidogyne* failed to break the *Fusarium* wilt resistance in *Chrysanthemum morifolium* Ram. cv. White Iceberg.

   Fassuliotis & Rau (1969) postulated two types of resistance to *Fusarium* wilt. In the first, a qualitative
high resistance is dependent on a basic incompatibility
of the host and the fungal pathogen that is unaffected by
nematode infection; the genetic basis of this interaction
has been examined by Sidhu & Webster (1974, 1977). The
second type of resistance is regarded as quantitative and
is readily influenced by nematode infection. Pitcher
(1965) refers to this type of breakdown in resistance as
a shift in the host-parasite equilibrium due to improved
nutrition for the fungus. Bergeson (1972) puts it simply
as "making a poor host a good host".

Other interactions

   Interactions between root-knot nematodes and mycorrhizal fungi are described in detail by Sikora in these
Proceedings. The following examples briefly indicate the
nature and variations of the interaction. Riffle (1973)
found that roots altered morphologically by mycorrhizal

symbionts appeared to be readily invaded by root-knot nematodes and supported giant cell development. Kellam & Schenck (1977) found there was less galling produced by *M.*
incognita in soybean in the presence of *Glomus macrocarpus*
that on roots without the fungus, and that the presence
of the nematode appeared to have little effect on the fungus. The endomycorrhizal *Gigaspora margarita* increased
the growth of cotton and nullified the stunting caused
by *M. incognita* in its absence (Roncadori & Hussey, 1977).

Some interactions have been reported between *Meloidogyne*
and virus infection, but these are difficult to interpret
because both organisms would seem to interact only *via*
the physiological processes of the plant. Ryder & Crittenden (1962) reported a synergistic interaction between root-
knot nematodes and tobacco ringspot virus in soybean. Bird
(1969) observed that *M. javanica* developed more rapidly
in roots of tomato plants infected with tobacco mosaic
virus, but although more nematode larvae entered the roots
of plants infected with tobacco ringspot virus they did
not develop at different rates from larvae in uninfected
plants. *Vigna sinensis* infected with cowpea mosaic virus
and exposed to *M. incognita* had fewer galls per g. root
than uninfected plants (Goswani *et al.*, 1974). On the contrary tobacco plants infected with tobacco ringspot virus
and exposed to *M. incognita* had more galls than healthy
plants with nematodes (Osores-Duran quoted *in* Powell,
1971b). The plants were also less stunted when inoculated
with both *M. incognita* and the virus than when the nematodes were present alone, and a similar effect was also
observed with *M. incognita* and cucumber mosaic virus.

Conclusions on complexes

Wounding by nematode penetration of roots may be responsible for increased susceptibility of plants to bacterial infections, but interactions with fungal pathogens
are obviously more complex. The observation that root-knot

nematodes must be present in the roots for 3 or 4 weeks before the plant is maximally predisposed to fungal infection points to physiological changes in host tissues as a major mechanism involved. This supposition is also supported by the observation that the predisposition to infection may be systemically translocated in the host from the site of nematode activity, as demonstrated by "split-root" experiments (Bowman & Bloom, 1966).

There is now ample evidence that the biochemical constituents of plant roots are substantially changed by *Meloidogyne* infection. Several authors have suggested that such changes manifested in galled tissues make them nutritionally favourable for the root-invading fungi and increase their rate of growth. Nematode infection may also quantitatively and qualitatively influence root exudates and so influence germination of resting spores of fungal pathogens in the rhizosphere, the attraction of their motile forms to the roots, and the development of the thallus on the root surface.

There is little doubt that root-knot nematodes can considerably influence the infection of roots by several fungi. However, in most *Meloidogyne*-fungi interrelationships there seem to be no benefits conferred on the nematodes and in most situations it would appear that nematodes are more often harmed than benefited by concomitant infections. For example, Powell & Nusbaum (1960) reported that giant cells were relatively more sensitive to *Phytophthora parasitica* than surrounding cells and were killed within 72 hr after invasion; and the development of *M. incognita acrita* on cabbage roots was drastically influenced by *Plasmodiophora brassicae* (Ryder & Crittenden, 1965). The influence of fungi on nematodes is probably competitive, as both compete for the same host substrate.

Interactions between *Meloidogyne* and microorganisms provide a fertile field for numerous types of investigations but the observation of Powell *et al.* (1971) may

serve as a cautionary note - "it is possible that the failure of disease control practices in certain situations
could be due to incomplete diagnosis of disease complexes,
resulting in incomplete or inappropriate control treatment.
It would seem unwise to give primary emphasis to root
diseases of plants caused by a single pathogen, when under
natural growing conditions, this is indeed rare".

References

Agarwal, D.K. and Goswani, B.D. (1973). Interrelationships between a
    fungus *Macrophomina phaseoli* (Maubl.) Ashby and root-knot nematode
    Meloidogyne incognita (Kofoid and White) Chitwood in soybean (*Gly-
    cine max* L.) Merill. *Proc.Indian Nat.Sc.Acad.B.*, **39**, 701-704.
Atkinson, G.F. (1892). Some diseases of cotton. *Bull.Ala.agric.Exp.
    Sta.*, **41**, 61-65.
Batten, C.K. and Powell, N.T. (1971). The *Rhizoctonia-Meloidogyne*
    disease complex in flue-cured tobacco. *J.Nematol.*, **3**, 164-169.
Bergeson, G.B. (1968). Evaluation of factors contributing to the path-
    ogenicity of *Meloidogyne incognita*. *Phytopathology* **58**, 49-53.
Bergeson, G.B. (1972). Concepts of nematode-fungus associations in
    disease complexes: a review. *Expl.Parasitol.*, **32**, 301-314.
Bergeson, G.B., Van Gundy, S.D. & Thomason, I.J. (1970). Effect of
    Meloidogyne javanica on rhizosphere microflora and *Fusarium* wilt
    of tomato. *Phytopathology* **60**, 1245-1249.
Bird, A.F. (1969). The influence of tobacco ringspot virus and tob-
    acco mosaic virus on the growth of *Meloidogyne javanica*. *Nematol-
    ogica* **15**, 201-209.
Bloom, J.R. and Burpee, L.L. (1973). Root-knot causes reduced root
    pressure in some tomato varieties. *Phytopathology* **63**, 199.
Bodrova, I.M. (1961). Peculiarities in the physiological processes
    and the influence of nutrients on the growth and development of
    plants infected with the root-knot nematode. *Tr.Vses.Inst.Zasch.
    Rast.*, **16**, 76-88.
Bowman, P. and Bloom, J.R. (1966). Breaking the resistance of tomato
    varieties to *Fusarium* wilt by *Meloidogyne incognita*. *Phytopathology*
    56, 871.
Brodie, B.B. and Cooper, W.G. (1964). Relation of parasitic nematodes
    to post-emergence damping-off of cotton. *Phytopathology* **54**, 1023-
    1027.
Buxton, E.W. (1957). Some effects of pea root exudates on physiologic
    races of *Fusarium oxysporum* fr. f. *pisi* (Linf.) Snyder & Hansen.
    Trans.Brit.Mycol.Soc., **40**, 145-154.
Cook, R.J. and Snyder, W.C. (1965). Influence of host exudates on
    growth and survival of germlings of *Fusarium solani* f. *phaseoli*
    in soil. *Phytopathology* **55**, 1021-1025.
Dropkin, V.H. (1972). Pathology of *Meloidogyne* galling, giant cell
    formation, effects on host physiology. *OEPP/EPPO Bull.*, **6**, 23-32.
Fassuliotis, G. and Rau, G.J. (1969). The relationship of *Meloidogyne
    incognita acrita* to the incidence of cabbage yellows. *J.Nematol.*,

1, 219-222.

Fawcett, H.S. (1931). The importance of investigations on the effects of known mixtures of organisms. *Phytopathology* 21, 545-550.

Fukudome, N. and Sakasegawa, Y. (1972). Interaction between root-knot nematode (*Meloidogyne incognita*) on the occurrence of Granville wilt on tobacco. *Proc.Assoc.Pl.Prot.Kyushu.*, 18, 100-102.

Golden, J.K. and Van Gundy, S.D. (1972). Influence of *Meloidogyne incognita* on root development by *Rhizoctonia solani* and *Thielaviopsis basicola* in tomato. *J.Nematol.*, 4, 225.

Goode, M.J. and McGuire, J.M. (1967). Relationship of root-knot nematodes to pathogenic variability in *Fusarium oxysporum* f.sp. *Lycopersici*. *Phytopathology* 57, 812.

Goswami, B.K., Singh, S. and Verma, V.S. (1974). Interaction of a mosaic virus with root-knot nematode *Meloidogyne incognita* in *Vigna sinensis*. *Nematologica* 20, 366-376.

Goswami, B.K., Singh, S. and Verma, V.S. (1976). Uptake and translocation of calcium and magnesium in tomato plants as influenced by infection with root-knot nematode *Meloidogyne incognita* and tobacco mosaic virus. *Nematologica* 22, 116-117.

Hiltner, L. (1904). Über neuere Erfahrungen und Probleme auf dem Gebiet der Bodenbakteriologie und unter besonderer Berücksichtigung der Grundungung und Brache. *Arb.Deut.Landwirsch.Ges.*, 98, 59-78.

Hunger, F.W.T. (1901). Een bacterie-ziekte der tomaat. *S Lands Plantentuin.Meded.*, 48, 1-57.

Ishibashi, N. and Shimizu, K. (1970). Gall formation by root-knot nematode, *Meloidogyne incognita* (Kofoid and White, 1919) Chitwood 1949, in grafted tomato plants and accumulation of phosphates on the gall tissues (Nematoda:Tylenchida). *Appl.Entomol.Zool.*, 5, 105-111.

Jenkins, W.R. and Coursen, B.W. (1957). The effect of root-knot nematodes, *Meloidogyne incognita acrita* and *M. hapla*, on *Fusarium* wilt of tomato. *Pl.Dis.Rptr.*, 41, 182-186.

Johnson, A.W. and Littrell, R.H. (1969). Effect of *Meloidogyne incognita*, *M. hapla* and *M. javanica* on the severity of *Fusarium* wilt of chrysanthemum. *J.Nematol.*, 1, 122-125.

Jones, J.P. and Woltz, S.S. (1972). Effect of amino acids on development of *Fusarium* wilt of resistant and susceptible tomato cultivars. *Proc.85th Ann.Mtg.Fla State Hort.Soc.Miami Beach,USA.*, 148-157.

Jones, J.P., Overman, A.J. and Crill, P. (1976). Failure of root-knot nematode to affect *Fusarium* wilt resistance of tomato. *Phytopathology* 66, 1339-1341.

Kellam, M.K. and Schenk, N.C. (1977). The effect of a vesicular-arbuscular mycorrhizal fungus, *Glomus macrocarpus*, on the amount of galling produced by *Meloidogyne incognita* on soybean. *Proc.Amer. Phytopath.Soc.Ann.Mtg. 1977. Abstr.No.192*

Knudson, L. (1920). The secretion of invertase by plant roots. *Am.J. Botany* 7, 371-379.

Kraft, J.M. (1974). The influence of seedling exudates on the resistance of peas to *Fusarium* and *Pythium* root-rot. *Phytopathology* 64, 190-193.

Lucas, G.B., Sasser, J.M. and Kelman, A. (1955). The relationship of root-knot nematodes to Granville wilt resistance in tobacco. *Phytopathology* 45, 537-540.

Majumdar, A. and Goswami, B.K. (1974). Studies on the interaction of a fungus, *Macrophomina phaseolina* (Maubl) Ashby and root-knot nematode *Meloidogyne incognita* (Kofoid & White) Chitwood in jute (*Corchorus capsularis* L.). *Labdev.J.Sc.Techn.India* 12B, 64-66.

Martin, W.J., Newson, L.D. and Jones, J.E. (1956). Relationship of nematodes to the development of *Fusarium* wilt in cotton. *Phytopathology* 46, 285-289.

Mayol, P.S. and Bergeson, G.B. (1970). The role of secondary invaders in *Meloidogyne incognita* infection. *J.Nematol.*, 2, 80-83.

McClellan, W.D,, Wilhelm, S. and George, A. (1955). Incidence of verticillium wilt in cotton not affected by root-knot nematodes. *Pl.Dis. Rptr.*, 39, 226-227.

Meléndez, P.L. and Powell, N.T. (1967). Histological aspects of the *Fusarium* wilt - root knot complex in flue-cured tobacco. *Phytopathology* 57, 286-292.

Mitchell, R. and Alexander, M. (1962). Microbiological processes associated with the use of chitin for biological control. *Proc. Soil Sci.Soc.Amer.*, 26, 556-558.

Moura, R.M. de., Echandi, E. and Powell, N.T. (1975). Interaction of *Corynebacterium michiganense* and *Meloidogyne incognita* on tomato. *Phytopathology* 65, 1332-1335.

Newsom, L.D. and Martin, W.J. (1953). Effects of soil fumigation on populations of parasitic nematodes, incidence of *Fusarium* wilt, and yield of cotton. *Phytopathology* 43, 292-293.

Nusbaum, C.J. and Chaplain, J.F. (1952). Reduction of the incidence of black shank in resistant tobacco varieties by soil fumigation. *Phytopathology* 42, 15.

Oteifa, B.A. and El-Gindi, D.M. (1962). Influence of *Meloidogyne javanica* on host nutrient uptake. *Nematologica* 8, 216-220.

Overman, A.J. and Jones, J.P. (1970). Effect of stunt and root-knot nematodes on *Verticillium* wilt of tomato. *Phytopathology* 60, 1306.

Owens, R.G. and Bottino, B.F. (1966). Changes in host cell wall composition induced by root-knot nematodes. *Contrib.Boyce Thompson Inst.Plant Res.*, 23, 171-180.

Owens, R.G. and Specht, H.N. (1966). Biochemical alterations induced in host tissues by root-knot nematodes. *Contrib. Boyce Thompson Inst.Plant Res.*, 23, 181-198.

Perry, D.A. (1963). Interaction of root-knot and *Fusarium* wilt of cotton. *Emp.Cott.Gr.Rev.*, 40, 41-47.

Pitcher, R.S. (1963). Interrelationships of nematodes and other pathogens of plants. *Helminth.Abstr.*, 34, 1-17.

Powell, N.T. (1963). The role of plant-parasitic nematodes in fungus diseases. *Phytopathology* 53, 28-34.

Powell, N.T. (1968). Disease complexes in tobacco involving interactions between *Meloidogyne incognita* and soil-borne fungal pathogens. *Proc.1st Int.Congr.Pl.Path.London.*, 153.

Powell, N.T. (1971a). Interactions between nematodes and fungi in disease complexes. *Ann.Rev.Phytopath.*, 9, 253-274.

Powell, N.T. (1971b). *In* "Plant Parasitic Nematodes." (Eds. B.M. Zuckerman, W.F. Mai & R.A. Rohde). 2, 119-136. Academic Press, New York.

Powell, N.T., Meléndez, P.L. and Batten, C.K. (1971). Disease complexes in tobacco involving *Meloidogyne incognita* and certain soil-borne fungi. *Phytopathology* 61, 1332-1337.

Porter, D.M. and Powell, N.T. (1967). Influence of certain *Meloidogyne* species on *Fusarium* wilt development in flue-cured tobacco. *Phytopathology* 57, 282-285.

Reynolds, H.W. and Hanson, R.G. (1957). *Rhizoctonia* disease of cotton in presence or absence of the cotton root-knot nematode in Arizona. *Phytopathology* 47, 256-261.

Riffle, J.W. (1973). Histopathology of *Pinus ponderosa* ectomycorrhizae infected with a *Meloidogyne* species. *Phytopathology* 63, 1034-1040.

Roncadori, R.W. and Hussey, R.S. (1977). A joint interaction between the endomycorrhizal fungus *Gigaspora margarita* and root-knot nematode on cotton. *Proc.Amer.Phytopath.Soc.Ann.Mtd.1977.*, Abstr No 195.

Rovira, A.D. (1965). In "Ecology of Soil-borne Plant Pathogens." (Eds. K.F. Baker & W.C. Snyder.) 571. Univ. Calif. Press, Berkeley.

Rumbos, I., Sikora, R.A. and Nienhaus, F. (1977). Rickettsia-like organisms in *Xiphinema index* Thorne & Allen found associated with yellows disease of grapevines. *Z.PflKrankh.Pfl Schutz.*, 84, 240-243.

Ryder, H.W. and Crittenden, H.W. (1962). Interrelationships of tobacco ringspot virus and *Meloidogyne incognita acrita* in roots of soybean. *Phytopathology* 52, 165-166.

Ryder, J.W. and Crittenden, H.W. (1965). Relationship of *Meloidogyne incognita acrita* and *Plasmodiophora brassicae* in cabbage roots. *Phytopathology* 55, 506.

Sasser, J.N., Lucas, G.B. and Powers, H.R. (1955). The relationship of root-knot nematodes to black shank resistance in tobacco. *Phytopathology* 45, 459-461.

Schroth, M.N. and Hildebrand, D.C. (1964). Influence of plant exudates on root infecting fungi. *Ann.Rev.Phytopath.*, 2, 101-132.

Schroth, M.N. and Snyder, W.C. (1961). Effect of host exudates on chlamydospore germination of the bean root rot fungus, *Fusarium solani* f. *phaseoli*. *Phytopathology* 51, 389-393.

Smith, A.L. (1948). Control of cotton wilt and nematodes with a soil fumigant. *Phytopathology* 58, 943-947.

Sidhu, G. and Webster, J.M. (1974). Genetics of resistance in the tomato to root-knot nematode-wilt-fungus complex. *J.Heredity* 65, 153-156.

Sidhu, G. and Webster, J.M. (1977). Predisposition of tomato to the wilt fungus (*Fusarium oxysporum lycopersici*) by the root-knot nematode (*Meloidogyne incognita*). *Nematologica* 23, 433-442.

Starr, J.L. and Mai, W.F. (1976). Effect of soil microflora on the interaction of three plant-parasitic nematodes with celery. *Phytopathology* 66, 1224-1228.

Stewart, R.N. and Schindler, A.F. (1956). The effect of some ecto-parasitic and endoparasitic nematodes on the expression of bacterial wilt in carnations. *Phytopathology* 46, 219-222.

Thomason, I.J., Erwin, D.C. and Garber, M.J. (1959). The relationship of root-knot nematode, *Meloidogyne javanica*, to *Fusarium* wilt in cowpea. *Phytopathology* 49, 602-606.

Van Andel, O.M. (1956). The importance of amino acids for the development of *Fusarium oxysporum* f. *lupini* Sn et H. in the xylem of lupins. *Acta.Bot.Neerl.*, 5, 280-286.

Van Gundy, S.D., Kirkpatrick, J.D. and Golden, J. (1977). The nature and role of metabolic leakage from root-knot nematode galls and

infection by *Rhizoctonia solan*. *J.Nematol.*, **9**, 113-121.
Walker, J.C. (1965). *In* "Ecology of Soil-borne Plant Pathogens." (Eds.
K.F. Baker, & W.C. Snyder.) 314-320. Univ. Calif. Press, Berkeley.
Wang, E.L.H. and Bergeson, G.B. (1974). Biochemical changes in root
exudate and xylem sap of tomato plants infected with *Meloidogyne
incognita*. *J.Nematol.*, **6**, 194-202.
Wang, E.L.H., Hodges, T.K. and Bergeson, G.B. (1975). *Meloidogyne in-
cognita*-induced changes in cell permeability of galled roots.
J.Nematol., **7**, 256-260.
Woltz, S.S. and Jones, J.P. (1968). Micronutrient effects on the *in
vitro* growth and pathogenicity of *Fusarium oxysporum* f.sp. *lycopersici*. *Phytopathology* **58**, 336-338.

Discussion

Triantaphyllou questioned the extent of the evidence of
Meloidogyne interactions with other pathogens obtained
from field situations. Taylor replied that investigations
in interrelationships were initiated because of Atkinson
(1892), Martin *et al.* (1956), Jenkins and Coursen (1957),
and others who observed an apparently synergistic interaction between the *Fusarium* wilt fungus and root-knot
nematodes on cotton. Since then many associations found
in field situations have been examined experimentally,
with attention being paid to the mechanism of the interaction, and particularly whether the nematodes predisposed the plants to fungal, bacterial or viral infection.
Mankau commented that virtually all of the laboratory investigations were made with very young plants, whereas
in the field the phenomenon of interaction occurs late,
with galled plants remaining relatively healthy and only
succumbing when they are also infected with the fungus.

Triantaphyllou referred to the view expressed that
bacterial infections do not favour the nematode but in
the case of the nitrogen-fixing *Rhizobium* there is evidence that root-knot nematodes multiply faster in the nodules. Lamberti commented that when there is a massive infection by the nematode it quickly results in rotting of
the roots and the production of *Rhizobium* nodules is consequently reduced.

Sikora observed that in view of the many pathogens

in the soil, those affecting plants are relatively few
and interactions between such organisms are rare. Taylor
commented that plant roots were constantly bathed in a
soup of micro-organisms and their infection was obviously
an exception rather than the rule; most plants displayed
a high level of resistance to infection. Interactions
between soil-borne organisms such as nematodes and fungi
were as rare as interactions between say insects and fungi, bacteria or viruses above ground.

Evans commented on the view that the primary damage
caused by root-knot nematodes is due to the blockage of
the vascular cylinder, and asked how experiments could be
done to simulate this. On a practical note Lamberti said
that the blockage of the xylem vessels was an important
aspect of nematode/host interaction because in the rainy
season little damage is encountered whereas in a dry
season root-knot damage can be extensive. Taylor referred
to the general view that the influence of the nematode
on the physiology of the whole plant results in the starvation of the roots by the establishment of syncytia which
divert the plant's energy to the nematode (see Bird in
these Proceedings. Orion (OEPP/EPPO Bull. No. 9, 67-71
(1973)) had proposed a scheme of host-root knot nematode
interrelationships in which at the site of infection, the
larva secretes substances which changed normal differentiation of the host cells into syncytia and xylem cells;
the giant cells are dependent on some metabolites from
the host-plant and in their absence giant cells do not
form and the nematode fails to mature normally. Annette
Kleineke said she had measured the permeability of roots
and found it to be higher in root galls than in roots
without nematodes. On the other hand Wang *et al.* (1975)
attributed an increase in nutrients released into the
rhizosphere by galled roots to a higher internal concentration of these nutrients, rather than to a significant
change in membrane permeability.

25 PREDISPOSITION TO *MELOIDOGYNE* INFECTION BY THE ENDOTROPHIC MYCORRHIZAL FUNGUS *GLOMUS MOSSEAE*

R.A. Sikora

Institut für Pflanzenkrankheiten, Universität Bonn, Federal Republic of Germany

Introduction

Predisposition can be defined as the tendency of treatments and conditions, acting before inoculation or before the introduction of the incitant, to affect susceptibility or resistance to biotic and abiotic pathogens (Yarwood, 1976). There are a multitude of reports demonstrating the existence of predisposition in experiments involving the interaction between root-knot nematodes and fungal pathogens. In most cases combined inoculation results in increased disease susceptibility, with the nematode acting as the predisposing agent.

Studies on predisposition in which root-knot nematodes and endotropic mycorrhizal fungi are involved are less common. Field observations have indicated that root-knot nematodes cause a reduction in mycorrhizal levels in the plant root system (Bird *et al.*, 1974; Schenck and Kinloch, 1974). Conversely, greenhouse tests have shown that plants preinoculated with endotrophic mycorrhizal fungi are less susceptible to root-knot attack than plants without mycorrhizae (Baltruschat *et al.*, 1973; Sikora & Schönbeck, 1975; Sikora, 1978); preinoculation allowed the slower growing fungal symbiont sufficient time to become established before nematode introduction.

The following observations are based on studies of

predispostion to *Meloidogyne* in plants preinoculated with
the endotrophic mycorrhizal fungus *Glomus mosseae*.

Predisposition

Population dynamics

Baltruschat *et al.* (1973) were the first to show that
plants preinoculated with *G. mosseae* were less susceptible
to root-knot infection. They reported a 75% reduction in
the number of *M. incognita* larvae that developed into
adults on tobacco plants preinoculated with *G. mosseae*,
compared with uninoculated controls. Sikora & Schönbeck
(1975) and Sikora (1978) also working with *G. mosseae* detected significant decreases in the number of larvae that
penetrated and developed to maturity on mycorrhizal tobacco, tomato and oat with *M. incognita*, and on carrot
with *Meloidogyne hapla*. Similar results were obtained by
Schenck *et al.* (1975) working with *Endogone macrocarpa*
and *E. heterogama* on soybean inoculated with *M. incognita*.
The results obtained in experimental work with root-knot
nematode and *G. mosseae* are presented in Fig. 1.

Attraction and penetration

Sikora (1978) demonstrated that the distribution of *M.
incognita* galls on mycorrhizal tomato plants was negatively correlated with *G. mosseae* levels in the root system. The majority of galls were formed at the periphery
of the root system where fungal levels were found to be
significantly lower than in the internal regions of the
root system. A significant decrease was detected in the
number of larvae that penetrated mycorrhizal tomato
roots, compared with nonmycorrhizal roots.

The results suggest that the nematode population decrease found in earlier experiments may be partially due
to changes in root attractiveness brought about by the
presence of the fungal symbiont in the root, or by the
inability of a significant number of larvae to penetrate

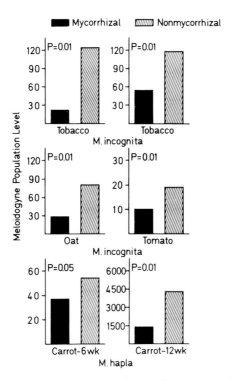

Fig. 1. Effect of *G. mosseae* on *M. incognita* and *M. hapla* population levels (from Sikora and Schönbeck (1975) and Sikora (1978)).

roots populated by the mycorrhizal fungus. Tests to determine whether root extracts from mycorrhizal plants repel or inactivate root-knot larvae gave negative results (Sikora & Schönbeck, 1975).

Nematode development

Meloidogyne incognita larval development was impeded in tomato plants inoculated with *G. mosseae* when measured 8 and 16 days after nematode introduction (Sikora, 1978). The slowdown in development may be responsible for the reduced size of galls often seen on mycorrhizal plants. Soil drenches with the amino acids arginine and citruline, both present in high concentrations in *G. mosseae* inoculated plants (Baltruschat & Schönbeck, 1975), caused significant reductions in *M. incognita* populations levels on tomato

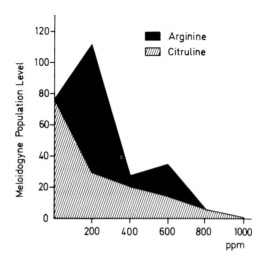

Fig. 2. Effect of arginine and citruline applied as a soil drench to 5 week old tomato, on *M. incognita* population levels 3 weeks after inoculation.
[1]Mean of 4 replicates, each with 500 larvae added 5 hours after amino acids.
[2]All citruline and the arginine treatments 400 to 1000 ppm significantly different (P = 0.01) from control.

(Fig. 2). The relationship between citruline concentration and nematode population density was negatively correlated (P = 0.01).

Giant cell development

Although *G. mosseae* has never been found in the galled tissue produced by *M. incognita*, it was often found in the root tissue surrounding the gall, and occasionally observed in the galls produced by *M. hapla* on carrot. Extensive histological studies have shown that although *G. mosseae* does not penetrate the giant cells produced by *M. incognita* on tomato, the presence of the fungus in the root retards giant cell formation. The giant cells produced on mycorrhizal tomato plants were smaller, fewer in number, and contained less nuclei than giant cells of the same age produced on nonmycorrhizal plants. In addition, the cytoplasm of the giant cells produced on mycorrhizal

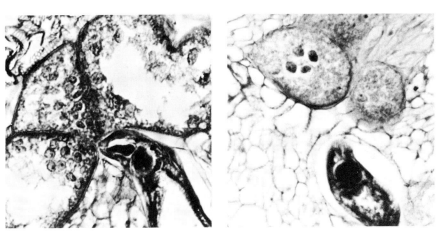

Fig. 3A. Section through 16 day old *M. incognita* gall on control tomato, showing advanced nematode development and highly vacuolated giant cells.
Fig. 3B. Section through a 16 day old *M. incognita* gall on mycorrhizal tomato, showing reduction in giant cell development and densely structured cell cytoplasm.

plants was densely structured while on controls it tended to be strongly vacuolated (Fig. 3A,B).

Discussion

It appears that the endotrophic mycorrhizal fungus *G. mosseae* predisposes plants to *Meloidogyne*, causing an increase in plant resistance. The main prerequisite for predisposition is the presence of the mycorrhizal fungus in the plant root system prior to nematode attack. The fungal symbiont may exert its influence on the nematode by
1) altering root attractiveness,
2) reducing larval penetration,
3) impeding larval development and
4) retarding giant cell formation.
These effects most probably result from complex physiological changes associated with the mycorrhizal infection and are not caused by direct competition between the two organisms. Attraction for example, might be adversely

influenced by changes in the chemical makeup of mycorrhizal root exudates. A hardening of the root tissue due to
increased lignin levels in the endo- and exodermis of myco-
corrhizal plants (Nehemiah, 1977) may limit nematode penetration and growth. The amino acids arginine and citruline
were found to arrest nematode development and possibly
increased phenolic and decreased auxin levels in the root
(Dehne, 1977) may impede larval growth and giant cell
formation.

The results suggest that preinoculation of transplants
with endotrophic mycorrhizal fungi could be a useful
factor in integrated nematode control.

References

Baltruschat, H., Sikora, R.A., and Schönbeck, F. (1973). Effect of
    VA-Mycorrhiza (*Endogone mosseae*) on the establishment of *Thielavi-
    opsis basicola* and *Meloidogyne incognita* in tobacco. *II Int.Cong.
    Plant Path.*, Minnesota, Nr. 0661.
Baltruschat, H. and Schönbeck, F. (1975). Untersuchungen über den
    Einfluß der endotrophen Mycorhiza auf den Befall von Tabak mit
    Thielaviopsis basicola. *Phytopath.Z.*, 84, 172-188.
Bird, G.W., Rich, J.R., and Glover, S.U. (1974). Increased endomyco-
    rrhizae of cotton roots in soil treated with nematicides. *Phyto-
    pathology* 64, 48-51.
Dehne, H.W. (1977). Untersuchungen über den Einfluß der Endotrophen
    Mycorrhiza auf die Fusarium welke an Tomate und Gurke. 150. Uni-
    versität Bonn und Universität Hannover.
Nehemiah, J. (1977). Untersuchungen über den Einfluss des Endotrophen
    Mycorrhizapilzes Glomus mosseae Gerd. & Trappe (*Endogone mosseae*
    Nicol. & Gerd.) auf *Zea mays*. 105. Universität Bonn.
Schenck, N.C. and Kinloch, R.A. (1974). Pathogenic fungi, parasitic
    nematodes and endomycorrhizal fungi associated with soybean roots
    in Florida. *Pl.Dis.Reptr.*, 58, 169-173.
Schenck, N.C., Kinloch, R.A., and Dickson, D.W. (1975). *In* "Endomyco-
    rrhiza." (Eds. F.E. Sanders, B. Mosse and P.B. Tinker.) 607-617.
    Academic Press, New York.
Sikora, R.A. and Schönbeck, F. (1975). Effect of vesicular-arbuscular
    mycorrhiza (*Endogone mosseae*) on the population dynamics of the
    root-knot nematodes (*Meloidogyne incognita*) and (*Meloidogyne hapla*)
    VIII Int. Plant Prot. Cong., Moscow, 5, 158-166.
Sikora, R.A. (1978). Einfluß der endotrophen mykorrhiza (*Glomus moss-
    eae*) auf das wirt-parasit-verhältnis von *Meloidogyne incognita* in
    tomaten. *Ztschr.Pflkrankh.u.Plschutz.*, 85, 197-202.
Yarwood, C.E. (1976). *In* "Physiological Plant Pathology." (Eds. R.
    Heitefuss and P.H. Williams.) 5, 703-718. Springer-Verlag, Berlin
    and New York.

F. Lamberti

*Laboratorio di Nematologia Agraria
del C.N.R., Bari, Italy*

The wide polyphagy of the more important species of root-knot nematodes means that the use of chemicals is likely to be the main method of control in many instances.

Fumigant nematicides

Soil fumigation is the most common measure used to achieve economical control in agricultural land, but to be effective it requires suitable conditions to eliminate interacting factors that may affect its application and effect (Thomason and McKenry, 1975).

1,3-D

Nematicides containing 1,3 dichloropropene (1,3-D) as the active ingredient are the most widely used (e.gs of trade names are Anguicid, D-D, Geofume, Geodden, Vidden, Vidden D, Nemagene, Nematox, Fumisol, Nemafume). Rates ranging from 200-300 l/ha of D-D or from 150-250 l/ha of Telone provide economical control of root-knot nematodes on several annual crops, especially for those remaining in the soil for only short periods (4-6 months).

In tomato fields infested with *Meloidogyne incognita*, in southern Italy, soil fumigation with 150 l/ha of Telone (93% 1,3-D) increased yields by 51% (Lamberti and Cirulli, 1970) and by 121% at the rate of 200 l/ha (Lamberti, 1971)

compared with the untreated plots, in dry summers. In 1972
(Lamberti, 1975), when it was a rainy summer, the same
dosage (200 1/ha) gave a yield increase of only about 34%.
To obtain similar results with D-D it is necessary
to apply 250-300 1/ha (50% of 1,3-D). In an experiment
done in the province of Lecce in 1974, tomato "Heinz 1350"
yielded 26, 36, 60, 68, and 89% more than the control in
plots treated respectively with 100, 200, 300, 400 and 500
1/ha (Lamberti, 1975). These results were improved, however, by following the initial preplanting fumigation of
200 1/ha of D-D with a foliar spray of oxamyl, at 2.5 1/ha
a.i. one month after transplanting; this gave a yield increase of 90%.

Investigations demonstrated the importance of transplanting nematode-free seedlings into fumigated soil but
also showed that even infested seedlings may yield well
in uninfested soil (Lamberti, 1975).

In glasshouse experiments, D-D applied at 200 1/ha increased the production of marketable tomatoes by 126% compared with untreated plants which were heavily infested
with *M. incognita* (Lamberti *et al.*, 1976). Under environmental conditions which prevented the nematode from building up large infestations, an application of 300 1/ha increased yields by 58%.

Applications of 500 1/ha D-D increased the yields of
cantaloupe by 238% in 1976 (Di Vito and Lamberti, 1978)
and of watermelon by 75 % in 1977 (Di Vito, unpublished)
compared with untreated controls infested with *M. incognita* .

Much less dramatic yield responses, of 25-34%, were
obtained, in southern Italy, with eggplant grown in soil
infested with *M. incognita* and treated before transplanting with 300 1/ha D-D (Lamberti, 1975).

Yield increases, but probably unimportant economically,
were obtained with pepper (Lamberti, 1975). This plant is
less sensitive to attack by *M. incognita*, mainly because

it is usually transplanted when the soil temperture is too
low for the activity of the nematode.

D-D applied at 500 l/ha increased root yields of sugarbeet by 183% in 1976 (Di Vito and Lamberti, 1977) in soil
heavily infested with *M. incognita*, while in 1977, D-D at
200 or 400 l/ha gave increases of 446 or 516% respectively,
compared with crops grown in untreated infested soil (Di
Vito *et al.*, 1977). It is thought that the most economical
dose of D-D for this crop, in southern Italy, is again 300
l/ha which has given increased yield of roots of 454% (Di
Vito *et al.*, 1977).

The response of tobacco to 1,3-D treatments can also
be exceptional in the presence of heavy and highly pathogenic populations of *M. incognita* (Fig. 1). Increases of
620 and 580% of fresh leaves were obtained in 1969 with
the application of 150 l/ha D-D or 100 l/ha Telone respectively (Lamberti, 1969). In 1971 substantial increases

Fig. 1. Growth of tobacco in soil infested with *Meloidogyne incognita*
following preplant treatment with 1,3 dichloropropane at 200 l/ha
(right) compared with untreated control (left).

(around 250% with respect to the control) of "Erzegovina" tobacco were obtained using D-D (300 1/ha) or Telone (150 1/ha) (Lamberti, 1972).

Because of the extreme susceptibility of levantine to-bacco to root-knot nematodes, farmers are advised to un-dertake chemical treatment of the soil every year if they wish to crop continuously. D-D applied one month before transplanting in a single application in 1970 or at diff-erent doses in 1970, 1971 and 1972 indicated the best yield increases were obtained with low doses repeated yearly (Lamberti, 1975).

A study of the factors affecting the yield responses of tobacco after fumigation with D-D have shown that the highest increases of yield with respect to the control are obtained when the crop is transplanted in the second half of April, 6-8 weeks after the application of the chemical. This treatment was more effective in controll-ing root-knot nematodes when residues of the previous crop were removed (Lamberti, 1975).

Soil fumigation with nematicides containing 1,3-D improved the quantity and quality of carnation flowers in soil infested with *M. incognita* and, again, D-D at the rates of 200 or 300 1/ha appears to be the best from the cultural and economical point of view (Lamberti, 1975).

A higher degree of nematode control is required when perennials are planted in order to avoid rapid reinfest-ation by the surviving inoculum. Raski and Lear (1962) apparently obtained complete kill of *M. incognita* down to 2 m with preplanting applications of either Telone or D-D at 1200 or 2000 1/ha respectively. However, after two growing seasons all the newly planted grapevines were heavily infested and galled.

In California, Abdalla *et al.* (1974) found in labor-atory tests that 100% kill of *M. incognita* required 2.5 ppm of 1,3-D and exposure for three days. In the field this concentration was reached only at 33 cm deep when

the chemical was applied at split depths: 1872 1/ha/81 cm
plus 468 1/ha/25.5 cm deep. However, nematodes were not
detected in soil samples collected to a depth of 91.5 cm
two years after treatments. In this case, a lower dosage
of the nematicide was effective because of the longer time
of exposure.

Various environmental factors affect the efficacy of
soil fumigation (Thomason and McKenry, 1975). Soil texture,
for instance, only indirectly affects the diffusion of
fumigants. The movement of the gas phase is restricted in
soil with moisture at field capacity, in soil rich in
organic matter, or when the structure is altered by com-
paction or deflocculation (Leistra, 1971; McKenry and
Thomason, 1974). When soil moisture is low the retention
of the fumigant may be insufficient to provide a lethal
exposure time.

Temperature is the most important ecological factor.
Low temperatures decrease the solubility of the chemical
limiting its diffusion and, conversely high temperatures
also limit diffusion by increasing volatilization (McKenry
and Thomason, 1974). Temperature also affects the rate of
hydrolysis of the chemical in the soil. Therefore it is
advisable to apply 1,3-D at 5-25°C (McKenry and Thomason,
1974), the optimal temperature range being 15-20°C (McKenry and Thomason, 1976).

The efficiency of chemical treatment of the soil can
be improved by taking these and various other factors into
account. For example, cultivation of the soil before application of the chemical can break up the layers which
may restrict its dispersal (Smelt et al., 1974; McKenry
and Thomason, 1976a); residues of a previous crop infested
with root-knot nematodes may increase soil populations to
the extent that vastly increased amounts of chemicals may
be required (McKenry et al., 1977); compacting the soil
surface after treatment results in an increased concentra-
tion-time of the product (Smelt et al., 1974; Thomason and

McKenry, 1975; Raski *et al.*, 1976); injection at two
depths increases the dynamics of the chemical in deeper
layers (Abdalla *et al.*, 1974; Raski *et al.*, 1976). Treatment applied along the replanting rows or localised to the
replanting spot only delays reinfestation (McKenry and
Thomason, 1976a) but by lowering the population of root-
knot nematodes to non-pathogenic levels this may give a
good start to new fruit plantings (Lownsbery *et al.*, 1968).

Massive and continuous applications of 1,3-D might
cause problems of soil and water contamination. The chemical or its degradation products can persist in the soil
for as long as six months but as non-dangerous concentrations (Roberts and Stoydin, 1976; Basile, unpublished)
and the increase of chlorine content induced by the treatments does not seem to be phytotoxic (Lamberti and Renzoni,
1974). No residues of the fumigant were found in the
water-bearing stratum below the treated soil and containing
the chemical (Basile, unpublished) nor in potato tubers
grown in treated soil (Roberts and Stoydin, 1976). However, when for some reason the chloroallyl alcohol resulting from the hydrolysis of 1,3-D is not biodegraded
(Castro and Belser, 1966; Belser and Castro, 1971; Van
Dijk, 1974) phytotoxicity may occur, especially when
young seedlings are transplanted in the treated soil
(Baines *et al.*, 1977).

A number of nematicides include 1,3-D in combination
with other products: Dorlone (dichloropropenes + ethylene
dibromide), Telone C, Teracide 15 or D-D-PIC (85% dichloropropenes + 15% chloropicrin), Telone PBC (80% dichloropropenes + 15% chloropicrin + 5% propargyl bromide), Vorlex, Di-Trapex or Trapex (80% dichloropropenes + 20%
methyl isothiocyanate)and Vorlex 201 (68% dichloropropenes + 17% methyl isothiocyanate + 15% chloropicrin).
Many of them are slightly more effective than 1,3-D alone
but also much more expensive. Those containing methyl
isothiocyanate are more effective because of their longer

persistence in the soil and the synergistic action of this
toxicant through the water phase (Greco and Lamberti,
1977). However, their increased beneficial effect to the
crops is due mainly to the fact that they control not only
nematodes but also pathogenic fungi and weeds. Therefore,
selection of a pesticide should always be related to the
particular problem which may be rather more conveniently
tackled by using compound products rather than 1,3-D alone.

Methyl bromide

Methyl bromide is generally very effective against
plant parasitic nematodes (Abdalla and Lear, 1975). Rates
of 450 - 600 kg/ha give excellent control of root-knot
nematodes (Raski *et al.*, 1976; McKenry and Thomason,
1976a) but it sometimes raises problems of side effects.
One of these is the high concentration of bromide in ed-
ible part of plants e.g. lettuce (Kempton and Maw, 1972;
Coosemans and Van Assche, 1976), tubers, tap roots or
strawberry fruits (Basile *et al.*, 1978). Nutritional de-
ficiences may also be apparent in crops newly planted in
treated soil due to the destruction of beneficial endotrophic mycorrhizae (Thomason and McKenry, 1975). Some
plants such as liliaceae (Lamberti, 1974) or carnation
(Kempton and Maw, 1974) are somewhat sensitive to methyl
bromide but it is possible to eliminate residues by thoroughly irrigating the treated soil before transplanting
(Vigodsky and Klein, 1976; Basile and Lamberti, 1977),
and this is economically feasible with high value cash
crops like carnation. Methyl bromide must, of course,
be applied by trained personnel which makes treatments
very expensive.

Ethylene dibromide

Ethylene dibromide, or 1,2-dibromoethane (EDB), is
another halogenated compound that gives very efficient
control of root-knot nematodes on both annual and perennial

crops. Application of 100-200 l/ha of EDB have proved
beneficial (Lownsbery *et al.*, 1968; Di Vito and Lamberti,
1977; Di Vito *et al.*, 1977; Di Vito and Lamberti, 1978).
The chemical is much less hydrolized than 1,3-D at higher
temperatures (Castro and Belser, 1968) and thus being
more persistent can substitute for 1,3-D compounds in
tropical regions where soil temperature are always above
25°C. However, because of this property it cannot be used
effectively in wet or cold (below 10°C) soils (McKenry
and Thomason, 1974).

1,2-Dibromo-3-chloropropane

Root-knot nematodes are killed by short exposures to
1,2-dibromo-3-chloropropane (DBCP) (Hodges and Lear, 1973).
However, it is little used as a preplanting treatment for
annuals or perennials as better results can be achieved
with less expensive chemicals such as 1,3-D or EDB, and
because its dispersion in the soil requires abundant irrigation and is greatly affected by the content of organic
matter (Johnson and Lear, 1968; 1969; O'Bannon and Tomerlin, 1975; O'Bannon *et al.*, 1975). DBCP is more frequently
used to control nematodes in established orchards, vineyards or rose plantings where it has been shown to increase yields and growth of young plants (Raski and
Schmitt, 1964; Lownsbery *et al.*, 1968, Johnson *et al.*,
1969). Such treatment decreases, but does not eliminate,
populations of root-knot nematodes in the soil, and because
the chemical does not penetrate the infested roots the endoparasitic stages of the nematode are largely unaffected.

Dazomet

Dazomet is a chemical that may give very good results
on annual crops but it has two limiting factors; namely
a very high and persistent phytotoxicity (Lamberti, 1972)
which requires at least 4-6 weeks interval between application and transplanting to eliminate it, and the

necessity for high moisture content of the soil because
water is required to hydrolyze it into compounds which
are toxic to the nematodes but harmless to plants. Dazomet
generally gives much better results when applied for
winter rather than for summer planting (Lamberti, 1975).

Metham sodium

This is a chemical that usually does not give good
results against root-knot nematodes even when applied in
very high dosages.

Non-fumigant nematicides

During the last twenty years several non-fumigant nematicides have become available. However, most of them are
still not well known and additional information is needed
on the mode of action, particularly those that are sys-
tematically translocated in plants, before proper use can
be made of them. They are mostly used on annual crops as
pre- or post-planting applications and occasionally in
established orchards and vineyards to reduce pre-existing
infestation. Their use in these latter cases is however
limited by their high cost.

Thionazin

The organic phosphorus compound thionazin or zinophos
(0,0-diethyl-0-2 pyrazinyl phosphorothioate) was the first
of the non-fumigant nematicides to be formulated. It is
only moderately effective against root-knot nematodes, but
this may depend on the plant species. On carnation, in
southern Italy, excellent protection against *M. incognita*
was obtained by spraying the foliage at 3-weekly intervals
(Lamberti, 1975). However, thionazin is toxic to several
plants especially if applied at high dosages (Vovlas and
Lamberti, 1974). Glasshouse experiments on the systemic
action of thionazin have shown that the chemical is translocated both upwards and downwards in the plants and

protection against nematode invasion persists for 18-20 days (Vovlas and Lamberti, 1976).

Aldicarb

Aldicarb (2-methyl-2(methylthio) propionaldehyde 0-(methylcarbamoyl) oxime) applied at 8 kg/ha (a.i.) before planting gave yield increases of about 200% compared with untreated controls in a tobacco field infested with *M. incognita* (Lamberti, 1972). The systemic action of aldicarb, at a concentration sufficient to inhibit root invasion by juveniles of *M. incognita*, persisted in the glasshouse for 18-20 days (Vovlas and Lamberti, 1976). At sublethal rates the chemical acts against root-knot nematodes by inhibiting egg hatching, migration and host invasion of infective stages and further development of second stage juveniles that had penetrated the roots (Cuany *et al.*, 1973, 1974; Hough and Thomason, 1975).

Carbofuran

Carbofuran (2,3-dihydro-2,2-dimethyl-7-benzofuranyl methylcarbamate) is another systemic chemical which in glasshouse tests prevented root invasion of tomato by *M. incognita* for 12 days (Vovlas and Lamberti, 1976). In general it is not very effective against root-knot nematodes. In formulation as a wettable powder, however, it gave satisfactory results when sprayed at monthly intervals on carnation plants planted in soil to which granular carbofuran has been incorporated at transplanting (Lamberti, 1975). Carbofuran may be toxic to tomato (Lamberti, 1971) and other plants in alkaline soils.

Oxamyl

Oxamyl is an oxime carbamate (S-methyl 1-(dimethyl-carbamoyl) N-((methylcarbamoyl) oxy) thioformimidate) which ha given excellent results on various crops, especially when the two formulations granular and liquid, were applied in combination (Lamberti, 1975). On tobacco, in a *M. incognita*-infested field, either soil or foliar applications of this chemical, or the two combined, more than doubled the production of fresh leaves compared with untreated controls (Lamberti, 1972).

The liquid formulation of oxamyl, especially when applied as a foliar spray, seems to be more effective than the granular. This is probably because in warmer climates the chemical in the soil is rapidly decomposed to non-nematoxic substances (Bunt, 1975; Roca *et al.*, 1975). Applied as a foliar spray, oxamyl is translocated basipetally to the roots where, under glasshouse conditions, it has been found to prevent root invasion for 25 days (Vovlas and Lamberti, 1976). Thus, foliar sprays applied at 2-3 weekly intervals should provide protection from nematode attacks during the whole of the critical period of growth of a crop.

Particularly efficient in controlling root-knot nematodes are the combined treatments of D-D preplanting, followed by one or two foliar applications of oxamyl some weeks after transplanting (Lamberti, 1975). Oxamyl also gives reasonably good results when used against pre-existing infestations in young plants (Lamberti, 1975).

Phenamiphos

Phenamiphos (ethyl 4-(methylthio)-m-tolyl-isopropyl phoramidate) is a systemic organo-phosphorus nematicide-insectide-miticide, with slight fungicidal action (Bunt, 1975; Lamberti and Zacheo, unpublished). The granular formulation is excellent against root-knot nematodes and because of its long persistence, even in warm soils, pre-planting applications (Bunt, 1975, Roca *et al.*, 1975) on annual crops have given yield increases equal or superior to those obtained with soil fumigation (Lamberti, 1972, 1975). The liquid formulation is less effective and is also highly phytotoxic if applied shortly before

transplanting (Lamberti, 1972). A paste of the chemical
brushed on the trunks of gardenia and fig gave good control of root-knot nematodes (Inserra and O'Bannon, 1974).

Prophos

Prophos or ethoprop (O-ethyl S,S-dipropylphosphorodi-
thioate) is a non-systemic nematicide with very little
persistence (Roca *et al.*, 1975). To obtain satisfactory
results with root-knot nematodes it is therefore necessary
to apply the chemical when juveniles are at peak populations in the soil. D'Errico and Di Maio (1977) increased
yields of tomato by more than 50% in soil infested with
M. incognita by applying prophos granules as a preplant
treatment, followed by a soil drench in June when they
had noted an outbreak of juvenile stages in the soil due
to the completion of the first new generation of the nematode.

Other nematicides

Other nematicides, such as acrylonitrile (propylene-
NH_3), disulfoton or thiodermethon (O,O-diethyl S-2 (ethyl-
thio)-ethyl phosphorodithioate), Kazide, (potassium
azide), methomyl (S-methyl-N-((methylcarbamoyl)-oxy)
thioacetamidate), Nema (potassium-N-hydroxymethyl-N-meth-
yldithiocarbamate), Nemamort (dichlorodiisopropyl ether)
and Thirpate (2,4-dimethyl-2-formyl-1-1,3-dithiolane oxime
methylcarbamate) have been tested for the control of root-
knot nematodes but the results have not always been satisfactory and often very erratic (Lamberti, 1972, 1973 and
1975). More investigations are needed to give better
evaluation of these chemicals.

Cultural methods of control

Sanitation of infested stocks

Sometimes one faces the problem of cleaning up nursery

stocks infested by root-knot nematodes. Attempts made in southern Italy of dipping bare roots of olive young trees in water suspensions of emulsions of various non-phytotoxic nematicides were unsuccessful in eradicating infestations (Lamberti and Di Vito, 1972; Vovlas *et al.*, 1975) although some success was achieved with higher concentrations of phenamiphos without damaging the plants (Lamberti and Di Vito, 1972).

It is doubtful if hot water dips would be successful as the high temperatures necessary to kill stages of the nematode deeply embedded in the root tissue would be lethal for the plants.

Physical methods

Physical treatments of the soil such as steam sterilization may give better results than those obtainable with applications of nematicides against root-knot nematodes (Lamberti *et al.*, 1976). Nevertheless, this practice cannot be considered as a suitable alternative to chemical control because of the need for specialized equipment and because of the high cost of energy. Therefore, the only alternative where chemicals are not available, or too costly as for example for low-cash crops, is the application of cultural practices.

Rotations

Because the most widespread and pathogenic species of root-knot nematodes are polyphagous it is very difficult to design rotations schemes which are effective in controlling the nematode and at the same time are economical.

M. naasi which has non-host plants among the cereals, and other species with few hosts, may easily be controlled by crop rotation. One or two years of oat or potato, or a one year fallow considerably decreased populations of the nematode (Franklin, 1973) and the following susceptible crop showed an increase in yield (Allen *et al.*, 1970).

M. arenaria and *M. hapla* present a more difficult problem and *M. incognita* and *M. javanica* are particularly
problematical because they reproduce on many weeds as
well as on a wide range of crops. Many authors agree that
good control of these species may be obtained if susceptible crops are spaced at intervals of three, four or even
five years (Di Muro, 1975). In southern Italy different
crops sequences have been tested in a three year rotat-
ional system in soil infested with *M. incognita*; yields
of tomato were slightly increased when grown in plots
that had grown susceptible plants only in winter (Lamberti
et al., unpublished). Similar results were obtained with
tobacco during a two year experiment (Lamberti *et al.*,
unpublished).

Time of planting

With knowledge of the biology of the nematode in each
region it is possible to obtain control by transplanting
early when soil temperatures are already good for development of the crop but not yet favourable for root invasion and reproduction of the nematode. This method is
very useful in regions where cold and warm seasons are
well delimited (temperate climates). In subtropical areas
where the change of season is very gradual or in tropical
places where such a change does not occur the practice
can be very risky. Nevertheless it has given some success
in southern Italy where *M. incognita* does not penetrate
roots of host plants when the soil temperature is below
18°C (Lamberti *et al.*, unpublished).

Removal of infested plants

It is important to reduce sources of inoculum by eliminating the residues of previous susceptible crops (Lamberti *et al.*, unpublished) and any weeds that may be host
for the nematode. The use of nematode-free planting stocks
is recommended not only to avoid the consequences of

nematode infestation of the crop, but also to avoid the possibility of bringing infestations to uncontaminated land. The use of resistant cultivars is an attractive and often economical solution to the problem (see elsewhere in these Proceedings) and may be used in combination with other control measures to provide a strategy for control referred today as integrated control.

References

Abdalla, N. and Lear, B. (1975). Lethal dosages of Methyl bromide for four plant-parasitic nematodes and the effect of soil temperature on its nematicidal activity. *Pl.Dis.Reptr.*, **59**, 224-228.

Abdalla, N., Raski, D.J., Lear, B. and Schmitt, R.V. (1974). Movement, persistence and nematicidal activity of a pesticide containing 1,3-Dichloropropene in soils treated for nematode control in replant vineyards. *Pl.Dis.Reptr.*, **58**, 562-566.

Allen, M.W., Hart, W.H. and Baghott, K. (1970). Crop rotation controls barley root-knot nematode at Tulelake. *Calif.Agric.*, **24**, 4-5.

Baines, R.C., Klotz, L.J. and De Wolfe, T.A. (1977). Some biocidal properties of 1,3-D and its degradation product. *Phytopathology* **67**, 936-940.

Basile, M., Di Vito, M. and Lamberti, F. (1978). Prove di lotta chimica contro *Ditylenchus dipsaci* su Fragola e residui di Bromo in fragole prodotte in terreno fumigato con Bromuro di metile. *Proc. Giornate Fitopatologiche, Acireale-Catania*, 389-396.

Basile, M. and Lamberti, F. (1977). Possibile impiego del bromuro di metile nella coltivazione del Garofano in provincia di Bari. *Riv.Ortoflorofrutticoltura It.*, **61**, 149-156.

Belser, N.O. and Castro, C.E. (1971). Biodehalogenation. The metabolism of the nematocides cis- and trans-3-chloroallyl alcohol by a bacterium isolated from soil. *J.Agric.Fd.Chem.*, **19**, 23-26.

Bunt, J.A. (1975). Effect and mode of action of some systemic nematicides. *Med.Landbou.Wageningen.*, **75**, 1-127.

Castro, C.E. and Belser, N.O. (1966). Hydrolysis of cis- and trans-1, 3-dichloropropene in wet soil. *J.Agric.Fd.Chem.*, **14**, 69-70.

Castro, C.E. and Belser, N.O. (1968). Biodehalogenation. Reductive dehalogenation of the biocides ethylene dibromide, 1,2-dibromo-3-chloropropane and 2,3-dibromobutane in soil. *Env.Sci. and Tech.*, **2**, 779-783.

Coosemans, J. and Van Assche, C. (1976). Investigations of the bromide concentration in Belgian greenhouse lettuce after methyl bromide disinfestation. *Med.Fac.Land.Gent.*, **41**, 1361-1369.

Cuany, A., Bergé, J.B. and Scotto La Massese, C. (1973). Observations sur l'emploi des nématicides endotherapiques contre le genre *Meloidogyne* en horticulture florale. *OEPP/EPPO Bull.*, **3**, 75-87.

Cuany, A., Bergé, J.B., Scotto La Massese, C. and Ritter, M. (1974). Considération sur les utilisations de l'Aldicarb et de l'Oxamyl

substances nématicides endotherapiques. *Phytiatrie-Phytopharmacie*
23, 199-210.

Dandria, D., Lamberti, F. and Vovlas, N. (1977). La lotta chimica
contro i nematodi galligeni del Pmoodoro a Malta. *Nematol.medit.*,
5, 127-131.

D'Errico, F.P. and Di Maio, F. (1977). Suggerimenti per una lotta
guidata contro *Meloidogyne incognita* su Pomodoro. *Nematol.medit.*,
5, 95-97.

Di Muro, A. (1975). Effetto di varie piante sui nematodi galligeni
del Tabacco in rotazioni biennali. *Ann.Ist.Sper.Tabacco* 2, 93-106.

Di Vito, M. and Lamberti, F. (1977). Prove di lotta chimica contro
i nematodi galigeni su Barbabietola da zucchero. *Nematol.medit.*,
5, 31-38.

Di Vito, M. and Lamberti, F. (1978). Prove di lotta chimica contro
i nematodi galligeni del Melone. *Proc.Giornate Fitopatologiche.*,
405-411.

Di Vito, M., Lamberti, F. and Carella, A. (1977). Ulteriori risultati di prove di lotta chimica contro *Meloidogyne incognita* su
Barbabietola da zucchero. *Nematol.medit.*, 5, 339-344.

Franklin, M.T. (1973). *Meloidogyne naasi*. *C.I.H. Descrip.Pl.par.Nem.*,
19, 4.

Goring, C.A. I., Laskowski, D.A., Hamaker, J.M. and Meikle, R.W.
(1975). *In* "Environmental Dynamics of Pesticides." (Eds. R. Hague
and V.H. Freed.) 135-172. Plenum Press, New York and London.

Greco, N. and Lamberti, F. (1977). Ricerca delle dosi ottimali di
alcuni nematocidi per la lotta contro *Heterodera carotae*. *Nematol.
medit.*, 5, 25-30.

Hodges, L.R. and Lear, B. (1973). Effect of time of irrigation on the
distribution of 1,2-Dibromo-3-chloropropane in soil after shallow
injfection. *Pestic.Sci.*, 4, 795-799.

Hough, A. and Thomason, I.J. (1975). Effect of Aldicarb on the behaviour of *Heterodera schachtii* and *Meloidogyne javanica*. *J.Nema-
tol.*, 7, 221-229.

Inserra, J.N. and O'Bannon, J.H. (1974). Systemic activity of Phe-
amiphos for control of *Meloidogyne arenaria* on *Gardenia jasmino-
ides* and *Ficus carica*. *Pl.Dis.Reptr.*, 58, 1075-1077.

Johnson, D.E. and Lear, B. (1968). Evaluating the movement of 1,2-
Dibromo-3-chloropropane through soil. *Soil Science* 105, 31-35.

Johnson, D.E. and Lear, B. (1969). The effect of temperature on the
dispersion of 1,2-dibromo-3-chloropropane in soil. *J.Nematol.*,
1, 116-121.

Johnson, D.E., Lear, B., Miyagawa, S.T. and Sciaroni, R.H. (1969).
Multiple application of 1,2-Dibromo-3-chloropropane for control
of nematodes in established rose plantings. *Pl.Dis.Reptr.*, 53,
34-37.

Kempton, R.S. and Maw, G.A. (1972). Soil fumigation with methyl bromide: bromide accumulation by lettuce plants. *Ann.app.Biol.*, 72,
71-79.

Kempton, R.S. and Maw, G.A. (1974). Soil fumigation with methyl bromide: the phytoxicity of inorganic bromide to carnation plants.
Ann.appl.Biol., 76, 217-229.

Lamberti, F. (1969). Prospettive di lotta menatocida sul tabacco in
provincia di Lecce. *Il Tabacco* 733, 8-10.

Lamberti, F. (1971). Prove di lotta chimica contro i nematodi galligeni del Pomodoro in Puglia. *Phytiatr.Photopharm.*, **3**, 140-146.

Lamberti, F. (1972). Chemical control of root-knot nematodes on tobacco in Apulia. *Meded.Fakul.Landbou.Gent.*, **37**, 790-797.

Lamberti, F. (1973). Yield responses in relation to chemical control of root-knot nematodes in Southern Italy. *OEPP/EPPO Bull.*, **3**, 55-66.

Lamberti, F. (1974). I nematodi parassiti delle liliacee da orto e loro controllo con particolare riferimento all'ambiente mediterraneo. *Proc.4th.Phytiatr.Phytopharm.*, 56-68.

Lamberti, F. (1975). Fumiganti e nematocidi sistemici nella lotta contro i fitoelminti ipogei. *Rep.S.I.F.*, 26.

Lamberti, F. and Cirulli, M. (1970). Prove preliminari di lotta a nematodi del gen. *Meloidogyne* su Pomodoro con due nematocidi sperimentali. *Italia agric.*, **107**, 721-723.

Lamberti, F., Dandria, D., Vovlas, N. and Aquilina, J. (1976). Prove di lotta contro i nematodi galligeni del Pomodoro da mensa in serra a Malta. *Colture Protette* **5**, 27-30.

Lamberti, F. and Di Vito, M. (1972). Sanitation of root-knot nematode infected olive-stocks. *Proc.III.Congr.Un.Fitopat.medit.*, 401-411.

Lamberti, F. and Renzoni, G. (1974). Chlorine content of soil treated for three years with halogenated fumigants. *Nematol.medit.*, **2**, 71-75.

Leistra, M. (1971). Diffusion of 1,3-Dichloropropene from a plane source in soil. *Pestic.Sci.*, **2**, 75-79.

Lownsbery, B.F., Mitchell, J.T., Hart, W.H., Charles, F.M., Gerdts, M.H. and Greathead, A.S. (1968). Responses to post-planting and preplanting soil fumigation in California peach, walnut, and prune orchards. *Pl.Dis.Reptr.*, **52**, 890-894.

McKenry, M.V. and Thomason, I.J. (1974). 1,3-Dichloropropene and 1,2 Dibromoethane compounds. *Hilgardia* **42**, 393-438.

McKenry, M.V. and Thomason, I.J. (1976). Dosage values obtained following pre-plant fumigation for perennials. *Pestic.Sci.*, **7**, 521-534.

McKenry, M.V. and Thomason, I.J. (1976a). Dosage values obtained following pre-plant fumigation for perennials. *Pestic.Sci.*, **7**, 535-544.

McKenry, M.V., Thomason, I.J. and Naylor, P. (1977). Dosage-response of root-knot nematode-infected grape roots to cis 1,3-Dichloropropene. *Phytopathology* **67**, 709-711.

O'Bannon, J.H., Tomerlin, A.T. and Rasmussen, G.K. (1975). Penetration of 1,2-Dibromo-3-chloropropane in a Florida soil. *J.Nematol.*, **7**, 252-255.

O'Bannon J.H. and Tomerlin, A.T. (1975). Efficacy of application and dispersion of 1,2-Dibromo-3-chloropropane in a Florida soil. *Soil and Crop Sci.Soc.Florida* **20**, 90-93.

Raski, D.J., Jones, N.O., Kissler, J.J. and Luvisi, D.A., (1976). Soil fumigation: one way to cleanse nematode-infested vineyard lands. *Calif.Agric.*, **30**, 4-7.

Raski, D.J. and Lear, B. (1962). Influence of rotation and fumigation on root-knot nematode populations on grape replants. *Nematologica* **8**, 143-151.

Raski, D.J. and Schmitt, R.V. (1964). Grapevine responses to chemical control of nematodes. *Am.J.En.Vitic.*, **15**, 199-203.

Roberts, T.R. and Stoydin, G. (1976). The degradation of (Z)- and
    (E)-1,3-Dichloropropenes and 1,2-Dichloropropane in soil. *Pestic.*
    Sci., **7**, 325-335.
Roca, F., Lamberti, F. and Siniscalco, A. (1975). Studi sulla persis-
    tenza di alcuni nematocidi granulari nella lotta contro i nematodi
    galligeni (*Meloidogyne* spp.). *Proc.Giornate Fitopatologiche* 265-
    269.
Smelt, J.H., Leistra, M., Sprong, M.C. and Nollen, H.M. (1974). Soil
    fumigation with Dichloropropene and Metham-Sodium: Effect of soil
    cultivation on dose pattern. *Pestic.Sci.*, **5**, 419-428.
Thomason, I.J. and McKenry, M.V. (1975). *In* "Nematode Vectors of
    Plant Viruses." (Eds. F. Lamberti, C.E. Taylor and J.W. Seinhorst)
    423-439. Plenum Press, London and New York.
Van Dijk, H. (1974). Degradation of 1,3-dichloropropenes in the soil.
    Agro-Ecosystems, **1**, 193-204.
Vigodsky, H. and Klein, L. (1976). Influence of methyl bromide soil
    fumigation on fungicidal efficacy and bromide residues. *Phyto-*
    parasitica **4**, 123-129.
Vovlas, N., Inserra, R.N. and Lamberti, F. (1975). Risanamento di
    piantonai di Arancio amaro, Olivo e Vite infestati da nematodi.
    Proc.Giornate Fitopatologiche, Torino, 271-277.
Vovlas, N. and Lamberti, F. (1974). Sensibilità di alcune colture
    erbacee alla Tionazina. *Inf.tore Fitopatol.*, **24**, 9-12.
Vovlas, N. and Lamberti, F. (1976). Studies on the systemic action
    of some chemicals in the control of root-knot nematodes. *Nematol.*
    medit., **4**, 111-113.

Discussion

The discussion indicated that acceptance of chemical
control of nematodes by farmers depended on the extent
of the problem, the availability of equipment for application of chemicals, and clear evidence of an economic
gain from treatments. Where a response to chemicals had
been demonstrated there was sometimes a tendency for
farmers to use overdoses in the hope of further improving
yields but then being discouraged because such treatments
appear uneconomic. On specific points, Lamberti stated
that bromide residues were not a problem following fumigation with methyl bromide; that organo-phosphorus compounds had made a considerable contribution to nematode
control, fenamiphos having been found particularly effective for many situations, although because of the suspic-
ion of undesirable residues the use of such compounds was
under surveillance by the Government. Chemical control
generally was more effective on irrigated land, partly

because a high humidity level of the air enables the infested plant to recover.

G. Fassuliotis

U.S. Vegetable Laboratory, Science and Education Administration, U.S. Department of Agriculture, Charleston, South Carolina, U.S.A.

Introduction

Root-knot nematodes, *Meloidogyne* spp., account for significant losses in yields of food, feed and fibre crops throughout the world. The most frequent species encountered worldwide are *M. incognita*, *M. javanica*, *M. arenaria*, and *M. hapla* (Sasser, 1977). These species have wide host ranges and are especially damaging to many important economic crops in the subtropical and tropical climates. They can also be important disease producing organisms in colder climates where high value crops are grown under glass.

Plant resistance is regarded as an extremely feasible method for controlling root-knot nematodes. It is an effective, economical, and environmentally safe means of reducing losses from diseases caused by these pests. Use of resistant plants enables the grower to control the root-knot disease without increasing production costs associated with the purchase of expensive chemicals, applicators, and in the numerous mechanical operations that go into the production of a crop. The benefits derived by growers using resistant tobacco 'NC 95' introduced in 1961 have been spectacular, saving millions of dollars (Moore *et al.*, 1962). Resistance is especially valuable in controlling root-knot nematodes in low value crops and minor crops which can increase crop yields equal to that obtained by soil fumigation (Good, 1972; Netscher and Manhoussin, 1973).

The importance of controlling root-knot nematodes through genetic means (resistance) has gained further significance since manufacturing restrictions were imposed on fumigants containing 1,2-dibromo-3-chloropropane because of their potential for causing cancer in some laboratory animals and sterility in man. Many nematology programmes which once emphasized control through chemical means have shifted research direction toward control through resistance. Breeding plants resistant to root-knot nematodes currently occupies a significant portion of the plant breeding programs at the U.S. Vegetable Laboratory, Charleston, South Carolina.

Successful employment of root-knot resistance requires the manipulation of genetic systems to transfer resistant genes from a resistant plant to a susceptible agronomic or horticulturally acceptable type. Significant accomplishments have been made over the past 60 years. The success achieved to date is reflected in Table I which shows over 235 cultivars for 15 major crops as having resistance to one or more *Meloidogyne* species. Most of these crops have resistance to M. *incognita* with the number of cultivars concentrated in tomato, tobacco, cotton, soybean, cowpea and sweet potato. This reflects both the past effort that has been placed in breeding for resistance against M. *incognita* and also the relative occurrence of resistance genes within these crops. Resistance to more than one species is found in tomato, cowpea, bermuda grass, peach, grapevine and lupin (Good, 1972).

Recent reviews by Hare (1965), Rohde(1965, 1972), Malo (1964), Kehr (1966), Hartmann (1976), Hunt *et al*. (1972), Sasser (1972a) and Singh *et al*. (1974) unfold the challenging problems and future opportunities which will keep plant breeders and nematologists occupied a long time.

TABLE I

Crops in which resistance to *Meloidogyne* spp. has been
selected or developed

Crop	*Meloidogyne* species	No. of cultivars
Tomato	*incognita, javanica, & arenaria*	65+
Cowpea	*incognita*	30
	javanica	4
	arenaria	4
Sweet potato	*incognita*	17
	javanica	3
	arenaria	7
Soybean	*incognita*	19
Tobacco	*incognita*	14+
Cotton	*incognita*	12
Pepper	*incognita*	3
	javanica	6
	arenaria	7
Peach	*incognita*	10
	javanica	4
	arenaria	3
Pole bean	*incognita*	8
Lima bean	*incognita*	4
Grapevine	*incognita, javanica, & arenaria*	4
Alfalfa	*incognita, javanica*	3
	& hapla	2
Lespedeza	*incognita*	3
Lupin	*javanica, hapla, & arenaria*	1
Bermuda grass	*incognita, javanica, hapla, & arenaria*	2
	Total	235+

This paper is not intended to review all of the crops
in which resistance to root-knot nematodes has been developed. It is a general review of the progress that has been
made, some of the problems that confront the nematologist
and the plant breeder, and the future needs of programmes
devoted to breeding for resistance.

Normal host reaction

The rhizosphere is a dynamic environment and a micro-
cosm of chemical, physical and biological activity. The
events occurring in this environment greatly influence
the host-parasite interactions that occur. After browsing
along the surface of the root to find a suitable location
for entrance, a root-knot nematode larva penetrates the
apical meristerm or other regions of the root and migrates
through the cortex until it reaches a region of differen-
tiating xylem where it begins to feed. It feeds by pierc-
ing cell walls with its stylet and discharging its salivary products from the stylet tip (Linford, 1937). Changes
occur in the root tissues within the first 24-48 hours
after penetration (Bird, 1962).

The cells within the vascular cylinder are modified
into giant cells on which the nematode must feed in order
to develop and reproduce. Under conditions for rapid
growth and development, practically all the larvae within
a host plant differentiate as females (Triantaphyllou,
1973). Under crowded conditions in which the supply of
available food to the larvae is limited, many second-
stage larvae develop into males.

Gall formation is a separate phenomenon from giant cell
development. It is the first macroscopic reaction to be
observed soon after larval invasion. Secretions from the
nematode cause a hypertrophy and hyperplasia of cortical
tissues and division of the pericycle. The female is
usually embedded in the gall and deposits her eggs within
the galled tissues or outside the root. The size of the
gall may vary, depending on the nematode species, host
species or cultivar and environmental conditions. For example, most bean lines are very susceptible to root-knot
nematodes and gall extensively. Some lines show no galling response to *M. incognita* (Fassuliotis *et al.*, 1970)
or to *M. javanica* (Bird, 1974) but giant cells develop
within the vascular cylinder and the nematodes develop

normally. The females protrude from the root and resemble immature *Heterodera* females (Bird, 1974).

Types of resistance

Resistance to root-knot nematodes may operate within the soil environment (preinfectional) or within the root environment (postinfectional). Preinfectional resistance, also referred to as passive, pre-existing or natural resistance (Wallace, 1973), operates before the nematode penetrates the surface of the root. This type of resistance may be associated with root exudates that either repel the infective stage larvae from the roots or are toxic to them (Rohde, 1972). Preinfectional resistance may be active against one nematode species but not against another. Sasser (1954) reported that *M. hapla* larvae do not readily penetrate oats and rye but that *M. incognita* larvae do. Goplen and Stanford (1959) found that *M. hapla* larvae failed to penetrate roots of a resistant selection of Vernal alfalfa but *M. javanica* entered readily. Plants exhibiting preinfectional resistance could properly be called non-hosts.

Postinfectional resistance is the most common type of resistance and is similar to active, provoked or induced resistance (Wallace, 1973). Resistance is manifested after the nematode penetrates the plant tissues. A host-parasite relationship is established that determines the fate of the host plant and the nematode.

Plants are categorized as being immune, resistant, susceptible, or tolerant to diseases. These are relative terms that represent a continuum of interactions between the nematode and the plant. An immune plant is one that is able to prevent infection with no disease expression, i.e. absolute resistance. A susceptible plant is one that cannot overcome or withstand the injurious effects of the nematode and permits it to develop to its fullest capacity. A resistant plant is a plant that cannot prevent

entrance by the parasite but is able to prevent, restrict
or retard its development. Tolerance refers to the ability
of a plant to endure invasion by the nematode without app-
reciable symptom expression or damage.

In a plant breeding programme the breeder is mainly
concerned with breeding plants that can withstand the
ravages of parasite attack, to enable them to grow, and
produce well.

Resistance reactions

Generally, larvae enter the roots of resistant plants
as readily as susceptible plants (Webster, 1975). Resistance reactions are manifested when some factors in the
host upsets the relationship between the host and nematode. Resistance is expressed by larvae failing to develop into adults, prolongation of the developmental cycle,
larval death with root tissues, alteration of sex ratio
toward maleness, or larval egress soon after penetration.

In two wild species of *Cucumis*, Fassuliotis (1970)
found that resistance was expressed by the development of
small giant cells which impeded the development of females
and shifted the sex ratio toward maleness. The host plants
failed to provide the nematode with an adeuqate supply of
nutrients for proper development. Reynolds *et al.* (1970)
observed that larvae readily entered the roots of resistant 'African' alfalfa but that after 7 days most of the
larvae left the root and the few that remained did not
develop.

Results of work by Fassuliotis and Dukes (1972) showed
that the galling reaction after penetration does not necessarily indicate successful development of the nematode.
Although the galling response is stimulated after attack
by *M. incognita* and *M. javanica* (Fassuliotis, unpublished),
the host fails to provide the necessary conditions for
normal parasite development.

The hypersensitive reaction is the most usual type of

resistance response. Hypersensitivity implies the death
of the host cells and the pathogen. This reaction occurs
within a few hours after penetration (Webster and Paulson,
1972). The browning reaction of the roots, which indicates
hypersensitivity (Fassuliotis *et al*., 1970), is due to
an accumulation of phenolic compounds in the tissues
around the infection sites (Milne *et al*., 1965; Pi and
Rohde, 1967).

Brodie *et al*. (1960) explored the nature of resistance
in cotton and concluded that resistance was associated
with severe root necrosis that killed whole sections of
root. This extreme sensitivity could cause considerable
stunting of the plant before new roots could develop to
offset the effects of the initial attack.

Resistance can be modified by the plant's genotype and
by environmental factors (Rohde, 1965). Soil temperature
is probably the most important environmental factor in
the expression of resistance. Holtzman (1965) found that
tomatoes were less resistant at 30 and 34.5°C than at
20 and 25°C. The detailed study by Dropkin (1969) showed
that the resistant tomato cv. Nematex grown at 28°C and
below, was highly resistant to *M. incognita acrita*. However, resistance diminished with each degree above 29°C
and at 33°C the seedlings were fully susceptible. Fig. 1
shows the effect of increased soil temperature on the expression of resistance in Atkinson tomatoes grown at 21,
26.7 and 32.2°C (Fassuliotis, unpublished). Similar temperature effects were observed on beans (Fassuliotis *et
al*., 1970) and sweet potato (Jatala and Russell, 1972).
However, not all resistant plants are affected by similarly increased temperatures. Dropkin (1969) also observed
that the resistance of the African horned cucumber, *Cucumis metuliferus* E. May increased as the temperature rose
from 28 to 32°C. He also found that resistance reactions
differed between cultivars within the same species. Two
contrary reactions were observed in lespedeza; with

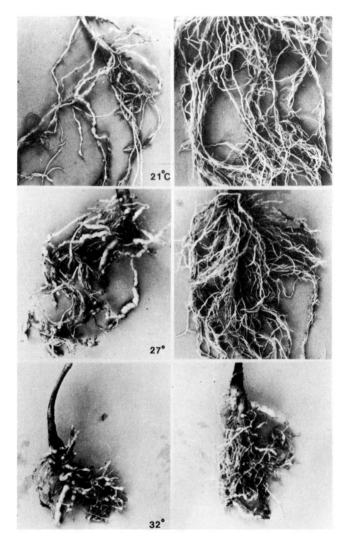

Fig. 1. The effect of increased soil temperatures on the expression of resistance in the root-knot nematode susceptible tomato cv. Homestead (left) and in the resistant cv. Atkinson (right).

increased temperature larval growth increased in 'Check No. 9' lespedeza but not in 'Alabama L7' (Table II).

The nematologist's role in a breeding programme

Breeding for root-knot nematode resistance involves the

TABLE II

Effect of increased temperature on host response and larval growth
of *Meloidogyne incognita acrita*[1]

	Temperature ^{0}C			
Host response	root necrosis %	larval growth %	root necrosis %	larval growth %
I Reduced necrosis & larval growth at high temp.				
'Anahu' tomato	53	11	1	43
II Low or no necrosis at either temp. & increase in larval growth at high temp.				
West India gherkin	0	4	0	46
'Auburn 56' cotton	7	1	7	28
'Check No. 9' lespedeza	1	0	0	25
III Necrosis & no larval growth at both temp.				
Crotolaria spectabilis	extensive	0	extensive	6
'African' alfalfa	84	4	58	2
IV No necrosis at either temp. & little larval growth				
'Alabama L7' lespedeza	0	24	0	30
V Increased resistance at high temp.				
African horned cucumber	2	19	25	0

[1]Adapted from Dropkin, V.H., 1969.

same basic principles as those used in breeding for resistance to other organisms. To be effective, root-knot resistance must be combined with desirable agronomic and horticultural characteristics of standard cultivars that already have resistance to one or more other diseases and
should be coordinated with the general improvement of the
crops concerned. The breeder should have an intimate
knowledge of the crop; the nematologist, in turn, must
have an intimate knowledge of the nematodes. My concept
of the nematologist's primary areas of responsibility in

a breeding programme consists of the following:
1) Accurately identify the root-knot nematode species
2) Establish pure cultures of nematode populations
3) Study the variability of the nematode populations and
their reaction to the resistant germplasm
4) Establish reliable methods for screening test plants
5) Seek sources of resistance
6) Define the mechanisms or nature of resistance
7) Support the breeder in technological problems that may
arise during routine screening of test plants.

Identification of the root-knot nematode species

Before initiating a programme of testing and breeding
the nematologist must determine which root-knot nematode
species are present in the geographical area of concern,
the distribution of the species and the species on which
emphasis should be placed in a breeding programme. Accurate identification is essential. The perineal pattern
of the saccate female is the principal character used
for determination of the species. However, because of
variability of the cuticular pattern, positive identification is often difficult, even by the "experts". To
overcome this, Sasser (1954, 1966, 1972b) proposed a number of host differentials to identify the species (see
these Proceedings). This system is useful also for identifying mixtures of species and for detecting pathogenic
variation within a species.

Establishing pure cultures

After a determination of the species has been made and
a decision on which root-knot nematode species and populations emphasis will be placed in a screening and breeding programme, it is important to establish pure cultures
of the species to be used as inoculum in screening tests.
While more than a single species may occur simultaneously
in a field, it is desirable to consider each species as

a separate problem for testing purposes. By screening
plants in mixed populations, valuable genetic material
with resistance to only one species and not to the other
may be discarded. Such breeding material could be of value
in areas where only one species predominates.

Pure cultures are started from a small number of specimens that have been positively identified. An egg mass
is removed from the root tissue and placed in a small
amount of water in a watch glass. A perineal pattern is
prepared of the corresponding female dissected from the
root. After microscopic identification of the species is
made with certainty, the egg mass is added to a 10 cm
pot of sterile soil in which a susceptible tomato seedling is planted. After 30-50 days the root system is examined for galling and nematode reproduction. The root-
knot nematode species, once again, is confirmed and the
infected roots are subdivided among other non-infected
tomato plants in order to increase the culture.

The culture is subsequently maintained in the greenhouse to provide a continuous source of inoculum to screen
large plant populations at any given time. The importance
of checking the population before beginning a test and
at the conclusion of a test cannot be over-emphasized in
order to insure that the roots of the test plants were
not infected by another root-knot nematode species.

Nematode variability

Breeding programmes cannot be expected to offer resistance to all populations of a *Meloidogyne* spp. and it
must be clear that the duration of resistance is limited.
How soon resistance will "break down" depends largely on
crop management practices and on the physiological variability of the nematode species. Disease organisms are
notorious for their ability to vary and the root-knot
nematodes are no exception (Sturhan, 1971).

Unquestionably, biotypes within a population exist

that differ in physiological characteristics. Successive
cropping with resistant cultivars can shift the distribution of frequency of a population from avirulence to virulence (Nelson, 1973). Riggs and Winstead (1959) in greenhouse experiments demonstrated that virulent populations
were selected from populations of M. *incognita* and M. *arenaria* that readily developed on tomato resistant to these
species. New pathogenic races of *Meloidogyne* spp. were
also reported by Sauer and Giles (1959) and Okamoto and
Mistui (1974) after continuous cropping with resistant
tomatoes; similar observations were made by Graham (1969)
on resistant tobacco. Giles and Hutton (1958) minimized
population shifts by planting resistant tomatoes with
susceptible tomatoes. After 5 years of continuous tomato
plantings they were able to retain resistance.

Populations of *Meloidogyne* spp. varying in pathogenicity are morphologically similar (Sturhan, 1971) and can
be detected only by differential host reactions (Sasser,
1954, 1966, 1972b). Variation of pathogenicity within a
species can limit the usefulness of a resistant cultivar.
Martin and Birchfield (1973) found that within M. *incognita* a population existed that was severely pathogenic
on soybeans but failed to mature on 'Centennial' sweet
potato, usually a very susceptible cultivar. On the other
hand, sweet potato cultivar La 4-73, resistant to Louisiana populations of M. *incognita*, was severely attacked
by a population from Maryland. Nishizawa (1974) reported
similar variability in Japanese populations of M. *incognita* on sweet potato. Populations of M. *incognita* have
been identified that attack and reproduce on the resistant NC-95 tobacco cultivar (Perry and Ziekus, 1972;
Phipps *et al.*, 1972; Reddy *et al.*, 1972; Sasser, 1972b;
Kirby *et al.*, 1975). Variability in host responses within
the root-knot nematodes collected from different parts of
the world was reported by Sasser (1972b).

The occurrence of pathotypes differing in host

specificity emphasizes the importance of screening plants
to more than one population of a given species in order
to broaden the usefulness of a resistant cultivar.

Screening methods

Techniques to differentiate resistant from susceptible
plants are as important to breeding for resistance as
are the sources of resistance. Success depends on the
ability to screen large numbers of plants with maximal
opportunity for infection. There are probably as many
techniques and modification as there are research workers
involved in screening. Laboratory, greenhouse and field
methods have been used to select resistant plants.

Infested soil or galled tissue are often used as inoculum in a screening test but they have certain disadvantages. Inoculum levels are difficult to control and they
are not biologically pure. The nematode may predispose the
plants to infection by other disease organisms present
in the inoculum (Powell, 1971).

Precise control of inoculum levels can be achieved by
using larvae or eggs as inoculum. Minton *et al.* (1966)
surface sterilized larvae collected in a mist chamber
with 0.001% 8-quinolinol sulphate for 30 min followed by
a rinse in tap water to eliminate fungal and bacterial
pathogens.

Hussey and Barker (1973) developed a method for releasing and collecting large numbers of eggs from infected roots utilizing NaOCl. This system is extremely satisfactory and has received wide acceptance among nematologists working with root-knot nematodes. Infected roots
which do not show any signs of rot are shaken vigorously
in 200 ml of 20% commercial bleaching solution (approximately 1% NaOCl) in a 1 litre container for 4 min. The
NaOCl solution is passed through a combination of 200/
500 mesh sieves. The eggs collected on the 500 mesh
sieve are rinsed with tap water to remove the residual

NaOCl. The roots are rinsed two more times with water to
remove additional eggs. Over a million eggs could be collected from one heavily infected tomato plant with this
method. The eggs are transferred to a stainless steel
pitcher. Stainless steel is preferred because the eggs
do not stick to the surface as they do to glass or plastic.
The concentration of eggs per ml is then calculated; the
volume of water is adjusted to dispense about 1,000 eggs
per ml. This method is simple and rapid. The eggs are
surface sterilized and the inoculum is easily standardized
for distribution around the root systems or for mixing
directly in the soil.

Laboratory methods   Dropkin *et al*. (1967) developed a
laboratory technique for primary screening of tomato. The
method was based on the assumption that the response of
small seedlings corresponded to the response of older
plants. Sterile seedlings with roots 1-2 cm in length
were transferred to agar slants of modified White's medium. They were inoculated with eggs suspended in 1% hyd-
oxyethylcellulose and rated for resistance 5 to 7 days
later based on the response of the seedlings to infection.

Fassuliotis and Corley (1967) utilized plastic seed
growth pouches as a container for screening. Seeds are
germinated on moist filter paper in petri dishes and
seedlings with root tips approximately 1 cm long are inserted through perforations in the trough. After the roots
have grown to 3 or 4 cm in length, they are inoculated
with a larval suspension in 0.5% carboxymethylcellulose.
Symptoms of infection are visible 2-3 days after inoculation and a high degree of accuracy for selecting resistant plants is obtained 5 days after inoculation which
corresponds to soil tests. Both techniques are useful for
routine screening of small-seeded plants since they permit
rapid disposal of the susceptible plants and resistant
selections could be transplanted to soil for further
testing. Milne (1966) screened *Nicotiana* seedlings for

resistance using the hypersensitive reaction as a criterion
of resistance. However, since the roots were killed and
stained with cotton blue lactophenol, the resistant plants
could not be saved for transplanting.

The rag doll approach was recently used for synchronous
inoculation of cotton seedlings with *M. incognita* larvae
(Carter *et al.*, 1977). The method can be applied for
screening tests as well.

Greenhouse methods   Most screening tests are done in the
greenhouse. Again, methodology will vary among workers.
Barrons (1939) found that bean and cowpea grown in artificially infested soil could be tested satisfactorily in
the seedling stage. Bailey (1940) reported similar results with tomato seeds planted in infested soil in the
greenhouse. Harrison (1960) transplanted tomato seedlings
into infested soil and left them until they were ready
for field planting. The plants were indexed, the susceptible plants were eliminated and resistant plants were
planted in the field.

Some plants such as the tomato and sweet potato can
be uprooted, evaluated for resistance and replanted;
others cannot. A difficulty encountered in breeding snap
beans for resistance to root-knot nematodes has been the
propagation of resistant plants following their evaluation for resistance (Fassuliotis *et al.*, 1970). The
plants are sensitive to root disturbance and will seldom
survive the shock of transplanting. Plant breeders must
then use remnant seed of resistant lines. This is time
consuming and does not ensure that the plants from the
remnant seed will be resistant. To avoid using remnant
seed, it is necessary to wait until the bean pods are
partially mature before evaluating the plants for resistance. But, by this time the plant is too old for back-
crossing or crossing. Root senescence and root rotting
organisms often cause cortical sloughing, making an accurate evaluation difficult.

   Wyatt and Fassuliotis (in preparation) developed a
technique for evaluating snap bean plants and propagating
resistant individuals for crossing and selfing. Seeds
were germinated and grown in paper towels until radicles
were 6 to 8 cm long. Seedlings were then placed in 10 cm
clay pots with the root tips protruding 1 cm through the
drain hole in the pot and filled with sterile soil. Holes
1 cm deep in the soil of a greenhouse bench were inoculated with root-knot nematode eggs and the clay pots were
placed over the holes. Roots that grew through the drain
hole became infected and were cut off for evaluation without permanently damaging the plant.

   Recently, I began using polystyrene planter flats,
specifically designed for growing transplants, to screen
large plant populations in a minimum amount of space.
Each flat or tray consists of 128 3.8 cm square inverted
pyramid-shaped root cells. They are filled with a mixture
of builders sand and soilless planting medium (2:1 v/v)
supplemented with a complete fertilizer. Each cell is
inoculated with 2000 to 5000 eggs and then planted with
either seed or seedling. The trays are open at the bottom
and set in aluminium T-rails, allowing the roots that grow
through the hole to be air pruned. An excellent root system grows within the cell. After 30 days, the plant is
lifted from the cell, washed of adhering growing medium
and evaluated. To date we have found this system effective
for evaluating tomatoes, beans and corn.

Field testing   Field plots provide sufficient land area
to screen large plant populations and opportunity to select desirable agronomic and horticultural types from
resistant plants. However, it cannot be recommended for
initial screening. Within field plots the soil may not be
uniformly infested and as a result escapes may erroneously
score as resistant. Escapes can be disastrous in a breeding programme. The breeder wastes time and money because
of saving and planting seed from escapes or using these

escapes as parental lines for further crossing.

However, where greenhouse facilities are limited, the
double row method (Ross and Brim, 1957) substantially reduces the risk of selecting escapes as resistant plants
since a row of susceptible check plants is planted only
15 cm from the test plants. This method was effectively
used by Sikora *et al.* (1973) to test the reaction of
nematode resistant tomato cultivars in India.

Rating schemes for resistance

Whatever method is used for testing plants for resistance, an accurate assessment of the plant's resistance
or susceptibility may be masked by several factors. The
level of inoculum is all too important. At low levels,
susceptible cultivars may show only very little of in-
termittent galling and at high levels resistant cultivars
may show some or even a moderate amount of galling (Hartmann, 1976). As discussed earlier, if soil temperatures
are high, resistance "breaks down" and no apparent differences show between resistant and susceptible cultivars.
Therefore, the selection of resistant plants requires
a basic understanding of the nematode, the host plant,
the types of interactions that may occur and the environmental relationships between the plant and the nematode
that may affect resistance.

Generally, the degree of resistance is indicative of
the ability of the nematode to reproduce on a host plant.
Because most susceptible plants infected with root-knot
nematodes develop galls, this response is often used as
the sole criterion for assessing the resistance of a
plant. Galling alone is not indicative of the nematode's
ability to reproduce (Golden and Schafer, 1958; Fassuliotis and Dukes, 1972; Fox and Miller, 1973) and an
evaluation based on this criterion can be misleading. The
index system used by most nematologists is based on a
1-5 scale in which 1 = no galls or reproduction,

2 = trace of galling and/or reproduction, 3 = light galling and/or reproduction, 4 = moderate galling and/or reproduction and 5 = abundant galling and/or reproduction.

Taylor (1971) presents a convenient classification
system than is based on nematode reproduction compared
to a susceptible control plant:

1) Susceptible: plants in which the reproduction is normal.

2) Slight resistant: reproduction is about 25% to 50% of
that on susceptible plants.

3) Moderately resistant: reproduction is 10% to 25% of
control.

4) Very resistant: reproduction is about 1% to 10% of
control.

5) Highly resistant: reproduction is less than 1% of
control.

6) Immunity: nematodes enter the roots but do not reproduce.

In order to be more objective, Fassuliotis (1967) used
female counts to determine resistance of *Cucumis* species
to infection by *M. incognita*. Roots were weighed and then
macerated in a mixture of nitric and chromic acid (Dropkin *et al.*, 1960) and females released from the roots
were counted and converted to number of females per gram
of root. Four arbitrary classes were used: 0-25 females
= most resistant, 26-50 = less resistant, 51-100 = less
susceptible, over 100 = most susceptible.

For more critical evaluation of advanced breeding
lines, the method described by Hussey and Barker (1973)
for extracting eggs is used in our laboratory. The number
of eggs recovered from the roots is expressed as an index
of reproduction. Index of reproduction is defined as the
number of eggs developing on a resistant cultivar expressed as a percentage of those developing on the susceptible cultivar. However, as with all systems there
are pitfalls to be wary of. The NaOCl solution does not

macerate the galls, and eggs within the galled tissue are
not released.

Sources of resistance

In a breeding programme for root-knot nematode resistance
we are concerned with two sources of genetic variability;
one is concerned with the variability within the crop
species and the other within the nematode. The variable
nature of the nematode has been presented. It is obvious
that to proceed with a breeding programme it is necessary
to have germplasm containing resistance-conferring genes.
There are two general sources of resistance: that found
within the crop species and that found in closely related
species. The transfer of resistance is greatly simplified
if the source of resistance can be found within commonly
grown and adapted cultivars in the locality where the
nematodes are a problem, i.e. cowpea, beans, sweet potato.
Simple intraspecific crosses, followed by sufficient back-
crossing, selfing, and screening for several generations
are all that is needed to obtain a new cultivar.

More frequently, sources of resistance are difficult
to find. Fassuliotis *et al.* (1970) found only two access-
ions from over 1100 plant introductions, cultivars and
breeding lines of beans, *Phaseolus vulgaris* L., with
sufficient resistance to include in a breeding programme.
Only after a search through over 270 lines of alfalfa,
were Stanford *et al.* (1958) able to find a percentage of
plants in Vernal and Hilmar alfalfa resistant to *M. hapla*.

When resistance cannot be found within the crop species,
attention must be turned to related wild species or genera. Wild species have been invaluable in providing breeders with root-knot resistance for inclusion in their breeders programmes. Shepherd (1974a, 1974b) developed a highly
resistant cotton cultivar resulting from a cross between
Gossypium barbadense L. and *G. hirsutum* L.

Lycopersicon peruvianum (L.) Mill. was the source of

resistance through which breeders were able to develop
resistant tomatoes. Because of sexual incompatibility
between the wild species and *L. esculentum* Mill., embryo
culture was used to develop the hybrid plant (Smith,
1944).

Although there has been much progress in selecting and
breeding root-knot resistance in many important crops,
sexually compatible germplasm has not been found in the
cucurbit cultivars (cucumber, melon, squash and pumpkin)
(Fassuliotis and Rau, 1963; Fassuliotis, 1971). However,
several wild species, *Cucumis anguria* L. *C. longipes*
Naud., *C. metuliferus* Naud. and *C. heptadactylus* Naud.
show resistance to *M. incognita* (Fassuliotis, 1967).
Attempts to produce a fruiting hybrid plant have yet to
be realized (Fassuliotis, 1877a, 1977b).

Eggplant is an important food crop in most warmer parts
of the world except in the United States where it is considered a minor crop. Resistance to *M. incognita* appears
to be non-existent within the species, *Solanum melongena*
L., but cv. Black Beauty appears to be tolerant (Birat,
1966; Alam *et al.*, 1974; Yadav *et al.*, 1975).

A source of resistance was found in *Solanum sisymbriifolium* Lam. (Choudhary *et al.*, 1969; Fassuliotis, 1973)
which may eventually provide the germplasm genes through
which resistance will be transferred into eggplant. Attempts to obtain a viable hybrid with conventional methods
have been unsuccessful.

In the United States, the Germplasm Institute of the
U.S. Department of Agriculture maintains stocks of seeds
which are made available to state and federal research
workers. The USDA maintains Regional Plant Introduction
Stations in the Northeastern (Geneva, New York), North
Central (Ames, Iowa), Southern (Experiment, Georgia),
and Western (Pullman, Washington) part of the U.S. Each
of these stations increases and maintains stocks of selected plants species. A partial list of the stocks is

TABLE III

Some crops maintained by the Regional Plant Introduction
Stations in the U.S.[1]

(Northeastern) Geneva, New York		(North Central) Ames, Iowa		(Southern) Experiment, Ga.		(Western) Pullman, Washington	
Pea	1500	Tomato	3579	Peanut	3816	Bean	6000
Perennial		Corn	2305	Sorghum	3423	Safflower	1317
clover	550	Alfalfa	969	Pepper	1784	Lentil	500
Trefoil	341	Bromegrass	617	Canta-		Lettuce	493
Onion	320	Cucumber	596	loupe	1626	Cabbage	263
Timothy	225			Cowpea	1212		

[1]From H.L. Hyland, (1975).

shown in Table III. In addition, basic stocks are maintained under long term storage in the National Seed Storage Laboratory, Fort Collins, Colorado (Hyland, 1975).

Future challenges

It is encouraging that over 235 cultivars for 15 major
crops have been reported as resistant to one or more root-
knot species. Many of the published lists (Singh *et al.*,
1974; Fassuliotis, 1976; Hartmann, 1976) should be used
for information only and selected cultivars should be
tested to local nematode populations. It is probable that
some cultivars reported as resistant will be susceptible.
The gene base in many of these crops is very narrow and
it is probable that resistant selections were made from
tests with only one or two nematode populations. Resistant
selections must be sought that are more temperature stable.
At higher soil temperatures, resistance becomes ineffective in bean, tomato, and sweet potato with no apparent
differences between resistant and susceptible cultivars.

Two crops which are widely grown and lacking resistance
to root-knot nematodes are the cucurbit vegetables and
eggplants. Attempts to hybridize them with other crop
species have been unsuccessful.

The concept of interspecific hybridization presents a
challenge and has been accomplished in tomato, cotton,

TABLE IV

Crops with resistance to *Meloidogyne* spp. developed through interspecific hybridization

Common name	Plant genus	Plant species crossed	*Meloidogyne* spp.	Reference
cotton	Gossypium	*G. hirsutum* (s) x *G. barbadense* (r)	*M. incognita*	Shepherd (1974a,b)
peach	Prunus	*P. persica* (s) x *P. davidiana* (r)	*M. incognita*	Havis *et al.*, (1950)
tobacco	Nicotiana	4n (*N. sylvestris* (s) x *N. tomentosiformis* (r) x *N. tabacum* (s)	*M. incognita*	Clayton *et al.*, (1958)
tomato	Lycopersicon	*L. esculentum* (s) x *L. peruvianum* (r)	*M. incognita* *M. javanica* *M. arenaria*	Smith, (1944)
vetch	Vicia	*V. sativa* (r) x *V. angustifolia* (s)	*M. incognita* *M. javanica*	Minton *et al.*, (1966)

r = resistant parent, s = susceptible parent

tobacco, peach and vetch to produce hybrids with resistance to one or more species of *Meloidogyne* (Table IV).

Currently, there is an even greater challenge to increase the germplasm for resistance to root-knot nematodes since one of the better nematicides has been removed from the market. The task will require innovative research. Embryo culture was used by Smith (1944) to obtain 3 hybrid plants from the cross *Lycopersicon esculentum* and *L. peruvianum*. The hybrid plants became the foundation from which all of the resistant tomato cultivars are currently derived (Thomason and Smith, 1957).

Results from some of our research indicate that it will be feasible to use embryo culture to obtain an interspecific hybrid between *Cucumis metuliferus* and *C. melo* L. Fassuliotis (1977a) observed that when these species were crossed with the wild species as the maternal parent, small fruits were produced but the ovules aborted and failed to produce viable seed. We have subsequently raised hybrid plantlets on synthetic culture media. But, transfer of these plantlets to soil and propagation to flowering plants has been elusive.

During the past decade there have been exciting developments in the field of tissue culture, especially in the field of protoplast culture and somatic hybridization (Bajal, 1974). Carlson *et al.* (1972) demonstrated the formation of an interspecific somatic cell hybrid between *Nicotiana glauca* Grah. and *N. langsdorfii* Schrank. This breakthrough has sparked much research activity in many laboratories to form other interspecific and intergenic somatic hybrids.

In our laboratory we are attempting to produce a somatic hybrid by fusion of eggplant protoplasts with *Solanum sisymbriifolium* protoplasts to obtain a root-knot resistant hybrid. The first step in this project was achieved with the successful regeneration of plants from undifferentiated callus of *S. sisymbriifolium* (Fassuliotis, 1975).

For somatic hybridization to be accomplished, protoplasts
of the two species must fuse and be induced to form a
common cell wall before mitotic divisions can proceed.
We have fused *S. sisymbriifolium* protoplasts but have
been unsuccessful in regenerating a new cell wall.

Many technical details have to be worked out before
somatic hybridization becomes a routine laboratory technique, but there is much optimism that the technique will,
in time, provide a powerful tool for crop improvement
(Bajaj, 1974; Carlson, 1975).

References

Alam, M., Mashkoor, K., Abrar, M., and Saxena, S.K., (1974). Reaction
of some cultivated varieties of egg plant, pepper, and okra to the
root-knot nematode, *Meloidogyne incognita*. *Ind.J.Nematol.*, 4,
64-68.

Bailey, D.M., (1940). The seedling test method for root-knot nematode
resistance. *Proc.Amer.Soc.Hort.Sci.*, 38, 573-575.

Bajaj, Y.P.S., (1974). Potentials of protoplast culture work in agriculture. *Euphytica* 23, 633-649.

Barrons, K.C., (1939). Studies on the nature of root-knot resistance.
J.Agric.Res., 58, 263-272.

Birat, R.B.S., (1966). Relative susceptibility of brinjal varieties to
Meloidogyne javanica (Treub, 1885) Chitwood, 1949. *Science and
Culture* 32, 192-193.

Bird, A.F., (1962). The inducement of giant cells by *Meloidogyne javanica*. *Nematologica* 8, 1-10.

Bird, A.F., (1974). Plant response to root-knot nematode. *Ann.Rev.
Phytopathology* 12, 69-85.

Brodie, B.B., Brinkerhoff, L.A. and Struble, F.B., (1960). Resistance
to the root-knot nematode, *Meloidogyne incognita acrita*, in Upland
cotton seedlings. *Phytopathology* 50, 673-677.

Carlson, P.S., (1975). Crop improvement through techniques of plant
cell and tissue culture. *Bioscience* 25, 747-749.

Carlson, P.S., Smith, H.H. and Dearing, R.D., (1972). Parasexual interspecific plant hybridization. *Proc.Nat.Acad.Sci.*, 69, 2292-2294.

Carter, W.W., Nieto, S. Jr., and Veech, J.A., (1977). A comparison
of two methods of synchronous inoculation of cotton seedlings with
Meloidogyne incognita. *J.Nematol.*, 9, 251-253.

Choudhary, B., Rajendran, R., Singh, B. and Verma, T.S., (1969).
Breeding tomato, brinjal and cowpea resistant to root-knot nematodes (*Meloidogyne* spp.). *Ind.Nematol.Symp.*, New Delhi, 46-47.

Clayton, E.E., Graham, T.W., Todd, F.A., Gaines, J.G. and Clark, F.A.
(1958). Resistance to the root-knot disease in tobacco. *Tobacco
Science* 2, 53-63.

Dropkin, V.H. (1969). The necrotic reaction of tomatoes and other
hosts resistant to *Meloidogyne:* Reversal by temperature.

Phytopathology **59**, 1632-1637.

Dropkin, V.H., Davis, D.W. and Webb, R.E. (1967). Resistance of
  tomato to *Meloidogyne incognita acrita* and to *M. hapla* (root-knot
  nematodes) as determined by a new technique. *Proc.Amer.Soc.Hort.
  Sci.*, **90**, 316-323.

Dropkin, V.H., Smith, W.L., Jr. and Myers, R.F. (1960). Recovery of
  nematodes from infected roots by maceration. *Nematologica* **5**,
  785-788.

Fassuliotis, G. (1967). Species of *Cucumis* resistant to the root-knot
  nematode, *Meloidogyne incognita acrita*. *Plant Dis.Reptr.*, **51**,
  720-723.

Fassuliotis, G. (1970). Resistance of *Cucumis* spp. to the root-knot
  nematode, *Meloidogyne incognita acrita*. *J.Nematol.*, **2**, 174-178.

Fassuliotis, G. (1971). Susceptibility of *Cucurbita* spp. to the root-
  knot nematode, *Meloidogyne incognita*. *Plant Dis.Reptr.*, **55**, 666.

Fassuliotis, G. (1973). Susceptibility of eggplant, *Solanum melongena*
  to root-knot nematode, *Meloidogyne incognita*. *Plant Dis.Reptr.*,
  57, 606-608.

Fassuliotis, G. (1975). Regeneration of whole plants from isolated
  stem parenchyma cells of *Solanum sisymbriifolium*. *J.Amer.Soc.Hort.
  Sci.*, **100**, 636-638.

Fassuliotis, G. (1976). Progress, problems and perspectives in breeding food crops for root-knot resistance. *Proc.Res.Plan.Conf.on
  Root-knot, Raleigh, N.C.*, 81-93.

Fassuliotis, G. (1977a). Self-fertilization of *Cucumis metuliferus*
  Naud. and its cross-compatibility with *C.melo* L. *J.Amer.Soc.Hort.
  Sci.*, **102**, 336-339.

Fassuliotis, G. (1977b). *In* "4th Annual Colloquim Plant Cell and
  Tissue Culture." Ohio State Univ., (in press).

Fassuliotis, G. and Dukes, P.D., (1972). Disease reactions of *Solanum melongena* and *S. sisymbriifolium* to *Meloidogyne incognita* and
  Verticillium albo-atrum. *J.Nematol.*, **4**, 222.

Fassuliotis, G. and Rau, G.J., (1963). Evaluation of *Cucumis* spp.
  for resistance to the cotton root-knot nematode, *Meloidogyne incognita acrita*. *Plant Dis. Reptr.*, **47**, 809.

Fassuliotis, G. and Corley, E.L., Jr. (1967). Use of seed growth
  pouches for root-knot nematode resistance tests. *Plant Dis.Reptr.*,
  51, 482-486.

Fassuliotis, G., Deakin, J.R. and Hoffman, J.C. (1970). Root-knot
  nematode resistance in snap beans: Breeding and nature of resistance. *J.Amer.Soc.Hort.Sci.*, **95**, 640-645.

Fox, J.A., and Miller, L.I., (1973). Comparison of gall and egg-mass
  indices of two races of *Meloidogyne incognita* on differential
  hosts. *Phytopathology* **63**, 801.

Giles, J.E., and Hutton, E.M. (1958). Combining resistance to the
  root-knot nematode, *Meloidogyne javanica* (Treub) Chitwood and
  fusarium wilt in hybrid tomatoes. *Austral.J.Agric.Res.*, **9**, 182-192.

Golden, A.M. and Shafer, T. (1958). Unusual response of *Hesperis
  matronalis* L. to root-knot nematodes (*Meloidogyne* spp.). *Plant
  Dis.Reptr.*, **42**, 1163-1166.

Good, J.M. (1972). Management of plant parasitic nematode populations.
  Proc.Ann.Tall Timbers Conf., Florida, Feb. 1972, 109-127.

Goplen, B.P. and Stanford, E.H., (1959). Studies on the nature of resistance in alfalfa to two species of root-knot nematodes. *Agron. J.* 51, 486-488.

Graham, T.W., (1969). A new pathogenic race of *Meloidogyne incognita* on flue-cured tobacco. *Tobacco Science* 43-44.

Hare, W.W., (1965). The inheritance of resistance of plants to nematodes. *Phytopathology* 55, 1162-1167.

Harrison, A.L., (1960).Breeding of disease resistant tomatoes with special emphasis on resistance to nematodes. *Proc.Campbel's Soup Pl. Sc.Sem., New Jersey,* 57-58.

Hartmann, R.W. (1976). Breeding for nematode resistance in vegetables. *SABRAO J.,* 8, 1-10.

Havis, L., Chitwood, B.G., Prince, V.E., Cobb, G.S. and Taylor, A.L., (1950). Susceptibility of some peach rootstocks to root-knot nematodes. *Plant Dis.Reptr.,* 34, 74-77.

Holtzmann, O.V., (1965). Effect of soil temperature on resistance of tomato to root-knot nematode (*Meloidogyne incognita*) *Phytopathology* 55, 990-992.

Hunt, O.J., Faulkner, L.R. and Pladen, R.N. (1972). *In* "Alfalfa Science and Technology." (Ed. C.H. Hanson). 355-370.

Hussey, R.S., and Barker, K.R. (1973). A comparison of methods of collecting inocula of *Meloidogyne* spp. including a new technique. *Plant Dis.Reptr.,* 57, 1025-1028.

Hyland, H.L. (1975). *In* "Crop Genetic Resources for Today and Tomorrow." (Eds. O.H. Frankel, J.G. Hawkeš.) 139-146. Cambridge University Press.

Jatala, P., and Russell, C.C. (1972). Nature of sweet potato resistance to *Meloidogyne incognita* and the effect of temperature on parasitism. *J.Nematol.,* 4, 1-7.

Kehr, A.E. (1966). *In* "Pest Control by Chemical, Biological, Genetic and Physical Means." *ARS Symp.,* 33-110.

Kirby, M.F., Dickson, D.W. and Smart, G.C., Jr. (1975). Physiological variation within species of *Meloidogyne* occurring in Florida. *Plant Dis.Reptr.,* 59, 353-356.

Linford, M.B. (1937). The feeding of the root-knot nematode in root tissue and nutrient solution. *Phytopathology* 27, 824-835.

Malo, S.E. (1964). A review of plant breeding for nematode resistance. *Proc.Soil and Crop Science Soc.Florida* 24, 354-365.

Martin, W.J., and Birchfield, W. (1973). Further observations of variability in *Meloidogyne incognita* on sweetpotatoes. *Plant Dis. Reptr.,* 57, 199.

Milne, D.L. (1966). Screening of *Nicotiana* plants for resistance to *Meloidogyne* spp. by the use of hypersensitive root reactions. *S. Afr.J.Agric.Sci.,* 9, 435-442.

Milne, D.L., Boshoff, D.N. and Buchan, P.W.W., (1965). The nature of resistance of *Nicotiana repanda* to the root-knot nematode, *Meloidogyne javanica. S.Afr.J.Agric.Sci.,* 8, 557-567.

Minton, N.A., Donnelly, E.D. and Shepherd, R.L. (1966). Reaction of *Vicia* species and F5 hybrids from *V. sativa* x *V. angustifolia* to 5 root-knot nematode species. *Phytopathology* 56, 102-107.

Moore, E.L., Jones, G.L. and Gwynn, G.R. (1962). Flue-cured tobacco variety NC 95 resistant to root-knot, black shank and the wilt diseases. *North Carolina Agr.Exp.Sta.Bull.,* 562.

Nelson, R.R. (1973). *In* "Breeding Plant for Disease Resistance." (Ed. R.R. Nelson.) 49-66. Pennsylvania State Univ. Press,

University Park and London.

Netscher, C., and Manhoussin, J.C., (1973). Results of an investigation of the comparative efficiency of a resistant tomato variety
and certain nematicides against *Meloidogyne javanica*. *Biologie*
21, 97-102.

Nishizawa, T., (1974). A new pathotype of *Meloidogyne incognita* breaking resistance of sweet potato and some trials to differentiate
pathotypes. *Jap.J.Nematology* 4, 37-42.

Okamoto, K., and Mistui, Y. (1974). Occurrence of a resistance breaking population of *Meloidogyne incognita* on tomato. *Jap.J.Nematol.*,
4, 32-36.

Perry, V.G. and Zeikus, J.A. (1972). Host variations among populations
of the *Meloidogyne incognita* group. *J.Nematol.*, 4, 231-232.

Phipps, P.M., Stipes, R.J. and Miller, L.I., (1972). A race of *Meloidogyne incognita* from *Albizzia julibrissin* parasitizes *Nicotiana
tabacum* 'NC 95'. *J.Nematol.*, 4, 232.

Pi, C.L. and Rhode, R.A., (1967). Phenolic compounds and host reaction in tomato to injury caused by root-knot and lesion nematodes.
Phytopathology 57, 344.

Powell, N.T. (1971). *In* "Plant Parasitic Nematodes." (Eds. B.M. Zuckerman, W.F. Mai, R.A. Rohde.) 2, 119-136. Academic Press, New York
and London.

Reddy, P.P., Setty, K.G., and Govindu, H.C., (1972). Susceptibility
of flue-cured tobacco variety 'NC 95' to southern root-knot nematode. *Mysore J.Agr.Sci.*, 6, 192-193.

Reynolds, H.W., Carter, W.W., and O'Bannon, J.H., (1970). Symptomless
resistance of alfalfa to *Meloidogyne incognita acrita*. *J.Nematol.*,
2, 131-134.

Riggs, R.D., and Winstead, N.N. (1959). Studies on resistance in
tomato to root-knot nematodes and on the occurrence of pathogenic
biotypes. *Phytopathology* 49, 716-724.

Rohde, R.A. (1965). The nature of resistance in plants to nematodes.
Phytopathology 55, 1159-1167.

Rohde, R.A. (1972). The expression of resistance in plants to nematodes. *Ann.Rev.Phytopathology* 10, 233-252.

Ross, J.P., and Brim, C.A. (1957). Resistance of soybeans to the
soybean cyst nematode as determined by a double-row method. *Plant
Dis.Reptr.*, 41, 923-924.

Sasser, J.N. (1954). Identification and host-parasitic relationships
of certain root-knot nematodes (*Meloidogyne* spp.). *Md.Agr.Exp.
Sta.Bull.*, A-77, 30.

Sasser, J.N., (1966). Behaviour of *Meloidogyne* spp. from various geographical locations on ten host differentials. *Nematologica* 12,
97-98.

Sasser, J.N. (1972a). Managing nematodes by plant breeding. *Proc.Ann.
Tall Timbers Conf., Florida, Feb. 1972*, 65-80.

Sasser, J.N., (1972b). Physiological variation in the genus *Meloidogyne* as determined by differential hosts. *OEPP/EPPO Bull.*, 6,
41-48.

Sasser, J.N. (1977). Worldwide dissemination and importance of the
root-knot nematodes, *Meloidogyne* spp. *J.Nematol.*, 9, 26-29.

Sauer, M.R., and Giles, J.E. (1959). A field trial with a root-knot
resistant tomato variety. *Irri.Res.Sta.Tech.Paper* 3.

Shepherd, R.L. (1974a). Breeding root-knot resistant *Gossypium hirsu-tum* L. using a resistant wild *G. barbadense* L. *Crop Sci.*, 14, 687-691.

Shepherd, R.L. (1974b). Transgressive segregation for root-knot nema-tode resistance in cotton. *Crop Sci.*, 14, 872-875.

Sikora, R.A., Sitaramaiah, K., and Singh, R.S. (1973). Reaction of root-knot nematode-resistant tomato cultivars to *Meloidogyne jav-anica* in India. *Plant Dis.Reptr.*, 57, 141-143.

Singh, B., Bhatti, D.S. and Singh, K., (1974). Resistance to root-knot nematodes (*Meloidogyne* spp.) in vegetable crops. *PANS*. 20, 58-67.

Smith, P.G. (1944). Embryo culture of a tomato species hybrid. *Proc.Amer.Soc.Hort.Sci.*, 44, 413-416.

Stanford, E.H., Goplen, B.P., and Allen, M.W. (1958). Sources of resistance of alfalfa to the northern root-knot nematode, *Meloidogyne hapla. Phytopathology* 48, 347-349.

Sturhan, D. (1971). *In* "Plant Parasitic Nematodes." (Eds. B.M. Zuckerman, W.F. Mai and R.A. Rohde.) 2, 51-71. Academic Press, New York and London.

Taylor, A.L. (1971). Introduction to research on plant nematology, an FAO guide to the study and control of plant-parasitic nema-todes. *FAO, UN, Rome. PL: CP/5-rev.l.*

Thomason, I.J. and Smith, P.G. (1957). Resistance in tomato to *Meloidogyne javanica* and *M. incognita acrita. Plant Dis.Reptr.*, 41, 180-181.

Triantaphyllou A.C. (1973). Environmental sex differentiation of nematodes in relation to pest management. *Ann.Rev.Phytopathology* 11, 441-462.

Wallace, H.R. (1973). Nematode ecology and plant disease. Crane, Russak & Co., New York, 228.

Webster, J.M. (1975). *In* "Advances of Parasitology." (Ed. B. Dawes.) 225-250. Academic Press, London.

Webster, J.M., and Paulson, R.E. (1972). An interpretation of the ultrastructural response of tomato roots susceptible and resis-tant to *Meloidogyne incognita* (Kofoid and White) Chitwood. *OEPP/EPPO Bull.*, 6, 33-39.

Wyatt, J.E., and Fassuliotis, G. (1978). Methods for screening snap beans for resistance to root-knot nematodes. (In manuscript).

Yadav, B.S., Nandwana, R.P, Lal, A. and Verma, M.K. (1975). Evaluation of certain brinjal varieties to the root-knot nematode *Meloidogyne incognita. Ind.J.Mycology and Plant Pathology* 5, 17.

Discussion

Ferris remarked that the largest percentage of the cultivars were resistant to *M. incognita*. Fassuliotis ack-nowledged that this was so. He pointed out that in the case of tomatoes, resistance was acquired through the wild species, *Lycopersicon peruvianum*, which showed vari-able resistance to populations of *M. incognita*, *M. arenaria* and *M. javanica*. After being hybridized with *L. esculentum* the degree of resistance was increased from the original resistant parent apparently by the addition of modifying factors transmitted by the susceptible parent.

He explained that nematologists interested in screening certain stocks of seeds for resistance to root-knot nematode can obtain these on an exchange basis by writing to Dr. George A. White, Chief, Germplasm Resources Laboratory, Agricultural Research Center-West, Beltsville, Maryland 20705.

Asked if there was any point in breeding for tolerance, Fassuliotis said this would be useful if no other path was available but it is difficult to assess. Field tests are useful back up information to the breeding programme and are carried out at a number of different stations throughout the United States.

M. Di Vito and F. Saccardo[1]

*Laboratorio di Nematologia Agraria del
C.N.R., Bari, Italy*

In Italy, the root-knot nematode *Meloidogyne incognita*
(Kofoid and White) Chitw. can cause considerable losses in
the yield of peppers, especially in the areas where the
crop is intensively cultivated. The use of nematicides is
not always feasible nor economically convenient and, if
used extensively, may give rise to problems of environmental pollution.

It is generally accepted that resistance is the most
desirable way of controlling parasites and Hare (1957)
and Langford *et al.* (1968) respectively have shown sources of resistance to *M. incognita*   *Capsicum frutescens*
L. and *C. annuum* L. We have tested the reaction of several lines of *Capsicum* spp. to attack by five Italian
populations of the nematode.

The results (Di Vito and Saccardo, 1978) indicate a
susceptible reaction in all commercial cultivars of *C. annuum*, in cv. INRA-372 and in a Californian line of *C. pen-
dulum* Wild, in a Californian line of *C. chacoense* Hunz.
and in cv. P.I.188478 of *C. chinense* Jacq. Two wild lines
of *C. annuum* (P.I.159237 and 159256 from Georgia) appeared to be tolerant to the nematode and *C. frutescens* cv.

[1]*Laboratorio Valorizzazione Colture Industriali del C.N.E.N., C.S.N.,
Casaccia, Roma, Italy.* Contribution No. 535.

Webb and the line Surrinam-4 of *C. chinense* showed resistance. A screening test of several lines of *C. frutescens*
has indicated that all of them are characterized by a
high level of resistance to *M. incognita*.

Differences in the infectivity of the five populations
of the nematode on either susceptible or resistant lines
were essentially similar to those reported by Di Vito and
Lamberti (1976) for tomato.

The preliminary results of a breeding programme have
shown the possibility of crossing *C. chinense* with *C. annuum*, but the latter appeared to be highly incompatible
with *C. frutescens* (Saccardo *et al.*, 1976; Saccardo and
Sree Ramulu, 1977).

Acknowledgement

This paper is contribution no. 535 of the Laboratorio
Valorizzazione Colture Industriali del C.N.E.N.

References

Di Vito, M. and Lamberti, F. (1976). Reazione di varietà di Pomodoro
   a popolazioni di *Meloidogyne* spp. in serra. *Nematol.medit.*, **4**,
   211-215.
Di Vito, M. and Saccardo, F. (1978). Risposta di linee e varietà di
   Capsicum spp. agli attacchi di *Meloidogyne incognita* in serra.
   Nematol.medit., **6**, 83-88.
Hare, W.W. (1957). Inheritance of resistance to root-knot nematodes
   in pepper. *Phytopathology* **47**, 455-459.
Langford, W.R., Corley, W.L., Massey, J.H. and Sowell, G. Jr. (1968)
   Catalogue of seed available at the Southern Regional Plant Intro-
   duction Station, Georgia Exp. Station.
Saccardo, F., Sree Ramulu, K. and Tomarchio, L. (1976). Isolation,
   trasnfer and induction of disease resistance in *Capsicum*. *Gen-
   etika* **8**, 247-254.
Saccardo, F. and Sree Ramulu, K. (1977). Cytogenetical investigations
   in the genus *Capsicum. 3rd Capsicum-EUCARPIA Meeting, Montfavet,
   France.* (in press).

O. Arrigoni

Istituto di Botanica, University of Bari, Bari, Italy

Introduction

It has long been known that the respiratory activity of plant tissues is increased after wounding or after attacks by pathogens. Because it occurs as a result of traumatic lesions, this respiration has been described by the term "wound respiration". A peculiar feature of wound respiration is that it is not mediated by cytochrome oxidase, but rather by another terminal oxidase which is resistant to cyanide: it is therefore known as "cyanide-resistant respiration" (Allen, 1954; Thimann, 1954; Uritani & Akazawa, 1959; Hackett *et al.*, 1960; Click & Hackett, 1963; Verleur & Uritani, 1965). To date, very little is known about its physiological significance. On the basis of various experimental data, we have been examining the hypothesis that cyanide-resistant respiration is a metabolic event that is closely related to the activation of biological defence mechanisms. Further, the data indicates that the development of cyanide-resistant respiration is conditioned by the presence of ascorbic acid (Vitamin C) in the cell. Thus, ascorbic acid can be considered as a factor of primary importance in the biological defence mechanisms of plants and animals.

Relation between ascorbic acid and cyanide-resistant respiration

Lycorine, an alkaloid extracted from Amarillidaceae, inhibits the biosynthesis of ascorbic acid (AA) in plants (Arrigoni *et al.*, 1975). We have demonstrated that AA in the cell is closely associated with the development of cyanide-resistant respiration.

Aerobically maintained potato tuber slices showed an increased uptake of oxygen within a day after cutting, from 20-30 µl O_2/h/g fresh weight to 100-150 µl O_2/h/g fresh weight; and the respiration became cyanide-resistant. During the period of aeration, the slices - while developing cyanide-resistant respiration - actively synthesized AA. The addition of lycorine inhibited AA biosynthesis and, at the same time, prevented the development of respiration. At 5 µm lycorine, the increase of respiration was almost completely inhibited. The degree of inhibition of AA biosynthesis correlates with the increased respiration, suggesting that AA is required to develop cyanide-resistant respiration. Further support for this conclusion is obtained from the evidence that lycorine inhibition in potato tuber slices is prevented by the administration of AA (Arrigoni *et al.*, 1976).

Because the rise of respiration depends on newly synthesized RNA and proteins (Click & Hackett, 1963), and because lycorine inhibits the increase of respiration while AA counteracts this effect, it was postulated that AA is required to synthesise some hydroxyproline-containing proteins linked to the development of cyanide-resistant respiration.

To test whether the elicitation of cyanide-resistant respiration is related to the biosynthesis of mitochondrial hydroxyproline-proteins, determinations of such proteins were compared in mitochondria isolated from aged slices of potato tubers (with cyanide-resistant respiration) and mitochondria prepared from slices which were without this alternate oxidase. The former contained three times as much hydroxyproline-proteins as the latter mitochondria. These findings support the supposition that the development of cyanide-resistant respiration is dependent on the synthesis of mitochondrial hydroxyproline-proteins (Arrigoni *et al.*, 1977c).

Experiments showed that the addition of lycorine during the ageing of potato tuber slices strongly inhibited the biosynthesis of hydroxyproline-proteins; furthermore, the inhibition induced in the slices by lycorine was almost completely prevented when AA was added to the slices during the ageing process. These results show that AA is required to carry out the *in vivo* synthesis of hydroxyproline-proteins (Arrigoni *et al.*, 1977a). Presumably hydroxyproline-proteins, such as collagen, are synthesized as the ordinary proline-containing polypeptide chains and subsequently some of the proline residues present in the chain are hydroxylated by means of the prolyl-hydroxylase; AA could be involved in the hydroxylation process.

Since AA is required in the synthesis of hydroxyproline-proteins and since the amount of these proteins rises in mitochrondria when cyanide-resistant respiration is present, it may be concluded that the relationship observed *"in vivo"* between AA and cyanide-resistant respiration is mediated by hydroxyproline-proteins synthesis in mitochondria.

Cyanide-resistant respiration is not peculiar to the plant cell, but also occurs in animal cells. It is known that the phagocytic process, which represents the main defence power of the organism, is accompanied in polymorphonuclear leukocytes by an increase of oxygen consumption which is proportional to the load of particles ingested. An interesting feature of this respiratory burst is its insensitivity to cyanide (Baldrige & Gerard, 1933; Sbarra & Karnovsky, 1959; Iyer *et al.*, 1961; Paul & Sbarra, 1968). Laboratory experiments showed that the respiratory

burst is controlled by ascorbic acid in polymorphonuclear
leukocytes, as is the case in plant mitochondria. In fact,
the respiratory burst of polymorphonuclear leukocytes
from normal guinea pigs was higher than that of leukocytes
from scorbutic animals. Leukocytes from scorbutic animals
also contained a lower amount of hydroxyproline-proteins.
This indicates a close relationship between the endogenous
level of AA and the biosynthesis of hydroxyproline-containing proteins (Arrigoni *et al.*, 1978). Consequently,
the development of cyanide-resistant respiration is
conditioned - both in the polymorphonucleates and in
plants - by the AA-dependent synthesis of hydroxyproline-
proteins.

I would point out that the amount of hydroxyproline-
proteins in the mitochondria of tomato roots increases in
nematode-resistant varieties when these are infested by
Meloidogyne incognita. Nematode infestation causes no increase of hydroxyproline-proteins in roots of tomato
varieties susceptible to *M. incognita*. From the data, it
appears that it is not the concentration of hydroxyproline-
proteins in healthy plants but the amount that the root
can produce under the stimulus of the parasite (Zacheo
et al., 1977) that is the important factor in determining
the level of resistance. It seems reasonable to suppose
that the synthesis of hydroxyproline-containing proteins
represents the metabolic response of the cell through
which, by developing a cyanide-insensitive respiration,
the latter initiates the biological mechanism to counteract the effects of the pathogen.

Respiratory pathway

Since oxygen consumption in cyanide-resistant respiration is not mediated by cytochrome oxidase, it presents
the question of what is the terminal oxidase of this new
electron transport pathway to oxygen? This has been the
subject of considerable discussion in the literature but

until now there has been little or no clarification.

From our own experimental data, it would seem possible to conclude that this respiratory pathway is via a NADPH-oxidase whose activity is conditioned by manganese. The NADPH-oxidase activity of mitochondria endowed with a high rate of cyanide-resistant respiration (mitochondria from aged slices of potato tubers) was 10 times greater than that of mitochondria devoid of this alternate oxidase (mitochondria from freshly cut slices). The addition of 0.2 mM $MnCl_2$ to mitochondria devoid of cyanide-resistant respiration produced, at pH 7.4, an approximately 9-fold increase in their NADPH-oxidase activity. Conversely, no stimulation was observed when $MnCl_2$ was added to mitochondria endowed with cyanide-resistant respiration. This would suggest that the NADPH-oxidase of mitochondria having cyanide-resistant respiration is a protein which contains manganese.

In the mitochondria of tissues that have been damaged or attacked by pathogens there are considerable changes of peroxidase and superoxide dismutase activity, together with the development of cyanide-resistant respiration. Both enzymes are considered to be of importance in the defence mechanism against infection or damage to tissues. Peroxidase activity has been shown to increase when tissues become infected with a wide variety of organisms (Evans, 1970; Hussey & Krusberg, 1970; Rautela & Payne, 1970; Seevers et al., 1971; Matsumo & Uritani, 1972; Maraite, 1973; Giebel, 1974; Ohguchi et al., 1974). A marked increase has been found in peroxidase activity of the mitochondria as cyanide-resistant respiration develops.

In mitochondria from aged potato tuber slices, with 27% cyanide-resistant respiration, peroxidase activity was 6 times greater than in mitochondria from freshly cut slices where respiration was via cytochrome oxidase. Similarly, in aged artichoke slices, when 32% of respiration was cyanide-resistant, there was a 10-fold increase in peroxidase activity compared with the fresh slices. Hence, NADPH-oxidase and peroxidase increase to the same extent during the development of cyanide-resistant respiration, suggesting that the two enzymes together form the terminal oxidase complex of cyanide-insensitive respiration.

Superoxidase dismutase catalyzes the reaction

$$O_2^- + O_2^- + 2H^+ \rightarrow H_2O_2 + O_2 \quad \text{(Fridovich, 1974; 1975)}$$

It should be noted that contrary to what happens with NADPH-oxidase and peroxidase, superoxide dismutase is considerably reduced as cyanide-resistant respiration develops; in mitochondria without such respiration, superoxide dismutase activity was found to be 7500 units/mg protein, whereas in those with 30% cyanide-resistant respiration such activity decreased to 2500 units/mg protein.

Biological defence mechanism

From the data discussion so far, the following set of reactions can be postulated as the basis of the biological defence mechanism in which oxidase complex and superoxide dismutase have a role to play:

Proposal of molecular events occurring in the biological defence mechanism

H_2O_2 *production:*

$$NADPH + (Enz-Mn)^{3+} \rightarrow NADP^\bullet + (Enz-Mn)^{2+} + H^+$$

NADPH-oxidase activity

$$NADP^\bullet + O_2 \rightarrow NADP^+ + O_2^-$$

$$O_2^- + (Enz-Mn)^{2+} + 2H^+ \rightarrow H_2O_2 + (Enz-Mn)^{3+}$$

O_2^- *production:*

Peroxidase activity

$$H_2O_2 + NADPH \rightarrow NADP^\bullet + (H_2O)$$

$$NADP^\bullet + O_2 \rightarrow NADP^+ + O_2^-$$

Oxidase complex

Pathogen inactivation regulated by superoxide dismutase

Killing mechanism by O_2^-	oxidation of functional groups of enzymes
	oxidation of phospholipids
	reduction of S-S bonds
	macromolecule and membrane injuries

The scheme includes an enzymatic reaction catalyzed by NADPH-oxidase which is represented here as a Mn-containing enzyme and accounts for hydrogen peroxide formation during NADPH oxidation. Hydrogen peroxide is the reactant in the next reaction catalyzed by peroxidase, while NADPH (or XH_2) is the hydrogen donor in the system: the formation of a free radical of NADPH (NADP$^•$) is postulated which is then able to react with O_2 to form the anionic radical of oxygen - the superoxide (O_2^-). Evidence that free radicals of nicotinamide coenzymes may be formed during the enzymatic oxidation of these coenzymes by peroxidase has already been presented (Yokota & Yamazaki, 1965).

According to the scheme, the role of cyanide-resistant respiration is to generate hydrogen peroxide, while the role of peroxidase is to generate the superoxide from which the process of pathogen inactivation is started. Because of the relatively long life of O_2^- compared with other radicals derived from oxygen (Rabani & Njelsen, 1969) superoxide can diffuse through the cell and cause damage. It has been shown that O_2^- is a highly toxic agent due to its strong oxidative capacity, and in this connection inactivation of many enzymes and damage to lipoprotein membranes have been reported (Green & Curzon, 1968; Jocelyn, 1970; Lavelle *et al.*, 1973; Zimmerman *et al.*, 1973).

It seems reasonable to conclude that as cyanide-resistant respiration increases, so does the production of hydrogen peroxide. Because of the increase in H_2O_2, the peroxidase (or the peroxidases) which at this time are

abundant in the cells, produce large amounts of superoxide.
Thus following an attack by a pathogen, the superoxide dismutase activity should diminish and thus allow the cells
to use adequate amounts of superoxide by which they will
be able to inactivate the pathogen. Evidence obtained in
our laboratory with regard to such reduced activity in
mitochondria during the development of cyanide-resistant
respiration substantiates this assumption.

At the initiation of the defence mechanism, however,
the tissues which are responsible for the inactivation
of the pathogen through the production of superoxide are
themselves liable to its toxic action. In fact, while
superoxide oxidates the phospholipids or the sulphydrilic
groups, or inactivates the enzymes and injures the membranes, it makes no distinction between the molecules of
the pathogen and those of the host cell, so that considerable damage can also occur in those cells that are engaged
in counteracting the pathogen. This accounts for the numerous modifications and necroses in the tissues surrounding the pathogen.

Again, on the basis of the proposed scheme, it might be
possible to indicate those factors which may be involved
in the determination of susceptibility, or resistance, of
a plant to a pathogen. For example, plants may differ in
their ability to develop a high rate of cyanide-resistant
respiration following an attack by a pathogen. The time
elapsing betwen the onset of the infection and the beginning of the respiratory rise may also differ; resistant
cultivars might be those which develop their cyanide-resistant respiration early, whereas the susceptible cultivars would be those which elicit such respiration only
very slowly and, possibly, to a minor extent, i.e. too
late for the plant to be able to counteract the pathogen's
action effectively.

It appears that cyanide-resistant respiration is the key
event in the plant's response to the pathogen. If ascorbic

acid controls the development of this respiration (Arrigoni *et al.*, 1978), then it also must be recognized as the factor of major importance in the defence mechanism (Arrigoni *et al.*, 1977b).

Moreover, as the action of ascorbic acid upon cyanide-resistant respiration in polymorphonuclear leukocytes and in plant mitochondria is mediated by the synthesis of hydroxyproline-containing proteins, we believe that our data may provide a link, at molecular level, between ascorbic acid and the biological defence mechanisms in both plants and animals.

References

Allen, P.J. (1954). Physiological aspects of fungus diseases of plants. *Ann.Rev.Plant Physiol.*, 5, 225-248.

Arrigoni, O., Arrigoni-Liso, R. and Calabrese, G. (1975). Lycorine as an inhibitor of ascorbic acid biosynthesis. *Nature* 256, 513-514.

Arrigoni, O., Arrigoni-Liso, R. and Calabrese, G. (1976). Ascorbic acid as a factor controlling the development of cyanide-insensitive respiration. *Science* 194, 332-333.

Arrigoni, O., Arrigoni-Liso, R. and Calabrese, G. (1977a). Ascorbic acid requirement for biosynthesis of hydroxyproline-containing proteins in plants. *FEBS* 81, 135-138.

Arrigoni, O., Calabrese, G., Liso, R. and Porcelli, S. (1977b). Acido ascorbico e meccanismi di bioresistenza nelle piante. *Ann.Inst. Sper.Ort.Salerno* 7, 1-16.

Arrigoni, O., De Santis, A., Arrigoni-Liso, R. and Calabrese, G. (1977c). The increase of hydroxyproline-containing proteins in Jerusalem artichoke mitochondria during the development of cyanide-insensitive respiration. *Biochem.Biophys.Res.Comm.*, 74, 1637-1641.

Arrigoni, O., De Santis, A., Arrigoni-Liso, R. and Calabrese, G. (1978). The relationship between Vitamin C and respiratory burst of Guinea pig polymorphonuclear leukocytes. *Nature* 259, (in press).

Baldridge, C.W. and Gerard, R.W. (1933). The extra respiration of phagocytosis. *Am.J.Physiol.*, 103, 235-241.

Click, R.E. and Hackett, D.P. (1963). The role of protein and nucleic acid synthesis in the development of respiration in potato tuber slices. *Proc.Natl.Acad.Sci.,USA* 50, 243-250.

Evans, J.J. (1970). Spectral similarities and kinetic differences of two tomato plant peroxidase isoenzymes. *Pl.Physiol.*, 45, 66-69.

Fridovich, I. (1974). Superoxide dismutase. *Advan.Enzymol.*, 41, 35-97.

Fridovich, I. (1975). Superoxide dismutases. *Ann.Rev.Biochem.*, 44, 147-159.

Giebel, J. (1974). Biochemical mechanism of plant resistance to nematodes: a review. *J.Nematol.*, 6, 175-184.

Green, A.R. and Curzon, G. (1968). Decrease of 5-hydroxytryptamine in the brain provoked by hydrocortisone and its prevention by allopurinol. *Nature* **220**, 1095-1097.

Hackett, D.P., Haas, D.W., Griffiths, S.K. and Niederpruen, D.J. (1960). Studies on development of cyanide-resistant respiration in potato tuber slices. *Plant Physiol.*, **35**, 8-19.

Hussey, R.S. and Krusberg, L.R. (1970). Histopathology of and oxidative enzyme patterns in Wando peas infected with two populations of *Ditylenchus dipsaci*. *Phytopathology* **60**, 1818-1825.

Iyer, G.J.N., Islam, M.F. and Quastel, J.H. (1961). Biochemical aspects of phagocytosis. *Nature* **192**, 535-541.

Jocelyn, P.C. (1970). The function of subcellular fractions in the oxidation of glutathione in rat liver homogenate. *Biochem.J.*, **117**, 951-956.

Lavelle, F., Michelson, A.M. and Dimitrijevic, L. (1973). Biological protection by superoxide dismutase. *Biochem.Biophys.Res.Comm.*, **55**, 350-357.

Maraite, H. (1973). Changes in polyphenoloxidases and peroxidas in muskmelon (*Cucumis melo* L.) infected by *Fusarium oxysporum* f.sp. *melonis*. *Physiol.pL.Path.*, **3**, 29-41.

Matsumo, H. and Uritani, I. (1972). Physiological behaviour of peroxidase isozymes in sweet potato root tissue injured by cutting or with black rot. *Pl.Cell Physiol.*, **13**, 1091-1101.

Ohguchi, T., Yamashita, Y. and Asada, Y. (1974). Isoperoxidases of Japanese radish root infected by downy mildew fungus. *Ann.Phytopath.Soc.Japan*, **40**, 419-426.

Paul, B. and Sbarra, A.J. (1968). The role of the phagocyte in host-parasite interactions. *Biochem.Biophys.Acta.*, **156**, 168-178.

Rabani, J. and Njelsen, S.O. (1969). Absorption spectrum and decay kinetics of O_2 and HO_2 in aqueous solutions by pulse radiolysis. *J.Phys.Chem.*, **73**, 3736-3745.

Rautela, G.S. and Payne, M.C. (1970). The relationship of peroxidase and ortho-diphenol oxidase to resistance of sugarbeet to *Cercospora* leaf spot. *Phytopathology* **60**, 238-245.

Sbarra, A.J. and Karnovsky, M.L. (1959). The biochemical basis of phagocytosis. *J.Biol.Chem.*, **234**, 1355-1362.

Seevers, P.M., Daly, J.M. and Catedral, F.F. (1971). The role of peroxidase isozymes in resistance to wheat stem rust disease. *Pl. Physiol.*, **48**, 353-360.

Thimann, K.V., Yokum, C.S. and Hackett, D.P. (1954). Terminal oxidases and growth in plant tissues. *Arch.Biochem.Biophys.*, **53**, 239-257.

Uritani, I. and Akazawa, T. (1959). *In* "Plant Pathology." (Eds. J.G. Horsfall and A.E. Dimond.) **1**, 349-390. Academic Press, New York.

Verleur, J.D. and Uritani, I. (1965). Respiratory activity of the mitochondrial fractions isolated from healthy potato tubers and from tuber tissue incubated after cutting or infection with *Ceratocystis fimbriata*. *Plant Physiol.*, **40**, 1008-1012.

Yokota, K. and Yamazaki, I. (1965). Reaction of peroxidase with NADH and NADPH. *Biochem.Biophys.Acta.*, **105**, 301-312.

Zacheo, G., Lamberti, F., Arrigoni-Liso, R. and Arrigoni, O. (1977). Mitochondrial protein-hdroxyproline content of susceptible and resistant tomatoes infected by *Meloidogyne incognita*. *Nematologica* **23**, 471-476.

Zimmermann, R., Flohe, L., Weser, U. and Hartmann, H.J. (1973). Inhibition of lipid peroxidation in isolated inner membrane of rat liver mitochondria by superoxide dismutase. *FEBS* **29**, 117-120.

Discussion

Asked about the link between hydroxy-proline proteins and superoxide formation, Arrigoni replied that as cyanide-resistant respiration increases there is an increase in hydroxy proline; this is paralleled by a marked increase in peroxidase activity which in turn generates superoxide. A start had been made on the measurement of cyanide-resistant respiration in whole plants; addition of ascorbic acid (Vitamin C) transformed nematode susceptible plants to resistant status. Bird observed that the addition of ascorbic acid in water culture should bring about a decrease in infectivity by root-knot nematodes.

Viglierchio asked how plant breeders could take advantage of these respiration mechanisms. Arrigoni replied there is a need to study carefully the synthesis of ascorbic acid in plants as the amount in the cell is a critically important factor, if his hypothesis is correct. The biosynthesis of ascorbic acid is important in plants because it is low in roots and therefore it needs to be synthesised in large amounts. Viglierchio observed that it therefore appears that the general resistance of plants to pathogens is at the basic level, which can increase or decrease, but in addition there are built in specific resistances. Arrigoni agreed that ascorbic acid is widespread in plant tissues and is part of a general mechanism of resistance, but that specific mechanisms are important. Teresa Zacheo observed that with nematode attacks the amount of ascorbic acid in the plant tissues decreases more in susceptible cultivars than in resistant ones.

THE ROLE OF ASCORBIC ACID IN THE DEFENCE MECHANISM OF PLANTS TO NEMATODE ATTACK

O. Arrigoni, G. Zacheo, R. Arrigoni-Liso,
T. Bleve-Zacheo and F. Lamberti

*Istituto di Botanica dell'Università and Laboratorio
de Nematologia Agraria del C.N.R., Bari, Italy*

Considerable evidence has been presented in the last 40 years that a high intake of ascorbic acid results in beneficial effects in many individuals affected by several different viral diseases e.g. the common cold (Lewin, 1976). The presence of a comparatively high concentration of vitamin C in leucocytes is indicative of the role of ascorbic acid (AA) in the mechanism of physiological defence (Rolli and Sherry, 1948). In fact, it has been recently reported that there is a relationship between vitamin C and respiratory burst of polymorphonuclear leucocytes of Guinea pigs and that this respiratory burst is a metabolic event of fundamental importance in phagocytosis (Cline, 1975).

Tonzig and Bracci (1951) have demonstrated that its application inhibits the formation of root nodules in *Pisum sativum* L. and we speculate that AA would probably be implicated in the defence mechanisms of plants as it is in animals. To test this hypothesis we experimentally varied the concentration of AA in tomato plants resistant and susceptible to *Meloidogyne incognita* and tested their reaction to attacks by the nematode.

The concentration of AA in plant roots was decreased by root application of an aqueous solution of lycorine, an alkaloid extracted from *Sternbergia lutea* Roem. & Schult,,

which has been proved to inhibit the synthesis of AA
(Arrigoni *et al.*, 1975; 1977).

The results of the experiments indicate that a decrease
in AA in plants induces a reduction in their resistance
to the root-knot nematode. In the roots of cv. Brecht the
rate of nematode reproduction was double in treated compared with untreated plants. Conversely, susceptible cultivars such as Roma VF and Marmande irrigated with a
water solution of 5 - 30 mM of AA reacted similarly to resistant cultivars, lowering penetration and reproduction
of the parasite (Arrigoni *et al.*, in preparation). The
amount of AA in susceptible plants was unaltered but in
resistant cultivars AA synthesis was always stimulated
by nematode attack.

Animals capable of synthetizing their own AA requirements have been shown to increase greatly its synthesis
when exposed to pathogens (Longenecker *et al.*, 1939). Our
data indicate a similarity in the defence mechanism of
plants and animals and a direct correlation between AA
and these defence mechanisms.

References

Arrigoni, O., Arrigoni-Liso, R. and Calabrese, G. (1975). Lycorine as
    an inhibitor of ascorbic acid biosynthesis. *Nature* 256, 513-514.
Arrigoni, O., Arrigoni-Liso, R. and Calabrese, G. (1977). Ascorbic
    acid requirement for biosynthesis of hydroxyproline-containing
    proteins in plants. *FEBS Letters* 82, 135-138.
Cline, M.J., (1975). The White Cell. Harvard University Press,
    Cambridge, Mass. and London, 564.
Lewin, S. (1976). *In* "Vitamin C: its Molecular Biology and Medical
    Potential." 1-4. Academic Press, London.
Longenecker, H.E., Musulin, R.R., Tully, R.H. and King, C.G. (1939).
    An acceleration of vitamin C synthesis and excretion by feeding
    known organic compounds in rats. *J.Biol.Chem.*, 129, 445-453.
Rolli, E.P. and Sherry, S. (1948). Adult survey and metabolism of
    vitamin C. *Medicine* 20, 251-261.
Tonzig, S. and Bracci, L. (1951). Ricerche sulla biologia dell'acido
    ascorbico. *Giorn.Bot.Ital.*, 58, 258-270.